THE TOP 40 THINGS YOU NEED TO KNOW FOR TOP SCORES IN CHEMISTRY

1. CHANGES

Understand and be able to identify the difference between physical and chemical changes.

See Chapter 1.

2. REACTIONS

Understand and be able to identify the difference between exothermic and endothermic reactions.

See Chapter 1.

3. MIXTURES

Know the differences between substances, mixtures, and the components of mixtures.

See Chapter 1.

4. GAS LAWS AND CALCULATIONS

Be able to use the gas laws to calculate moles, pressure, volume, and mass of a sample of gas at various temperatures and conditions.

See Chapter 2.

5. MATTER

Be able to name the changes in phases of matter and identify them on a heating/cooling curve.

See Chapter 2.

6. SUBATOMIC PARTICLES

Understand the properties of the subatomic particles and how they allow isotopes to exist.

See Chapter 3.

7. ELECTRON CONFIGURATION

Be able to provide the electron configuration of an element given the number of electrons.

See Chapter 3.

8. MOLECULES

Know how to distinguish between the various hybridization states and the shapes of molecules that can be formed.

See Chapter 3.

9. THE OCTET RULE

Understand the octet rule and how it allows atoms and ions to be stable.

See Chapter 3.

10. GROUPING AND THE PERIODIC TABLE

Know the properties and names of various groups/families within the periodic table.

See Chapter 4.

11. METALS, NONMETALS, AND THE PERIODIC TABLE

Know the properties and locations of the metals and nonmetals.

See Chapter 4.

12. TRENDS AND THE PERIODIC TABLE

Know the trends for electronegativity, ionization energy, and atomic radius across the periodic table.

See Chapter 4.

13. BONDS

Be able to distinguish between the various intramolecular bonds: covalent (polar vs. nonpolar), ionic, network covalent, hydrogen, coordinate covalent, metallic, dispersion/Van der Waals, and molecule-ion attraction.

See Chapter 5.

14. SIGMA AND PI BONDS

Be able to tell the difference between sigma and pi bonds and be able to locate them within a molecule.

See Chapter 5.

15. COMPOUNDS

Be able to name ionic and covalent compounds using both traditional methods and the stock method.

See Chapter 5.

16. CHEMICAL EQUATIONS

Know how to balance and classify chemical equations.

See Chapter 6.

17. CALCULATIONS OF COMPOUNDS

Be able to calculate percent hydration and percent composition of a compound.

See Chapter 6.

18. SOLUBILITY RULES

Understand how to use solubility rules to predict the products of a reaction and write net ionic equations.

See Chapter 6.

19. SOLUTIONS

Know how to calculate the concentration of a solution.

See Chapter 6.

20. THE MOLE

Understand how to use the mole to calculate the number of liters a gas will occupy, the number of molecules present, the mass of a sample, and the number of moles of another substance in a reaction.

See Chapter 6.

21. POTENTIAL ENERGY DIAGRAMS

Be able to draw and label a potential energy diagram.

See Chapter 7.

22. HEAT

Know how to use a potential energy diagram and Hess's Law to calculate heat involved in reactions.

See Chapter 7.

23. RATE OF REACTION

Be able to determine how to change the rate of reaction.

See Chapter 8.

24. EQUILIBRIUM

Be able to determine how changing conditions changes the point of equilibrium of a reaction.

See Chapter 8.

25. PRODUCTS AND REACTANTS

Understand how to use K_{eq} and K_{sp} values to find concentrations of products and reactants.

See Chapter 8.

26. SPONTANEOUS REACTIONS

Know how to determine if a reaction will be spontaneous.

See Chapter 8.

27. ACIDS AND BASES

Understand the various operational and conceptual methods for defining acids and bases.

See Chapter 9.

28. K_a

Understand what K_a can tell us about an acid or a base.

See Chapter 9.

29. MOLARITY AND pH

Know how to calculate the molarity and pH of an acid or a base solution.

See Chapter 9.

30. OXIDATION NUMBERS

Be able to determine the oxidation numbers for the elements in a compound.

See Chapter 10.

31. OXIDIZING AND REDUCING AGENTS

Know how to identify the substances that had a change in oxidation number and identify which serve as an oxidizing or a reducing agent in a half-reaction.

See Chapter 10.

32. REDOX REACTIONS

Be able to balance both simple and complex redox reactions.

See Chapter 10.

33. VOLTAIC AND ELECTROLYTIC CELLS

Know how to determine the reactions that occur within voltaic and electrolytic cells.

See Chapter 10.

34. PREFIXES

Understand the prefixes used in organic chemistry so as to know the number of carbon atoms present in a molecule.

See Chapter 11.

35. FUNCTIONAL GROUPS IN ORGANIC CHEMISTRY

Be able to distinguish between the various functional groups in organic chemistry which contain oxygen, nitrogen, and double and triple bonds.

See Chapter 11.

36. RADIATION

Be able to identify the risks and benefits of using radiation.

See Chapter 12.

37. NUCLEAR EMANATIONS

Know the differences between various types of nuclear emanations.

See Chapter 12.

38. HALF-LIFE AND MASS CALCULATIONS

Be able to calculate the half-life of an isotope or the mass of a radioactive sample after a certain period of time.

See Chapter 12.

39. LABORATORY TECHNIQUES

Understand safe and general laboratory techniques.

See Chapter 13.

40. LABORATORY CALCULATIONS

Know how to make all necessary calculations pertaining to experiments carried out in the laboratory.

See Chapter 13.

McGRAW-HILL's

SAT

SUBJECT TEST

CHEMISTRY

McGRAW-HILL's

SAT
SUBJECT TEST

CHEMISTRY

Third Edition

Thomas A. Evangelist

Assistant Principal, Supervision of Science
New York City Department of Education

New York / Chicago / San Francisco / Lisbon / London / Madrid / Mexico City
Milan / New Delhi / San Juan / Seoul / Singapore / Sydney / Toronto

McGRAW-HILL's SAT Subject Test Chemistry

Copyright © 2012, 2009 by The McGraw-Hill Companies, Inc. All rights reserved. Printed in the United States of America. Except as permitted under the Copyright Act of 1976, no part of this publication may be reproduced or distributed in any forms or by any means, or stored in a database or retrieval system, without the prior written permission of the publisher.

3 4 5 6 7 8 9 10 11 12   QVS/QVS   1 9 8 7 6 5 4 3

ISBN 978-0-07-176875-7
MHID 0-07-176875-0

e-ISBN 978-0-07-176876-4
e-MHID 0-07-176876-9

Library of Congress Control Number: 2011914988

This publication is designed to provide accurate and authoritative information in regard to the subject matter covered. It is sold with the understanding that neither the author nor the publisher is engaged in rendering legal, accounting, futures/securities trading, or other professional service. If legal advice or other expert assistance is required, the services of a competent professional person should be sought.

> —*From a Declaration of Principles jointly adopted by a Committee of the American Bar Association and a Committee of Publishers*

McGraw-Hill books are available at special quantity discounts for use as premiums and sales promotions, or for use in corporate training programs. To contact a representative, please e-mail us at bulksales@mcgraw-hill.com.

SAT is a registered trademark of the College Entrance Examination Board, which was not involved in the production of, and does not endorse, this product.

Dedicated to:

My loving and supportive family,
My students who are worth all the effort,
My friends who have been there over the years,
My teammates who put it all on the line game after game,
My colleagues who have been behind me 100 percent,
Everyone who has taught me something over the past 30-plus years.

Let's never stop learning. . . .

CONTENTS

McGRAW-HILL's

SAT
SUBJECT TEST
CHEMISTRY

PART I

INTRODUCTION TO THE SAT CHEMISTRY TEST

All About the SAT Chemistry Test

THE SAT SUBJECT TESTS

What Are the SAT Subject Tests?

The SAT Subject Tests (formerly called the SAT II tests and the Achievement Tests) are a series of college entrance tests that cover specific academic subject areas. Like the better-known SAT test, which measures general verbal and math skills, the SAT Subject Tests are given by the College Entrance Examination Board. Colleges and universities often require applicants to take one or more SAT Subject Tests along with the SAT.

SAT Subject Tests are generally not as difficult as Advanced Placement tests, but they may cover more than is taught in basic high school courses. Students usually take an SAT Subject Test after completing an Advanced Placement course or an Honors course in the subject area.

How Do I Know if I Need to Take SAT Subject Tests?

Review the admissions requirements of the colleges to which you plan to apply. Each college will have its own requirements. Many colleges require that you take a minimum number of SAT Subject Tests—usually one or two. Some require that you take tests in specific subjects. Some may not require SAT Subject Tests at all.

When Are SAT Subject Tests Given, and How Do I Register for Them?

SAT Subject Tests are usually given on six weekend dates spread throughout the academic year. These dates are usually the same ones on which the SAT is given. To find out the test dates, visit the College Board Web site at www.collegeboard.com. You can also register for a test at the Web site. Click on the tabs marked "students" and follow the directions you are given. You will need to use a credit card if you register online. As an alternative, you can register for SAT Subject Tests by mail using the registration form in the SAT Registration Bulletin, which should be available from your high school guidance counselor.

How Many SAT Subject Tests Should I Take?

You can take as many SAT Subject Tests as you wish. According to the College Board, more than one-half of all SAT Subject Test takers take three tests, and about one-quarter take four or more tests. Keep in mind, though, that you can take only three tests on a single day. If you want to take more than three tests, you'll need to take the others on a different testing date. When deciding how many SAT Subject Tests to take, base your decision on the requirements of the colleges to which you plan to apply. It is probably not a good idea to take many more SAT Subject Tests than you need. You will probably do better by focusing only on the ones that your preferred colleges require.

Which SAT Subject Tests Should I Take?

If a college to which you are applying requires one or more specific SAT Subject Tests, then of course you must take those particular tests. If the college simply requires that you take a minimum number of SAT Subject Tests, then choose the test or tests for which you think you are best prepared and likely to get the best score. If you have taken an Advanced Placement course or an Honors course in a particular subject and done well in that course, then you should probably consider taking an SAT Subject Test in that subject.

When Should I Take SAT Subject Tests?

Timing is important. It is a good idea to take an SAT Subject Test as soon as possible after completing a course in the test subject, while the course material is still fresh in your mind. If you plan to take an SAT Subject Test in a subject that you have not studied recently, make sure to leave yourself enough time to review the course material before taking the test.

What Do I Need on the Day of the Test?

To take an SAT Subject Test, you will need an admission ticket to enter the exam room and acceptable forms of photo identification. You will also need two number 2 pencils. Be sure that the erasers work well at erasing without leaving smudge marks. The tests are scored by machine, and scoring can be inaccurate if there are smudges or other stray marks on the answer sheet.

Any devices that can make noise, such as cell phones or wristwatch alarms, should be turned off during the test. Study aids such as dictionaries and review books, as well as food and beverages, are barred from the test room.

THE SAT CHEMISTRY TEST

What Is the Format of the SAT Chemistry Test?

The SAT Chemistry test is a one-hour exam consisting of 85 multiple-choice questions. According to the College Board, the test measures the following knowledge and skills:

- Familiarity with major chemistry concepts and ability to use those concepts to solve problems
- Ability to understand and interpret data from observation and experiments and to draw conclusions based on experiment results
- Knowledge of laboratory procedures and of metric units of measure
- Ability to use simple algebra to solve word problems
- Ability to solve problems involving ratio and direct and inverse proportions, exponents, and scientific notation

The test covers a variety of chemistry topics. The following chart shows the general test subject areas, as well as the approximate portion of the test devoted to each subject.

SAT Chemistry Subject Areas

Subject Area	Approximate Percentage of Exam
1. Structure of Matter	25%
2. States of Matter	15%
3. Reaction Types	14%
4. Stoichiometry	12%
5. Equilibrium and Reaction Rates	7%
6. Thermodynamics	6%
7. Descriptive Chemistry	13%
8. Laboratory	8%

When you take the SAT Chemistry test, you will be given a test booklet that includes a periodic table of the elements. The table will show only the element symbols, atomic numbers, and atomic masses. It will not show electron configurations or oxidation numbers. You may not use your own reference tables or a calculator.

What School Background Do I Need for the SAT Chemistry Test?

The College Board recommends that you have at least the following experience before taking the SAT Chemistry test:

- One-year chemistry course at the college preparatory level
- One-year algebra course
- Experience in the chemistry laboratory

How Is the SAT Chemistry Test Scored?

On the SAT Chemistry test, your "raw score" is calculated as follows: You receive one point for each question you answer correctly, but you lose one-quarter point for each question you answer incorrectly. You do not gain or lose any points for questions that you do not answer at all. Your raw score is then converted into a scaled score by a statistical method that takes into account how well you did compared to others who took the same test. Scaled scores range from 200 to 800 points. Your scaled score will be reported to you, to your high school, and to the colleges and universities that you designate to receive it.

Scoring scales differ slightly from one version of the test to the next. The scoring scales provided after each practice test in this book are only samples that will show you your approximate scaled score.

When Will I Receive My Score?

Scores are mailed to students approximately three to four weeks after the test. If you want to find out your score a week or so earlier, you can do so for free by accessing the College Board Web site or for an additional fee by calling (866)756-7346.

How Do I Submit My Score to Colleges and Universities?

When you register to take the SAT or SAT Subject Tests, your fee includes free reporting of your scores to up to four colleges and universities. To have your scores reported to additional schools, visit the College Board Web site or call (866)756-7346. You will need to pay an additional fee.

SAT CHEMISTRY QUESTION TYPES

The SAT Chemistry test consists entirely of multiple-choice questions. Most are the regular five-answer-choice format that you will be familiar with from taking other standardized tests. Some, however, have special formats that do not appear on other tests and that you need to be aware of. The College Board calls these formats "classification sets" and "relationship analysis questions." Review the following examples before you tackle the Diagnostic Test.

Regular Multiple-Choice Questions

On the SAT Chemistry test, most of the questions are in the regular five-answer-choice format that is used on standardized tests such as the SAT. Here is an example:

1. Which oxidation half reaction below demonstrates conservation of mass and charge?
    (A) $Mg^{2+} + 2e^- \rightarrow Mg$
    (B) $Cl^{1-} + 1e^- \rightarrow Cl_2$
    (C) $2Ag^{1+} \rightarrow 2Ag + 1e^-$
    (D) $Mg \rightarrow Mg^{2+} + 2e^-$
    (E) $F_2 + 2e^- \rightarrow 2F^{1-}$

The correct answer is choice D. Note that with this question, as with many other questions on the test, you can find the correct answer by using the process of elimination. The half reactions shown in choices A, B, and E are all reduction half reactions, so those choices can be eliminated. Both remaining choices, C and D, show oxidation and a loss of electrons. But choice C does not demonstrate conservation of charge and mass; if it did, there would have to be two electrons on the left side of the reaction. So the correct answer must be choice D.

You will see a variation of this basic format in which you are offered three choices indicated by the Roman numerals I, II, and III. Your task is to decide which combination of the three choices answers the question. Here is an example:

2. Which of the following indicates an acidic solution?
   I.   Litmus paper turns blue.
   II.   Phenolphthalein turns pink.
   III. Hydronium ion concentration is greater than hydroxide ion con-
         centration.
         (A) I only
         (B) II only
         (C) III only
         (D) I and II only
         (E) I, II, and III

The correct answer is choice C. First, review the choices. Choices I and II indicate a basic solution. If they were acidic, then the solutions would be red
for litmus and clear for phenolphthalein. Only choice III holds true for an
acidic solution. In an acidic solution the concentration of hydronium ions ex-
ceeds that of hydroxide ion concentration.

Classification Sets

In a classification set, you are given five answer choices lettered A through E.
The choices may be chemistry principles, substances, numbers, equations,
diagrams, or the like. The choices are followed by three or four numbered
questions. Your task is to match each question with the answer choice to
which it refers. Here are sample directions for a classification set, followed
by a sample of this question format.

Directions: Each of the following sets of lettered choices refers to the numbered formulas or
statements immediately below it. For each numbered item, choose the one lettered choice that
fits it best. Then fill in the corresponding oval on the answer sheet. Each choice in a set may be
used once, more than once, or not at all.

Questions 3–5:
        (A)   Coordinate covalent bonding
        (B)   Ionic bonding
        (C)   Nonpolar covalent bonding
        (D)   Metallic bonding
        (E)   Hydrogen bonding

3. HF

4. N_2

5. KI

3. The correct answer is choice E. The bond between the atoms of hydrogen and fluorine is a polar covalent bond, a choice that is not present in
the choices above. Now look at the bonding between the molecules of HF.
HF can exhibit dipole forces between its molecules, yet another choice that
is not present. HF can, however, exhibit hydrogen bonding, a choice that
is present.

4. The correct answer is choice C. Nitrogen gas has no difference in electro-negativity between the nitrogen atoms. The two nitrogen atoms will form a nonpolar covalent bond. The type of bonding present between the molecules of nitrogen gas will be dispersion forces, the forces present between nonpolar molecules.

5. The correct answer is choice B. Potassium iodide is formed from a metal, potassium, and a nonmetal, iodine. The type of bonding that forms between metals and nonmetals is ionic bonding.

Relationship Analysis Questions

Relationship analysis questions are probably not like any question type that you have seen before. Each question consists of two statements labeled I and II with the word BECAUSE between the two statements. For each question, you have three tasks. You must:

- Determine if statement I is true or false.
- Determine if statement II is true or false.
- Determine if statement II is the correct explanation for statement I.

On the answer sheet, you will mark true (T) or false (F) for each statement, and you will mark "correct explanation" (CE) ONLY if statement II is a correct explanation of statement I. Here are sample directions for this kind of question, followed by two examples and a sample of a correctly marked answer sheet.

Directions: Each question below consists of two statements. For each question, determine whether statement I in the leftmost column is true or false and whether statement II in the rightmost column is true or false. Fill in the corresponding T or F ovals on the answer sheet provided. Fill in the oval labeled "CE" only if statement II correctly explains statement I.

| 101 HCl is an Arrhenius acid | BECAUSE | HCl is a proton donor. |
| 102 Water is a polar molecule | BECAUSE | the dipole forces in a molecule of water will counterbalance each other and cancel out. |

101. T, T, CE   HCl will donate a proton (hydronium ion) when it reacts. This classifies it as an Arrhenius acid.

102. T, F   Because of the bent shape of a water molecule, the dipole forces in the molecule will not counterbalance or cancel out. This is what causes a water molecule to be a polar molecule.

Here is how you would mark these answers on the answer sheet:

	I	II	CE*
101	● (F)	● (F)	●
102	● (F)	(T) ●	○

How to Use This Book

The SAT Chemistry test covers a very large amount of material, and your preparation time may be short. That is why it is important to use your study time wisely. This book provides a comprehensive review of everything you need to know for the test, and it has been organized to make your study program practical and efficient. It will help you:

- Identify the chemistry topics that you most need to focus on.
- Familiarize yourself with the test format and test question types.
- Review all the basic chemistry you need to know for the test.
- Check your progress with questions at the end of each review chapter.
- Practice your test-taking skills using sample tests.

The following four-step study program has been designed to help you make the best use of this book.

STEP 1   TAKE THE DIAGNOSTIC TEST

Once you have read through this chapter, start your preparation program by taking the Diagnostic Test. This test is carefully modeled on the real SAT Chemistry test in terms of format, types of questions, and topics tested. Take the Diagnostic Test under test conditions and pay careful attention to the one-hour time limit. When you complete the test, score yourself using the scoring information at the end of the test. Then read through the explanations to see which test topics gave you the most trouble. Look for patterns. Did you miss questions in one or two specific subject areas? Did specific question formats give you trouble? When did you need to guess at the answer? Use your results to identify the topics and question types that were most difficult for you. Once you know your chemistry strengths and weaknesses, you'll know which subjects you need to focus on as you review for the test.

STEP 2   REVIEW THE TEST TOPICS

This book provides a full-scale review of all the topics tested on the SAT Chemistry test. Once you have identified the topics that give you the most trouble, review the relevant chapters. You do not need to work through the review chapters in the order in which they appear. Skip around if you like, but remember to focus on the topics that gave you the most trouble on the Diagnostic Test.

Each review chapter ends with practice problems that you can use to see how well you have mastered the material. If you get a problem wrong, go back into the chapter and reread the section that covers that particular topic.

Make a study schedule. If you have the time, plan to spend at least two weeks or so working your way through the review chapters. Be sure to set aside enough time at the end of your schedule to take the practice tests at the end of the book. However, if you do not have much time before the test, you may want to shorten your review time and focus instead entirely on the practice tests.

STEP 3   BUILD YOUR TEST-TAKING SKILLS

As you work through the examples and review questions in each review chapter, you'll become familiar with the kinds of questions that appear on the SAT Chemistry test. You'll also practice the test-taking skills essential for top scores. These include:

- The ability to recall and comprehend major concepts in chemistry and to apply them to solve problems
- The ability to interpret information gained from observations and experiments
- The ability to make inferences from experimental data, including data presented in graphs and tables

STEP 4   TAKE THE PRACTICE TESTS

Once you have completed your review of all the SAT Chemistry topics, get ready for the real exam by taking the four practice tests at the back of this book. When you take each test, try to simulate actual test conditions. Sit in a quiet room, time yourself, and work through as much of the test as time allows. The tests are ideal for practice because they have been constructed to be as much like the real test as possible. The directions and practice questions are very much like those on the real test. You'll gain experience with the test format, and you'll learn to pace yourself so that you can earn the maximum number of points in the time allowed.

Each test will also serve as a review of the topics tested because complete explanations are provided for every question. The explanations can be found at the end of each test. If you get a question wrong, you'll want to review the explanation carefully. You may also want to go back to the chapter in this book that covers the question topic.

Each review chapter ends with practice problems that you can use to see how well you have mastered the material. If you get a problem wrong, go back into the chapter and reread the section that covers that particular topic.

At the end of each test you'll also find scoring information. Calculate your raw score, then use the table provided to find your approximate scaled score. The scaling on the real test may be slightly different, but you'll get a good idea of how you might score on the actual test.

Strategies for Top Scores

When you take the SAT Chemistry test, you'll want to do everything you can to make sure you get your best possible score. That means studying right, building good problem-solving skills, and learning proven test-taking strategies. Here are some tips to help you do your best.

STUDY STRATEGIES

- **Get to know the format of the exam.** Use the practice tests in this book to familiarize yourself with the test format, which does not change from year to year. That way, you'll know exactly what to expect when you see the real thing on test day.

- **Get to know the test directions.** If you are familiar with the directions ahead of time, you won't have to waste valuable test time reading them and trying to understand them. The format and directions used in the practice exams in this book are modeled on the ones you'll see on the actual SAT Chemistry exam.

- **Get to know what topics are covered.** Get to know what specific topics are covered on the exam. You'll find all of them in the review material and practice exams in this book. The following are the "hot topics" on the SAT Chemistry exam:

- Structure and theories of the atom
- Periodic trends and the chemical and physical properties of the elements
- Bonding between atoms and molecules
- Molecular geometries
- Nuclear chemistry
- Kinetic molecular theory and the gas laws
- Liquids, solids, and changes of phase
- Concentration
- Solubility and precipitation of compounds
- Electrolytes and conductivity
- Solution chemistry
- Colligative properties
- Redox reactions and reactions of acids and bases
- Mole relationships and Avogadro's number
- Molecular and empirical formulas
- Percent composition by mass
- Limiting and excess reagents
- Rates of reaction
- Le Châtelier's Principle
- Equilibrium constant expressions for gases and slightly soluble salts
- Energy in chemical and physical changes
- Hess's Law
- Entropy and Gibbs free energy
- Organic and environmental chemistry
- Laboratory safety, procedures, skills, and setups

- **Study hard.** If possible, plan to study for at least an hour a day for two weeks before the test. You should be able to read this entire book and complete all five practice exams during that time period. Be sure to write notes in the margins of the book and paraphrase what you read. Make study cards from a set of index cards. Those cards can "go where you go" during the weeks and days before the test. If you are pressed for time, focus on taking the five practice exams, reading the explanations, and reviewing the particular topics that give you the most trouble.

PROBLEM-SOLVING STRATEGIES

- **Solve problems in whatever way is easiest for you.** There are usually several ways to solve any problem in chemistry and arrive at the correct answer. For example, when converting units some students prefer to use a dimensional analysis whereas others prefer to set up a proportion.

Do what is easiest for you. Remember that the SAT exam is all multiple choice. That means that no one is going to be checking your work and judging you by which solution method you chose. So solve the problem any way you like.

- **Build good problem-solving skills.** When you tackle SAT Chemistry problems, try following this three-step process:

1. When you first read a question, make a list of the given values and variables and the units for the variables.
2. Ask yourself, "What do I have and what do I need to get?" The link between what you have and what you need to get is either an equation that you should be familiar with or certain specific steps to follow to solve particular types of problems.
3. Solve the problem and see if the answer makes sense. For example, if you know that one variable should be much larger than another, make sure your answer reflects that relationship. You'll see how this works with many of the problems in this book.

- **Make sure you know what the question is asking.** The questions on the SAT Chemistry test are not deliberately designed to trick you, but it is still important that you look closely at each one to make sure you know what it is asking. If a question asks which compound has the lowest hydrogen ion concentration, don't pick the answer choice with the highest concentration. Pay special attention to questions that include the words NOT or EXCEPT. You may want to circle these words to make sure you take them into account as you choose your answer.

TEST-TAKING STRATEGIES

- **Answer all the easy problems first, then tackle the harder ones.** Keep in mind that the test is only one hour long. There isn't much time to spend trying to figure out the answers to harder problems, so skip them and come back to them later. There are three reasons why you should do this. The first reason is that every question counts the same in the scoring of the exam. That means that you are better off spending time answering the easier questions, where you are sure to pick up points. The second reason to skip past harder questions is that later on in the test you might come to a question or a set of answer choices that jogs your memory and helps you to go back and answer the question you skipped. The third reason is that by answering the easier questions, you'll build your confidence and get into a helpful test-taking rhythm. Then when you go back to a question you skipped, you may find that it isn't as hard as you first thought.

- **Use the process of elimination.** Keep in mind that on the SAT Chemistry test, like any other multiple-choice test, the answer is right in front of you. Try eliminating answer choices that you know are incorrect. Often this can help you select the correct answer.

- **If you must guess, make an educated guess.** The SAT has a one-quarter-point penalty for wrong answers to discourage random guessing. So if you have absolutely no idea how to answer a question, you are better off skipping it entirely. However, you may be able to eliminate one or more answer choices. If you can do that, you can increase your odds of guessing the correct answer. If you can make this kind of educated guess, go ahead. If you guess correctly, you'll earn another point.

- **Be wary of answer choices that look familiar but are not correct.** Sometimes in the set of answer choices there will be one or more wrong answers that include familiar expressions or phrases. You might be tempted to pick one of these choices if you do not work out the problem completely. That is why it is important to work through each problem thoroughly and carefully to make sure that you pick the correct answer choice.

- **You don't have to answer every question.** If you do not know the answer to a question and cannot eliminate any answer choices, skip it and go on. It is better to do that than to risk losing one-quarter of a point for a wrong answer. If you have time at the end of the test, you can return to skipped questions and try to make an educated guess. But you do not have to answer every question to get a good score.

TIPS FOR TEST DAY

- **Don't panic!** Once test day comes, you're as prepared as you're ever going to be, so there is no point in panicking. Use your energy to make sure that you are extra careful in answering questions and marking the answer sheet.

- **Use your test booklet as scratch paper.** Your test booklet is not going to be reused by anyone when you're finished with it, so feel free to mark it up in whatever way is most helpful to you. Circle important words, underline important points, write your calculations in the margins, and cross out wrong answer choices.

- **Be careful when marking your answer sheet.** Remember that the answer sheet is scored by a machine, so mark it carefully. Fill in answer ovals completely, erase thoroughly if you change your mind, and do not make any stray marks anywhere on the sheet. Also, make sure that the answer space you are marking matches the number of the question you are answering. If you skip a question, make sure that you skip the corresponding space on the answer sheet. Every 5 or 10 questions, check the question numbers and make sure that you are marking in the right spot. You may want to mark your answers in groups of 5 or 10 to make sure that you are marking the answer sheet correctly.

- **Be especially careful when marking the answers to Chemistry Test questions 101–115.** On the SAT Chemistry test, questions 101 through 115 comprise a special section. The questions in this section have their own format. (For more about these questions, see the section of this book titled "SAT Chemistry Question Types.") On the answer sheet, you must mark your answers to these questions in a special section labeled "Chemistry" at the lower left corner. Make sure that you locate this answer sheet section and mark the answers to these questions in the proper place.

- **Watch the time.** Keep track of the time as you work your way through the test. Try to pace yourself so that you can tackle as many of the 80 questions as possible within the one-hour time limit. Check yourself at 10- or 15-minute intervals using your watch or a timer.

- **Don't panic if time runs out.** If you've paced yourself carefully, you should have time to tackle all or most of the questions. But if you do run out of time, don't panic. Make sure that you have marked your answer sheet for all the questions that you have answered so far. Then look ahead at the questions you have not yet read. Can you answer any of them quickly, without taking the time to do lengthy calculations? If you can, mark your answers in the time you have left. Every point counts!

- **Use extra time to check your work.** If you have time left over at the end of the test, go back and check your work. Make sure that you have marked the answer sheet correctly. Check any calculations you may have made to make sure that they are correct. Take another look at any questions you may have skipped. Can you eliminate one or more answer choices and make an educated guess? Resist the urge to second-guess too many of your answers, however, as this may lead you to change an already correct answer to a wrong one.

PART II
DIAGNOSTIC TEST

DIAGNOSTIC TEST

▰ ANSWER SHEET

This Diagnostic Test will help you measure your current knowledge of chemistry and familiarize you with the structure of the SAT Chemistry test. To simulate exam conditions, use the answer sheet below to record your answers and the tables in Appendix 2, 3, and 4 of this book to obtain necessary information.

If there are more answer spaces than you need, leave the extra spaces blank.

1. Ⓐ Ⓑ Ⓒ Ⓓ Ⓔ	21. Ⓐ Ⓑ Ⓒ Ⓓ Ⓔ	41. Ⓐ Ⓑ Ⓒ Ⓓ Ⓔ	61. Ⓐ Ⓑ Ⓒ Ⓓ Ⓔ
2. Ⓐ Ⓑ Ⓒ Ⓓ Ⓔ	22. Ⓐ Ⓑ Ⓒ Ⓓ Ⓔ	42. Ⓐ Ⓑ Ⓒ Ⓓ Ⓔ	62. Ⓐ Ⓑ Ⓒ Ⓓ Ⓔ
3. Ⓐ Ⓑ Ⓒ Ⓓ Ⓔ	23. Ⓐ Ⓑ Ⓒ Ⓓ Ⓔ	43. Ⓐ Ⓑ Ⓒ Ⓓ Ⓔ	63. Ⓐ Ⓑ Ⓒ Ⓓ Ⓔ
4. Ⓐ Ⓑ Ⓒ Ⓓ Ⓔ	24. Ⓐ Ⓑ Ⓒ Ⓓ Ⓔ	44. Ⓐ Ⓑ Ⓒ Ⓓ Ⓔ	64. Ⓐ Ⓑ Ⓒ Ⓓ Ⓔ
5. Ⓐ Ⓑ Ⓒ Ⓓ Ⓔ	25. Ⓐ Ⓑ Ⓒ Ⓓ Ⓔ	45. Ⓐ Ⓑ Ⓒ Ⓓ Ⓔ	65. Ⓐ Ⓑ Ⓒ Ⓓ Ⓔ
6. Ⓐ Ⓑ Ⓒ Ⓓ Ⓔ	26. Ⓐ Ⓑ Ⓒ Ⓓ Ⓔ	46. Ⓐ Ⓑ Ⓒ Ⓓ Ⓔ	66. Ⓐ Ⓑ Ⓒ Ⓓ Ⓔ
7. Ⓐ Ⓑ Ⓒ Ⓓ Ⓔ	27. Ⓐ Ⓑ Ⓒ Ⓓ Ⓔ	47. Ⓐ Ⓑ Ⓒ Ⓓ Ⓔ	67. Ⓐ Ⓑ Ⓒ Ⓓ Ⓔ
8. Ⓐ Ⓑ Ⓒ Ⓓ Ⓔ	28. Ⓐ Ⓑ Ⓒ Ⓓ Ⓔ	48. Ⓐ Ⓑ Ⓒ Ⓓ Ⓔ	68. Ⓐ Ⓑ Ⓒ Ⓓ Ⓔ
9. Ⓐ Ⓑ Ⓒ Ⓓ Ⓔ	29. Ⓐ Ⓑ Ⓒ Ⓓ Ⓔ	49. Ⓐ Ⓑ Ⓒ Ⓓ Ⓔ	69. Ⓐ Ⓑ Ⓒ Ⓓ Ⓔ
10. Ⓐ Ⓑ Ⓒ Ⓓ Ⓔ	30. Ⓐ Ⓑ Ⓒ Ⓓ Ⓔ	50. Ⓐ Ⓑ Ⓒ Ⓓ Ⓔ	70. Ⓐ Ⓑ Ⓒ Ⓓ Ⓔ
11. Ⓐ Ⓑ Ⓒ Ⓓ Ⓔ	31. Ⓐ Ⓑ Ⓒ Ⓓ Ⓔ	51. Ⓐ Ⓑ Ⓒ Ⓓ Ⓔ	71. Ⓐ Ⓑ Ⓒ Ⓓ Ⓔ
12. Ⓐ Ⓑ Ⓒ Ⓓ Ⓔ	32. Ⓐ Ⓑ Ⓒ Ⓓ Ⓔ	52. Ⓐ Ⓑ Ⓒ Ⓓ Ⓔ	72. Ⓐ Ⓑ Ⓒ Ⓓ Ⓔ
13. Ⓐ Ⓑ Ⓒ Ⓓ Ⓔ	33. Ⓐ Ⓑ Ⓒ Ⓓ Ⓔ	53. Ⓐ Ⓑ Ⓒ Ⓓ Ⓔ	73. Ⓐ Ⓑ Ⓒ Ⓓ Ⓔ
14. Ⓐ Ⓑ Ⓒ Ⓓ Ⓔ	34. Ⓐ Ⓑ Ⓒ Ⓓ Ⓔ	54. Ⓐ Ⓑ Ⓒ Ⓓ Ⓔ	74. Ⓐ Ⓑ Ⓒ Ⓓ Ⓔ
15. Ⓐ Ⓑ Ⓒ Ⓓ Ⓔ	35. Ⓐ Ⓑ Ⓒ Ⓓ Ⓔ	55. Ⓐ Ⓑ Ⓒ Ⓓ Ⓔ	75. Ⓐ Ⓑ Ⓒ Ⓓ Ⓔ
16. Ⓐ Ⓑ Ⓒ Ⓓ Ⓔ	36. Ⓐ Ⓑ Ⓒ Ⓓ Ⓔ	56. Ⓐ Ⓑ Ⓒ Ⓓ Ⓔ	76. Ⓐ Ⓑ Ⓒ Ⓓ Ⓔ
17. Ⓐ Ⓑ Ⓒ Ⓓ Ⓔ	37. Ⓐ Ⓑ Ⓒ Ⓓ Ⓔ	57. Ⓐ Ⓑ Ⓒ Ⓓ Ⓔ	77. Ⓐ Ⓑ Ⓒ Ⓓ Ⓔ
18. Ⓐ Ⓑ Ⓒ Ⓓ Ⓔ	38. Ⓐ Ⓑ Ⓒ Ⓓ Ⓔ	58. Ⓐ Ⓑ Ⓒ Ⓓ Ⓔ	78. Ⓐ Ⓑ Ⓒ Ⓓ Ⓔ
19. Ⓐ Ⓑ Ⓒ Ⓓ Ⓔ	39. Ⓐ Ⓑ Ⓒ Ⓓ Ⓔ	59. Ⓐ Ⓑ Ⓒ Ⓓ Ⓔ	79. Ⓐ Ⓑ Ⓒ Ⓓ Ⓔ
20. Ⓐ Ⓑ Ⓒ Ⓓ Ⓔ	40. Ⓐ Ⓑ Ⓒ Ⓓ Ⓔ	60. Ⓐ Ⓑ Ⓒ Ⓓ Ⓔ	80. Ⓐ Ⓑ Ⓒ Ⓓ Ⓔ

Chemistry   *Fill in oval CE only if II is correct explanation of I.

	I	II	CE*		I	II	CE*
101.	Ⓣ Ⓕ	Ⓣ Ⓕ	◯	109.	Ⓣ Ⓕ	Ⓣ Ⓕ	◯
102.	Ⓣ Ⓕ	Ⓣ Ⓕ	◯	110.	Ⓣ Ⓕ	Ⓣ Ⓕ	◯
103.	Ⓣ Ⓕ	Ⓣ Ⓕ	◯	111.	Ⓣ Ⓕ	Ⓣ Ⓕ	◯
104.	Ⓣ Ⓕ	Ⓣ Ⓕ	◯	112.	Ⓣ Ⓕ	Ⓣ Ⓕ	◯
105.	Ⓣ Ⓕ	Ⓣ Ⓕ	◯	113.	Ⓣ Ⓕ	Ⓣ Ⓕ	◯
106.	Ⓣ Ⓕ	Ⓣ Ⓕ	◯	114.	Ⓣ Ⓕ	Ⓣ Ⓕ	◯
107.	Ⓣ Ⓕ	Ⓣ Ⓕ	◯	115.	Ⓣ Ⓕ	Ⓣ Ⓕ	◯
108.	Ⓣ Ⓕ	Ⓣ Ⓕ	◯				

DIAGNOSTIC TEST

Time: 60 minutes

Part A

<u>Note:</u> Unless otherwise stated, for all statements involving chemical equations and/or solutions, assume that the system is in pure water.

<u>Directions:</u> Each of the following sets of lettered choices refers to the numbered formulas or statements immediately below it. For each numbered item, choose the one lettered choice that fits it best. Then fill in the corresponding oval on the answer sheet. Each choice in a set may be used once, more than once, or not at all.

<u>Questions 1–4</u>

(A) $CH_3CH_2CH_2OH$

(B) $CH_3CH(Br)CH(Br)CH_3$

(C) $CH_3CH_2CH(Br)CH_3$

(D) $CH_3CH_2CH_2COOH$

(E) $CH_3CH_2CH_2CH_3$

$CH_3CH=CHCH_3 + H_2 \rightarrow (1) + Br_2 \rightarrow (2) + HBr$

$+ Br_2$

(3)

4. Which compound would most likely turn litmus paper to a red color?

<u>Questions 5–8</u>

(A) Heisenberg Uncertainty Principle

(B) Pauli Exclusion Principle

(C) Schrödinger Wave Equation

(D) Hund's rule

(E) Bohr model of the hydrogen atom

5. No two electrons can have the same quantum number because they must have opposite spins.

6. We cannot know the exact location of an electron in space.

7. The electrons will occupy an orbital singly, with parallel spins, before pairing up.

8. The energy changes that an electron may undergo are quantized.

<u>Questions 9–12</u>

(A) H_2

(B) CO_2

(C) H_2O

(D) NaCl

(E) CH_2CH_2

9. Contains just one sigma bond.

10. Has a bond formed from the transfer of electrons.

11. Has an atom that is sp hybridized.

12. Is a polar molecule.

<u>Questions 13–16</u>

(A) F

(B) Li

(C) Fe

(D) He

(E) Si

13. Shows both the properties of both metals and non-metals.

14. Has the greatest ionization energy.

15. Has the greatest electronegativity.

16. Has colored salts that will produce colored aqueous solutions.

GO ON TO THE NEXT PAGE

(A) $NaC_2H_3O_2$

(B) $HC_2H_3O_2$

(C) KCl

(D) NH_3

(E) HCl

a salt that will undergo hydrolysis to form a basic olution.

Will form a coordinate covalent bond with a hydro-nium ion.

9. Is a strong acid.

Questions 20–22

(A) $q = mc\Delta T$

(B) $q = H_v m$

(C) $P_1V_1 = P_2V_2$

(D) $D = m / V$

(E) $K = C + 273$

20. Can be used to find the mass of an irregularly shaped solid.

21. Boyle's Law.

22. Used to find energy gained or lost during a particular phase change.

Questions 23–25

(A) Alpha particle

(B) Beta particle

(C) Neutron

(D) Gamma ray

(E) Positron

23. Has the greatest mass.

24. Has the greatest positive charge.

25. Has the same mass and charge as an electron.

GO ON TO THE NEXT PAGE

DIAGNOSTIC TEST—*Continued*

Part B

Directions: Each question below consists of two statements. For each question, determine whether statement I in the left most column is true or false and whether statement II in the rightmost column is true or false. Fill in the corresponding ⸍ or F ovals on the answer sheet provided. Fill in the oval labeled "CE" only if statement II correctly explains statement I.

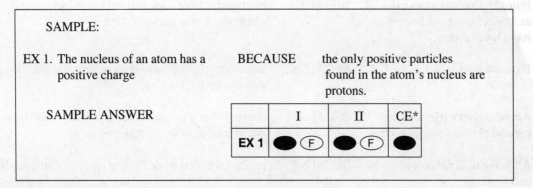

SAMPLE:

EX 1. The nucleus of an atom has a positive charge   BECAUSE   the only positive particles found in the atom's nucleus are protons.

SAMPLE ANSWER

	I	II	CE*
EX 1	● Ⓕ	● Ⓕ	●

	I		II
101.	^{12}C is an isotope of ^{14}C	BECAUSE	the nuclei of both atoms have the same number of neutrons.
102.	Ne is an inert gas	BECAUSE	Ne has a complete octet in its valence shell.
103.	A solution with a pH of 5 is less acidic than a solution with a pH of 8	BECAUSE	a solution with a pH of 5 has 1,000 times more hydronium ions than the solution with a pH of 8.
104.	A reaction with a positive ΔH is considered to be exothermic	BECAUSE	an exothermic reaction has more heat released than absorbed.
105.	A voltaic cell spontaneously converts chemical energy into electrical energy	BECAUSE	a voltaic cell needs an externally applied current to work.
106.	K is considered to be a metal	BECAUSE	when K becomes an ion its atomic radius increases.
107.	At equilibrium the concentration of reactants and products remain constant	BECAUSE	at equilibrium the rates of the forward and reverse reactions are equal.
108.	Powdered zinc will react faster with HCl than one larger piece of zinc of the same mass	BECAUSE	powdered zinc has less surface area than one larger piece of zinc of the same mass.

GO ON TO THE NEXT PAGE ➤

	pound with the mula C_4H_{10} can compounds	BECAUSE	n-Butane and 2-methylpropane are isomers that have the molecular formula of C_4H_{10}.
	2.4 liters of He will same volume as one H_2 (assume ideal gases)	BECAUSE	one mole or 22.4 liters of any gas at STP will have the same mass.
	gen molecules can exist olids, liquids, or gases at om temperature	BECAUSE	as nonpolar molecules are considered by increasing mass the dispersion forces between them increases.
	Hydrocarbons will dissolve in water	BECAUSE	substances that have the same polarity are miscible and can dissolve each other.
	Ammonia has a trigonal pyramidal molecular geometry	BECAUSE	ammonia has a tetrahedral electron pair geometry with three atoms bonded to the central atom.
14.	$AlCl_3$ is called aluminum trichloride	BECAUSE	prefixes are used when naming covalent compounds.
115.	When a Li atom reacts and becomes an ion, the Li atom can be considered to be a reducing agent	BECAUSE	the Li atom lost an electron and was oxidized.

GO ON TO THE NEXT PAGE

DIAGNOSTIC TEST—*Continued*

Part C

Directions: Each of the multiple-choice questions or incomplete sentences below is followed by five answers or completions. Select the one answer that is best in each case and then fill in the corresponding oval on the answer sheet provided.

Question 26

26. 117 grams of NaCl are dissolved in water to make 500 mL of solution. Water is then added to this solution to make a total of one liter of solution. The final molarity of this solution will be

   (A) 4 M
   (B) 2 M
   (C) 1 M
   (D) 0.5 M
   (E) 0.585 M

Questions 27–28 refer to a molecule of 2-butyne, CH_3–C≡C–CH_3.

27. How many pi bonds can be found in this molecule?

   (A) 1
   (B) 2
   (C) 4
   (D) 6
   (E) 10

28. How many atoms lie in a straight line in this molecule?

   (A) 10
   (B) 8
   (C) 6
   (D) 4
   (E) 2

29. A solution of a weak acid, HA, has a concentration of 0.100 M. What is the concentration of hydronium ion and the pH of this solution if the K_a value for this acid is 1.0×10^{-5}?

   (A) 1.0×10^{-3} M and pH = 11
   (B) 1.0×10^{-6} M and pH = 6
   (C) 1.0×10^{-4} M and pH = 8
   (D) 3.0×10^{-4} M and pH = 4
   (E) 1.0×10^{-3} M and pH = 3

30. Given the reaction at STP: Mg(s) + 2HCl(aq) → $MgCl_2$(aq) + H_2(g)

   At STP, how many liters of H_2(g) can be produced from the reaction of 12.15 grams of Mg with excess HCl(aq)?

   (A) 2.0 liters
   (B) 4.0 liters
   (C) 11.2 liters
   (D) 22.4 liters
   (E) 44.8 liters

31. A student performed a titration using 2.00 M HCl to completely titrate 40.00 mL of 1.00 M NaOH. If the initial reading on the buret containing HCl was 2.05 mL, what will the final reading of this buret be?

   (A) 82.05 mL
   (B) 42.05 mL
   (C) 20.00 mL
   (D) 10.00 mL
   (E) 22.05 mL

32. Which of the following was not a conclusion of Rutherford's gold foil experiment?

   (A) The atom is mainly empty space.
   (B) The nucleus has a negative charge.
   (C) The atom has a dense nucleus.
   (D) Alpha particles can pass through a thin sheet of gold foil.
   (E) All of the above are correct regarding the gold foil experiment.

33. In a reaction the potential energy of the reactants is 40 kJ/mol, the potential energy of the products is 10 kJ/mol and the potential energy of the activated complex is 55 kJ/mol. What is the activation energy for the reverse reaction?

   (A) 45 kJ/mol
   (B) −30 kJ/mol
   (C) 15 kJ/mol
   (D) 35 kJ/mol
   (E) −55 kJ/mol

GO ON TO THE NEXT PAGE

34. Which reactions would form at least one solid precipitate as a product? Assume aqueous reactants.

     I. $AgNO_3 + NaCl \rightarrow NaNO_3 + AgCl$
     II. $Pb(NO_3)_2 + 2KI \rightarrow PbI_2 + 2KNO_3$
     III. $2NaOH + H_2SO_4 \rightarrow Na_2SO_4 + 2H_2O$

     (A) I only
     (B) II only
     (C) III only
     (D) I and II only
     (E) II and III only

Questions 35 and 36 refer to the following reaction at equilibrium:

$$2A(aq) + B(aq) \longleftrightarrow 3C(aq) + D(s)$$

35. The mass action equation is written as

     (A) $K_{eq} = \dfrac{[C]^3}{[A]^2[B]}$

     (B) $K_{eq} = \dfrac{[C]^3 + [D]}{[A]^2 + [B]}$

     (C) $K_{eq} = \dfrac{[A]^2 + [B]}{[C]^3}$

     (D) $K_{eq} = \dfrac{[A]^2 - [B]}{[C]^3 - [D]}$

     (E) $K_{eq} = \dfrac{[C]}{[A] + [B]}$

36. If the equilibrium constant for the reverse reaction is 9.0×10^{-4}, what is the equilibrium constant for the forward reaction?

     (A) 3.0×10^{-2}
     (B) -3.0×10^{-2}
     (C) -9.0×10^{-2}
     (D) $1 / 9.0 \times 10^{-4}$
     (E) $1 / -9.0 \times 10^{-2}$

Questions 37–47

37. A compound's composition by mass is 50% S and 50% O. What is the empirical formula of this compound?

     (A) SO
     (B) SO_2
     (C) S_2O
     (D) S_2O_3
     (E) S_3O_4

38. What percentage of the total mass of $KHCO_3$ is made up by nonmetallic elements?

     (A) 17%
     (B) 83%
     (C) 61%
     (D) 20%
     (E) 50%

39. Which aqueous solution is expected to have the highest boiling point?

     (A) $0.2\ m\ CaCl_2$
     (B) $0.2\ m\ NaCl$
     (C) $0.1\ m\ AlCl_3$
     (D) $0.2\ m\ CH_3OH$
     (E) $0.2\ m\ NaC_2H_3O_2$

40. Which of the following solids are known to undergo sublimation?

     I. CO_2
     II. I_2
     III. Naphthalene

     (A) I only
     (B) II only
     (C) I and II only
     (D) II and III only
     (E) I, II, and III

41. Which one of the following demonstrates a decrease in entropy?

     (A) Dissolving a solid into solution
     (B) An expanding universe
     (C) Burning a log in a fireplace
     (D) Raking up leaves into a trash bag
     (E) Spilling a glass of water

42. Which of the following substances is/are liquid(s) at room temperature?

     I. Hg
     II. Br_2
     III. Si

     (A) I only
     (B) II only
     (C) I and II only
     (D) II and III only
     (E) I, II, and III

GO ON TO THE NEXT PAGE

43. Which of the following would be considered to be unsafe in a laboratory setting?
    (A) Using a test tube holder to handle a hot test tube
    (B) Tying one's long hair back before experimenting
    (C) Wearing open-toe shoes
    (D) Pouring liquids while holding the reagent bottles over a sink
    (E) Working under a fume hood

44. A sample of a gas at STP contains 3.01×10^{23} molecules and has a mass of 20.0 grams. This gas would
    (A) have a molar mass of 20.0 grams/mole and occupy 11.2 liters
    (B) occupy 22.4 liters and have a molar mass of 30.0 grams/mol
    (C) occupy 22.4 liters and have a molar mass of 20.0 grams/mol
    (D) have a molar mass of 40.0 grams/mole and occupy 33.6 liters
    (E) have a molar mass of 40.0 grams/mole and occupy 11.2 liters

45. Given the reaction: $Ca(s) + Cl_2(g) \rightarrow CaCl_2(s)$
    When 80 grams of Ca (molar mass is 40 grams) is reacted with 213 grams of Cl_2 (molar mass is 71) one will have
    (A) 40 grams of Ca in excess
    (B) 71 grams of Cl_2 in excess
    (C) 293 grams of $CaCl_2$ formed
    (D) 133 grams of $CaCl_2$ formed
    (E) 113 grams of $CaCl_2$ formed

46. A student performed an experiment to determine the solubility of a salt at various temperatures. The data from the experiment can be seen below:

Trial	Temperature in °C	Solubility in 100 grams of water
1	20	44
2	30	58
3	40	67
4	50	62
5	60	84

Which trial seems to be in error?
    (A) 1
    (B) 2
    (C) 3
    (D) 4
    (E) 5

47. Given the following reaction at equilibrium: $3H_2(g) + N_2(g) \longleftrightarrow 2NH_3(g)$ + heat energy

    Which of the following conditions would shift the equilibrium of this reaction so that the formation of ammonia is favored?
    (A) Increasing the pressure on the reaction
    (B) Heating the reaction
    (C) Removing hydrogen gas from the reaction
    (D) Adding more ammonia to the reaction
    (E) Removing nitrogen gas from the reaction

Questions 48–49 refer to these gases:

H_2 Ne   Kr H_2S Cl_2 F_2 Ar

48. Given equal conditions, which gas from above is expected to have the greatest density?
    (A) H_2
    (B) Ne
    (C) Ar
    (D) H_2S
    (E) Cl_2

49. Given equal conditions, which gas from above is expected to have to have the greatest rate of effusion?
    (A) H_2
    (B) Ar
    (C) Kr
    (D) F_2
    (E) Cl_2

Questions 50–70

50. Ideal gases
    (A) have forces of attraction between them
    (B) are always linear in shape
    (C) never travel with a straight line motion
    (D) have molecules that are close together
    (E) have low masses and are spread far apart

GO ON TO THE NEXT PAGE

51. Which substance will combine with oxygen gas to produce a greenhouse gas?

    (A) Na
    (B) Si
    (C) H_2
    (D) Ne
    (E) C

52. Which general formula below represents that of an organic ester?

    (A) R—OH
    (B) R—COOH
    (C) R—O—R
    (D) R—COO—R
    (E) R—CO—R

53. When an alkaline earth metal, M, reacts with oxygen the formula of the compound produced will be

    (A) M_2O
    (B) MO
    (C) M_2O_3
    (D) MO_2
    (E) M_3O_4

54. A catalyst can change the

    (A) heat of reaction and the potential energy of the reactants
    (B) heat of reaction and the time it takes the reaction to proceed
    (C) activation energy of the reverse reaction and the potential energy of the activated complex
    (D) potential energy of the reactants and the time it takes the reaction to proceed
    (E) activation energy of the forward reaction and the potential energy of the products

55. A neutral atom has a total of 17 electrons. The electron configuration in the outermost principal energy level will look closest to

    (A) $1s^2 2s^2 2p^5$
    (B) $3s^5 3p^2$
    (C) $s^2 p^5$
    (D) $s^2 p^8 d^7$
    (E) sp^7

56. Given a 22.4-liter sample of helium gas is at STP. If the temperature is increased by 15 degrees Celsius and the pressure changed to 600 torr, what would the new volume of the gas sample be?

    (A) $\dfrac{(760)(22.4)(15)}{(273)(600)}$

    (B) $\dfrac{(273)(600)(288)}{(760)(22.4)}$

    (C) $\dfrac{(760)(22.4)(15)}{(600)}$

    (D) $\dfrac{(760)(22.4)(288)}{(273)(600)}$

    (E) $\dfrac{(273)(600)}{(760)(22.4)(288)}$

57. Which of the following are correct about the sub-atomic particles found in $^{37}Cl^{1-}$?

    I. 21 neutrons
    II. 17 protons
    III. 16 electrons

    (A) II only
    (B) III only
    (C) I and II only
    (D) I and III only
    (E) II and III only

58. A hydrated blue copper(II) sulfate salt with a formula of $YCuSO_4 \cdot XH_2O$ is heated until it is completely white in color. The student who performed the dehydration of this salt took note of the mass of the sample before and after heating and recorded it as follows:

    Mass of the hydrated salt = 500 grams

    Mass of the dehydrated salt = 320 grams

    What is the value of "X" in the formula of the hydrated salt?

    (A) 1
    (B) 2
    (C) 4
    (D) 5
    (E) 10

GO ON TO THE NEXT PAGE

59. Which of the following oxides can dissolve in water to form a solution that would turn litmus indicator red in color?

    (A) MgO
    (B) K_2O
    (C) CO_2
    (D) ZnO
    (E) H_2O

60. The process in which water vapor changes phase to become a liquid is called

    (A) deposition
    (B) sublimation
    (C) vaporization
    (D) fusion
    (E) condensation

61. What is the value for ΔH for the reaction: $N_2O_4 \rightarrow 2NO_2$?

    $(2NO_2 \rightarrow N_2 + 2O_2 \quad \Delta H = -16.2 \text{ kcal})$

    $(N_2 + 2O_2 \rightarrow N_2O_4 \quad \Delta H = +2.31 \text{ kcal})$

    (A) + 13.89 kcal
    (B) + 18.51 kcal
    (C) + 37.42 kcal
    (D) − 13.89 kcal
    (E) − 18.51 kcal

62. A liquid will boil when

    (A) enough salt has been added to it
    (B) the vapor pressure of the liquid is equal to the atmospheric or surrounding pressure
    (C) the vapor pressure of the liquid reaches 760 mm Hg
    (D) conditions favor the liquid's molecules to be closer together
    (E) it has been brought up to a higher elevation

63. A conductivity experiment is set up with a light bulb and five beakers of 0.1 M solutions of the substances below. Which solution would allow the bulb to glow the brightest?

    (A) $C_6H_{12}O_6$
    (B) HCl
    (C) SiO_2
    (D) $HC_2H_3O_2$
    (E) CH_3OH

64. Which of the following represents a correctly balanced half reaction?

    (A) $Cl_2 + 2e^- \rightarrow Cl^{1-}$
    (B) $2e^- + Fe \rightarrow Fe^{2+}$
    (C) $O_2 \rightarrow 2e^- + 2O^{2-}$
    (D) $Al^{3+} \rightarrow Al + 3e^-$
    (E) $2H^+ + 2e^- \rightarrow H_2$

65. A student prepares for an experiment involving a voltaic cell. Which of the following is needed the least to perform the experiment?

    (A) Buret
    (B) Salt bridge
    (C) Strip of zinc metal
    (D) Copper wire
    (E) Solution of zinc sulfate

66. When the equation: ___ C_3H_8 + ___ O_2 → ___ CO_2 + ___ H_2O is balanced using the lowest whole number coefficients, the coefficient before the O_2 will be

    (A) 1
    (B) 2.5
    (C) 5
    (D) 10
    (E) 13

67. Which nuclear equation below demonstrates beta decay?

    (A) $^{238}U \rightarrow {}^{234}Th + X$
    (B) $^1H + X \rightarrow {}^3H$
    (C) $^{14}N + X \rightarrow {}^{17}O + {}^1H$
    (D) $^{234}Pa \rightarrow {}^{234}U + X$
    (E) None of the above demonstrates beta decay.

68. Which of these processes could be associated with the following reaction: $2H_2O \rightarrow 2H_2 + O_2$?

    I.   Electrolysis
    II. Neutralization
    III. Decomposition

    (A) I only
    (B) III only
    (C) I and III only
    (D) I and II only
    (E) II and III only

GO ON TO THE NEXT PAGE ➤

69. The following reaction occurs in a beaker: $Ag^{1+}(aq)$ + $Cl^{1-}(aq) \longleftrightarrow AgCl(s)$. If a solution of sodium chloride were added to this beaker

    (A) the solubility of the sodium chloride would decrease
    (B) the reaction would shift to the left
    (C) the concentration of silver ions in solution would increase
    (D) the solubility of the silver chloride would decrease
    (E) the equilibrium would not shift at all

70. How many atoms are represented in the equation $Pb(NO_3)_2 + 2KI \rightarrow PbI_2 + 2KNO_3$?

    (A) 5
    (B) 12
    (C) 13
    (D) 18
    (E) 26

STOP

IF YOU FINISH BEFORE TIME IS CALLED, GO BACK AND CHECK YOUR WORK.

ANSWERS AND EXPLANATIONS TO DIAGNOSTIC TEST

1. E This reaction shows the addition of hydrogen gas to an alkene. In this case the product will be the alkane named butane.

2. C This reaction is called a substitution reaction. A hydrogen atom has been lost from the alkane and a halogen has been added. A hydrogen halide is also a byproduct.

3. B This, again, demonstrates the addition of a diatomic molecule to an alkene. The bromine atoms, just like the hydrogen atoms, will be added to the carbon atoms that had the double bond.

4. D Litmus will become red in color when mixed with acids. The organic acid is characterized by the group R—COOH.

5. B The Pauli Exclusion Principle states that no two electrons can have the same four quantum numbers. This makes it impossible for two electrons in the same orbital to have the same spin.

6. A The uncertainty principle states that it is impossible to know simultaneously the momentum of an electron and its location in space.

7. D Hund's rule states that the electrons fill orbitals so that there is the maximum number of parallel spins. For example, this is why the valence electrons for a nitrogen atom are configured

$$\underset{2s}{\uparrow\downarrow} \; \underset{2p_x}{\uparrow} \; \underset{2p_y}{\uparrow} \; \underset{2p_z}{\uparrow} \quad \text{and not} \quad \underset{2s}{\uparrow\downarrow} \; \underset{2p_x}{\uparrow\downarrow} \; \underset{2p_y}{\uparrow\downarrow} \; \underset{2p_z}{\uparrow}$$

8. E In the Bohr model of the atom an electron may orbit the atom's nucleus only at certain radii corresponding to certain energies. He called this the quantum theory of the atom.

9. A A sigma bond is a single bond that is formed from the overlap of s orbitals. A diatomic hydrogen molecule falls under this category.

10. D Ionic bonds are formed when a metal loses electrons to a nonmetal.

11. B sp hybridization can be found at an atom that has formed a triple bond with a single bond or that has formed two double bonds. There are two double bonds found around the carbon atom in carbon dioxide.

12. C Water, because of its shape (bent) and polar bonds, has a dipole moment and is a polar molecule.

13. E Semimetals (or metalloids) show the properties of both metals and nonmetals.

14. D Noble (or inert) gases are very stable because of their complete outermost principal energy levels. It takes a greater amount of energy to remove an electron from these stable elements.

15. A Because of its size and nuclear charge, fluorine has a great electronegativity.

16.   C   Transition elements are famous for their multiple oxidation states and colors. Compounds containing iron are usually reddish in color (hemoglobin and rust are examples).

17.   A   Hydrolysis is the "opposite" of neutralization. The acid and base that would need to react to form sodium acetate are $NaOH$ and $HC_2H_3O_2$. Sodium hydroxide is a strong base and acetic acid is a weak acid and their combination results in a basic solution.

18.   D   Ammonia has a free pair of electrons that can be shared with a hydronium ion to form a bond. When one atom donates both electrons to a bond, a coordinate covalent bond is formed.

19.   E   HCl ionizes completely and releases all of its hydronium ions. These hydronium ions are what contribute to the acidity of a solution.

20.   D   By placing an irregularly shaped solid into a graduated cylinder with a known volume of water, the water is displaced by a certain volume. If the density of the solid is known, then the mass of the object can be calculated. If the mass of the solid is known, then the density of the object can be calculated.

21.   C   Boyle's Law describes the inverse relationship between pressure and volume.

22.   B   The energy required to vaporize a liquid already at its boiling point can be calculated by multiplying the mass of the liquid by its heat of vaporization.

23.   A   Alpha particles have a mass of four atomic mass units because they contain two protons and two neutrons.

24.   A   While both the positron and the alpha particle have positive charges, the charge of the positron is 1+ while the charge of the alpha particle is 2+.

25.   B   A beta particle has essentially no mass and a 1– charge, the same as an electron.

101.   T, F   Isotopes are the same element with a different mass number because of their different number of neutrons.

102.   T, T, CE The Noble or inert gases have a complete outermost principal energy level. As a result, they are less inclined to react with other substances.

103.   F, T   A pH of 5 is more acid than a pH of 8. Each number on the pH scale indicates a 10-fold difference in the number of hydronium ions present. Three pH units indicates a 1,000-fold difference in acidity.

104.   F, T   Endothermic reactions have a negative value for the heat of reaction. Exothermic reactions will release more heat than they absorb.

105.   T, F   A voltaic cell will generate its own electrical current without using any outside current. Electrolytic cells are not spontaneous and require an external current.

106.   T, F   Metals tend to lose electrons and form ions that have a smaller ionic radius.

107.   T, T, CE   When the forward and reverse reaction rates are equal there will be no change in the concentrations or quantities of the reactants or products.

108.   T, F   Powdered substances have more surface area than larger pieces of the same substance. This will allow them to react faster.

109.   T, T, CE   Isomers are compounds with the same molecular formula but different structures. The compounds presented in this problem are the two isomers that can exist for C_4H_{10}.

110.   T, F   One mole of an ideal gas at STP will have the same number of molecules present and occupy the same volume. Because different gases have different molar volumes, the masses of the sample will be different.

111.   T, T, CE   Fluorine and chlorine exist as gases while bromine is a liquid and iodine is a solid at room temperatures. These differences in phases can be explained by the increase in dispersion (or Van der Waals) forces as these nonpolar molecules become greater in mass.

112.   F, T   Substances will dissolve in one another if they have the same polarity. Hydrocarbons are generally nonpolar and they will not dissolve in water which is polar.

113.   T, T, CE   Ammonia has a tetrahedral electron pair geometry. When three of the four electron pairs around the central atom are bonded to three other atoms, the resulting shape of the molecule will be trigonal pyramidal.

114.   F, T   The correct name of this compound is aluminum chloride. Prefixes are not used when naming ionic compounds. Prefixes are used when naming covalent compounds.

115.   T, T, CE   Lithium is a metal and will lose electrons, undergoing the process of oxidation. Any substance that has been oxidized can be considered to be a reducing agent.

26.   B   117 grams of NaCl divided by its molar mass of 58.5 gives us two moles of NaCl. The molarity is calculated by dividing two moles by 0.500 L ($M = $ moles/liters) which gives a concentration of four molar. Using the equation $M_1V_1 = M_2V_2$, if the volume of the solution is doubled the concentration of the solution will be halved. This gives a final answer of 2 M NaCl(aq).

27.   B   In this molecule there is one triple bond. The first bond of the triple bond is a sigma bond, the other two bonds of the triple bond are called pi bonds. All the other bonds in the molecule are single bonds and are called sigma bonds.

28. D   There are two sp hybridized carbon atoms in this molecule. This type of hybridization gives rise to a linear shape in the molecule, which includes the atoms to the left and right (labeled L and R below) of the sp hybridized carbon atoms. Therefore, the four carbon atoms in this molecule will lie in a straight line.

$$\begin{array}{ccc} L & sp & R \\ C-C & \equiv & C-C \\ L & sp & R \end{array}$$

29. E   Start with the equation, $K_a = \dfrac{[H^+][A^-]}{[HA]}$

Substitute the numbers from the problem

$$1.0 \times 10^{-5} = \frac{[H^+][A^-]}{[0.100\ M]}$$

and $[H^+][A^-] = 1.0 \times 10^{-6}$. Because the amount of $[H^+]$ is the same as $[A^-]$, you have the same values multiplying each other. Now you can write, $x^2 = 1.0 \times 10^{-6}$. Solving you get x (or $[H^+]$) = $1.0 \times 10^{-3}\ M$. This concentration of hydronium ion will yield a pH of 3 when you use the equation pH = $-\log[H^+]$.

30. C   12.15 grams of Mg is 0.5 moles of Mg (molar mass is 23). 0.5 moles of Mg will give 0.5 mole of hydrogen gas because the magnesium and hydrogen gas are used and formed in a 1:1 ratio. 0.5 moles of hydrogen gas (or any gas at STP) will occupy 11.2 L.

31. E   To calculate molarities and volumes used during a titration use the formula $M_aV_a = M_bV_b$. Substituting you find $(2.00\ M)(V_a) = (1.00\ M)(40.00\ mL)$. The volume of HCl used is 20.00 mL. Because of the way a buret is set up and used, if the initial reading of the meniscus was 2.05 mL and 20.00 mL of HCl were used, the final buret reading will be 22.05 mL.

32. B   The nucleus of an atom does not have a negative charge because of the protons and neutrons that are located there.

33. A   To find the activation energy for a reaction first find the potential energy (PE) of the activated complex (55 kJ/mol). Then the PE of either the reactants or products is subtracted from the PE of the activated complex. Because you are looking at the reverse reaction in this case, subtract the PE of the products. This gives an answer of 45 kJ/mol.

34. D   All chlorides and iodides are soluble except for those chemically combined with silver, mercury, and lead. These exceptions will form precipitates. Salts containing sodium or nitrate ion will be soluble in water and not form precipitates.

35.   A   When writing a mass action equation, remember the rule, "Products over reactants, coefficients become powers." Be sure that the equation demonstrates multiplication and division of the concentrations. Finally, never include the concentrations of solids or liquids in the equation. The concentrations are substituted with the value of "1."

36.   D   In a reverse reaction the products are now the reactants and the reactants are now the products. This would reciprocate the value mass action equation. This is demonstrated by $1/9.0 \times 10^{-4}$.

37.   B   Because this compound is composed of 50% S and 50% O by mass you might rush to give SO as the answer. But, because oxygen has half the mass of sulfur, twice as many oxygen atoms will be in the empirical formula. This leads to an answer of SO_2.

38.   C   The total molar mass of this compound is 100. The nonmetals, one hydrogen atom, one carbon atom, and three oxygen atoms have a total mass of 61. Therefore, the nonmetals make up 61% of the total mass.

39.   A   Boiling and freezing point are affected most by a solution that contains the greatest number of particles. Because calcium chloride yields three ions, a 0.2 molal solution will, in effect, act as if it were 0.6 molal. This is greater than any of the other choices.

40.   E   Dry ice (solid carbon dioxide), iodine crystals, and naphthalene (mothballs) all sublime and change from solid to gas phase without any apparent liquid phase.

41.   D   Entropy is disorder, chaos, or randomness. Raking up leaves is a more orderly process.

42.   C   Bromine is a brown-orange liquid at room temperature. Mercury, because it is a liquid, is able to be used in a thermometer and expand and contract with changes in temperature.

43.   C   Spills can flow off table tops in the laboratory. Open-toe shoes leave toes exposed to these spills.

44.   E 3.01×10^{23} molecules is 0.5 moles of a substance. Because you are dealing with a gas at STP, the volume will be 11.2 liters, leaving choices A and E. If 0.5 mole of the substance has a mass of 20.0 grams then the mass of one mole (the molar mass) will be 40.0 grams.

45.   B   80 grams of Ca is equivalent to two moles of Ca. Two moles of Ca will react with two moles of Cl_2 because the mole ratio from the balanced equation is 1:1. Two moles of Cl_2 have a mass of 142 grams. This leaves 71 grams of the Cl_2 gas in excess.

46.   D   As the temperature of a solute increases, the solubility of the salts dissolved increases as well. Trial 4 shows a decrease in solubility from that of trial 3.

47.   A   An increase in pressure favors a smaller volume (Boyle's Law). On the left side of the equation there are four "volumes" of hydrogen and nitrogen gas (add the coefficients). On the right side there are two "volumes" of ammonia gas. An increase in pressure will shift the equilibrium to the side with fewer "volumes" and favor the formation of ammonia.

48.   E   Gases with greater masses will have greater densities. Cl_2 gas has the greatest molar mass of these choices. You could also calculate the density of the gases by using the equation: D_{gas} = molar mass / 22.4 liters.

49.   A   Graham's Law of Effusion states that at the same temperature and pressure, gases diffuse at a rate inversely proportional to the square roots of their molecular masses. What this translates to is that lighter (less dense) gases travel faster than heavier (more dense) gases.

50.   E   Gases behave ideally when they have low masses and are spread out as far as possible. This is best achieved at low pressures and high temperatures.

51.   E   Carbon and oxygen can react to form carbon dioxide, a greenhouse gas.

52.   D   The functional groups given as choices are alcohol, carboxylic acid, ether, ester, and ketone, respectively.

53.   B   Alkaline earth metals are in group 2 of the periodic table. They will form ions with a charge of 2+. Oxides will have a charge of 2−. This means that alkaline earth metals will react with oxygen in a 1:1 ratio. An example is CaO.

54.   C   Catalysts will speed up a reaction by lowering the PE of the activated complex and, therefore, lower the activation energy of the forward and reverse reactions.

55.   C   The atom in question is an atom of chlorine. Seventeen electrons will fill with a configuration of $1s^2 2s^2 2p^6 3s^2 3p^5$.

56.   D   This question involves many factors and needs to be done in steps. The experiment starts at STP, meaning that P_1 = 760 torr and T_1 = 273 K. The initial volume is 22.4 liters, so V_1 = 22.4 L. The final pressure, P_2 is 600 torr. The final temperature is 15 degrees Celsius above 273 K. A rise of 15 degrees Celsius is an equal rise in the Kelvin temperature (K = C + 273). So the final Kelvin temperature, T_2 = 288 K. Be sure to use the Kelvin scale and not the Celsius scale. Next use the combined gas law: $\dfrac{P_1 V_1}{T_1} = \dfrac{P_2 V_2}{T_2}$

Substitution gives: $\dfrac{(760\ torr)(22.4\ L)}{(273\ K)} = \dfrac{(600\ torr)(V_2)}{(288\ K)}$

Cross-multiplication and division to solve for the new volume gives: $\dfrac{(760)(22.4)(288)}{(273)(600)}$

While the question did not ask you to solve for the value of the final volume, you can predict that it is greater than 22.4 L because the temperature increased (volume increases) and the pressure decreased (volume increases again).

57.   A   This isotope of chlorine ion has 17 protons (atomic number), 20 neutrons (mass number minus atomic number), and 18 electrons (gained one electron).

58.   D   180 grams of water were lost due to the dehydration of the salt. This means 10 moles of water were lost; 320 grams of $CuSO_4$, or 2 moles of the salt remained. According to these calculations the formula should be $2CuSO_4 \bullet 10H_2O$. Because ionic compounds are written as empirical formulas, the lowest ratio is $CuSO_4 \bullet 5H_2O$ and the coefficient in front of the water has a value of 5.

59.   C   Litmus will turn red in acidic solutions (blue in basic solutions). Nonmetal oxides, such as carbon dioxide, will dissolve in solution to form acidic solutions.

60.   E   Condensation is the changing from gas phase to a liquid phase.

61.   A   Start by reversing the first reaction to put the NO_2 on the right:

$$N_2 + 2O_2 \rightarrow 2NO_2 \qquad \Delta H = +16.2 \text{ kcal}$$

Next reverse the second equation to get the N_2O_4 on the left:

$$N_2O_4 \rightarrow N_2 + 2O_2 \qquad \Delta H = -2.31 \text{ kcal}$$

Add up the heats of reaction: +13.89 kcal

62.   B   Boiling points can change with changes in atmospheric pressure (or surrounding pressure). When the vapor pressure of the liquid is equal to the atmospheric pressure or surrounding pressure, the liquid will boil.

63.   B   Strong acids are also strong electrolytes and will be better conductors of electricity. HCl is a very strong acid. Acetic acid is a weak acid and will allow the bulb to glow dimly. Glucose and methanol are covalently bonded compounds and will not form ions to carry an electrical current. Silicon dioxide is sand and will not dissolve in water to form ions that can carry a current.

64.   E   Half reactions must show conservation of mass and charge. The hydrogen half reaction shows two atoms of hydrogen entering and leaving the equation. It also shows a total charge of zero entering and leaving the reaction as well.

65.   A   Burets are excellent for performing titrations. They are generally not used in experiments involving voltaic cells.

66.   C   First balance the number of carbon atoms by changing the coefficient before the CO_2: $C_3H_8 + O_2 \rightarrow 3CO_2 + H_2O$

Next balance the number of hydrogen atoms by changing the coefficient before the water: $C_3H_8 + O_2 \rightarrow 3CO_2 + 4H_2O$

Finally, balance the number of oxygen atoms and the final balanced equation will be: $C_3H_8 + 5O_2 \rightarrow 3CO_2 + 4H_2O$.

67.   D   A beta particle can be represented by the symbol $^{0}_{-1}e$. The reaction: $^{234}_{91}Pa \rightarrow$ $^{234}_{92}U + ^{0}_{-1}e$ shows the mass numbers to add up to be the same on both sides of the equation and the atomic numbers (nuclear charges) to add up to be the same on both sides as well. All masses and nuclear charges have been conserved.

68.   C   The Hoffmann apparatus is used to break down water into hydrogen and oxygen gases via the use of a battery or external current. This process is called electrolysis. The reaction can also be classified as a decomposition reaction because the water is being broken down into simpler substances.

69.   D   Adding NaCl to the solution increases the amount of chloride ions in solution. This will shift the equilibrium to the right and form more solid silver chloride. If more solid silver chloride is precipitating then it is not dissolving into the solution, meaning that the solubility of the silver chloride has decreased.

70.   E   The quantities of atoms are as follows: Pb = 2, N = 4, O = 12, I = 4, and K = 4. The total number of atoms is 26.

DIAGNOSTIC TEST

■ SCORE SHEET

Number of questions correct: _____

Less: 0.25 × number of questions wrong: _____

(Remember that omitted questions are not counted as wrong.)

Equals your raw score: _____

Raw Score	Test Score		Raw Score	Test Score		Raw Score	Test Score		Raw Score	Test Score		Raw Score	Test Score
85	800		63	710		41	570		19	440		−3	300
84	800		62	700		40	560		18	430		−4	300
83	800		61	700		39	560		17	430		−5	290
82	800		60	690		38	550		16	420		−6	290
81	800		59	680		37	550		15	420		−7	280
80	800		58	670		36	540		14	410		−8	270
79	790		57	670		35	530		13	400		−9	270
78	790		56	660		34	530		12	400		−10	260
77	790		55	650		33	520		11	390		−11	250
76	780		54	640		32	520		10	390		−12	250
75	780		53	640		31	510		9	380		−13	240
74	770		52	630		30	500		8	370		−14	240
73	760		51	630		29	500		7	360		−15	230
72	760		50	620		28	490		6	360		−16	230
71	750		49	610		27	480		5	350		−17	220
70	740		48	610		26	480		4	350		−18	220
69	740		47	600		25	470		3	340		−19	210
68	730		46	600		24	470		2	330		−20	210
67	730		45	590		23	460		1	330		−21	200
66	720		44	580		22	460		0	320			
65	720		43	580		21	450		−1	320			
64	710		42	570		20	440		−2	310			

<u>Note:</u> This is only a sample scoring scale. Scoring scales differ from exam to exam.

PART III
CHEMISTRY TOPIC REVIEW

CHAPTER 1
MATTER AND ENERGY

IN THIS CHAPTER YOU WILL LEARN ABOUT...

Matter
Substances
Chemical and Physical Properties
Energy
Types of Energy
Endothermic and Exothermic Reactions

MATTER

Chemistry is the study of *matter,* that is, anything that takes up space and has mass. *Mass* (a measure of the number of particles in an object) should not be confused with weight, which is the influence of gravity on mass. Because gravitational forces can differ, an object that has the same mass on Earth as it does on the moon will have a lower weight on the moon because there is less of a gravitational pull on the moon. Because matter has mass and takes up a certain *volume* (space), the *density* for any variety of matter can be calculated using the equation: D = m/V. Units typically used to calculate density are grams (for mass) and milliliters (mL for volume). Sometimes the unit cubic centimeters (cm³) is used instead of milliliters. These volumes are equivalent. You will see an exception to the general rule when calculating the density for gases later in this book.

PROBLEM: What is the density of a solid cube that has a length of 2.0 cm on each side and weighs 6.0 grams?

> *Solution:* Because the length of each side of this cube is 2.0 cm, the volume will be the length × width × height. This is (2.0 cm)(2.0 cm)(2.0 cm) or 8 cm³. The mass is 6.0 grams. Substitute into the equation D = m/V and you get D = 6.0 grams/8.0 cm³. The density of this solid is 0.75 grams/cm³.

Think about this: *If you had 1,000 grams of feathers and 1,000 grams of the metal lead, which one would you use to fill your pillow? Even with both samples having the same mass, it is hard to imagine having much fun at a pillow fight if the pillow were filled with lead!*

SUBSTANCES

A *substance* can be defined as any variety of matter with identical properties and composition. Substances are classified as either *elements* or *compounds*. Elements cannot be broken down chemically, whereas compounds can be broken down chemically. An element is made up of a particular *atom*, the basic building block of matter. Compounds are formed from the bonding of two or more elements. Consider the reaction: $CH_4 + 2O_2 \rightarrow 2H_2O + CO_2$. The reaction shows the elements carbon, hydrogen, and oxygen in different compounds. The chemical equation also shows how the compounds change over the course of the reaction. Although the compounds on the left of the arrow are not the same as those to the right of the arrow, the elements in the reaction are still carbon, hydrogen, and oxygen.

Mixtures are the results of the combination of elements and/or compounds. In a mixture:

- The substances are not chemically combined (each substance retains its properties).
- The ratios of substances can vary.
- The substances can be separated into the original elements and/or compounds.

Mixtures can be classified as *homogeneous* (the same throughout) or *heterogeneous* (not the same throughout). An example of a homogeneous mixture is homogenized milk. You do not have to shake milk before using it because all samples of homogenized milk will be the same. Solutions are homogeneous mixtures of a *solute* dissolved in a *solvent* and can be represented by a substance followed by the symbol (aq) to show that the substance has formed a homogeneous mixture with water (an aqueous *solution*). An example of a heterogeneous mixture is a mixture of sand in water. The sand is sure to settle to the bottom of the container no matter how much you stir the mixture.

PROBLEM: Classify the following as elements, compounds, or mixtures: salt water, water, argon, methane, and iron.

> *Solution:* Salt water is a homogeneous mixture of water and NaCl. Water is a compound made up of hydrogen and oxygen. Argon is an element. Methane is a compound made up of hydrogen and carbon. Iron is an element.

Think about this: *One simple mixture to consider is a salad. You can vary the amounts/ratios of the vegetables you put in a salad. If you do not like a particular vegetable, that vegetable can easily be pulled out of the mixture. A salad is a heterogeneous mixture. Each portion you eat is likely to contain a different combination of vegetables. Finally, the vegetables in a salad retain their properties. Imagine a carrot and a piece of lettuce combining to form a new substance!*

CHEMICAL AND PHYSICAL PROPERTIES

All substances have physical and chemical properties. *Physical properties* are the observable and measurable properties of substances. These include phase (solid, liquid, or gas), color, odor, density, boiling or melting point. *Chemical properties* are the properties observed when a substance reacts with other substances. Chemical changes result in substances with different physical properties. For example, when iron (a gray, solid metal) reacts with oxygen gas (an odorless, colorless gas) the result is iron oxide or rust (a solid that is orange-red in color). You could also note the changes in the density, melting points, and boiling points of the iron, oxygen, and iron oxide.

PROBLEM: Classify the following as physical or chemical changes: burning a piece of paper, smashing a piece of chalk, melting an ice cube, and the rusting of an iron nail.

Solution: Burning paper changes the paper's chemical composition. Smashing chalk is a physical change; the chalk is still the same chalk, only in smaller pieces. Melting an ice cube does not change the composition of the water; it's a physical change. Rusting an iron nail is a chemical change; the iron is now an iron oxide.

ENERGY

Chemistry is defined as the study of matter, but *energy* plays an important role in chemistry. Energy is defined as the ability to do work. Energy is conserved, that is, it is not created or destroyed. This means that the amount of energy lost by one system is always equal to the amount of energy gained by another. Energy can also be converted from one form to another. For example, a toaster or a hairdryer converts electrical energy into heat energy.

The units used for measuring amounts of energy are the *joule* or the *calorie*. Most people are more familiar with the term calorie. This should not be a problem because the simple relationship between the two units is that one calorie is equal to 4.18 joules. This ratio is helpful in setting up problems that ask for a conversion of one unit to another.

TYPES OF ENERGY

Energy is found in many forms. As mentioned above, energy can exist as heat or electricity. Other forms of energy are sound, light, chemical energy, and nuclear energy. Probably the two most important forms of energy in chemistry are *potential energy* and *kinetic energy*. Potential energy is stored energy. A good example is someone holding up a hammer, ready to strike a nail. The hammer has the potential of falling onto the head of the nail. As the hammer is swung downward and moves through the air, the potential energy is converted to kinetic energy, or moving energy. Because all of the energy is conserved, the potential energy stored in the hammer is turned into kinetic energy. As the hammer is lifted to strike the nail again, the movement of the hammer upward becomes the potential energy stored for the next strike on the head of the nail. One thing to remember in chemistry is that nature prefers a lower energy state. This rule will be noted again and again in this chemistry review.

ENDOTHERMIC AND EXOTHERMIC REACTIONS

Energy may also be absorbed or released in a reaction. When more energy is released than absorbed, the reaction is said to be *exothermic*. When more energy is absorbed than released, the reaction is said to be *endothermic*. A potential energy diagram can be used to graph these changes as in Figure 1.1.

Exothermic—more energy is released than absorbed.

$$A + B \rightarrow C + D + Heat$$

Reactants   Products

Endothermic—more energy is absorbed than released.

$$A + B + Heat \rightarrow C + D$$

Reactants   Products

Notice that energy is always absorbed and released in a reaction. The relative amounts are what cause the reaction to be endothermic or exothermic. Also, it takes energy to start the reaction. This is called the *activation energy* (E_a).

Finally, take note of the difference in the energy of the *reactants* and *products*. The change in the energy of the reactants or products is called the *heat of reaction*. This is designated by the symbol ΔH. This symbol stands for the change in heat energy or *enthalpy*. The simple way to remember how to calculate the change in enthalpy is to use this mnemonic device:

$$\Delta H = \text{potential energy of the products minus the}$$

$$\text{potential energy of the reactants}$$

or

$$\Delta H = PEP - PER \text{ (delta H is equal to "pepper")}$$

When ΔH has a positive value, the reaction is said to be endothermic (enters heat). When ΔH has a negative value, the reaction is said to be exothermic (exits heat).

PROBLEM: In a reaction the potential energy of the reactants is 150 joules/mole and the potential energy of the products is 400 joules/mole. What is the heat of reaction for this reaction? Does this demonstrate an endothermic or exothermic process?

Solution: Use PEP – PER! ΔH = 400 joules/mole – 150 joules/mole = +250 joules/mole. Because the sign is positive the reaction is endothermic.

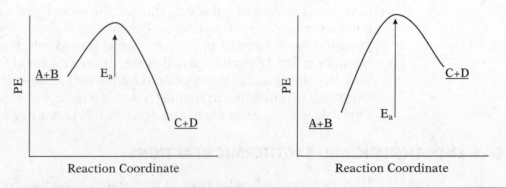

Figure 1.1 Two Potential Energy Diagrams

CHAPTER REVIEW QUESTIONS

1. Which substance can be decomposed chemically?

   (A) Ammonia
   (B) Iron
   (C) Neon
   (D) Hydrogen
   (E) Fluorine

2. Which units could be used to express the amount of energy absorbed or released during a chemical reaction?

   (A) Degree and gram
   (B) Torr and mmHg
   (C) Gram and liter
   (D) Calorie and joule
   (E) Meter and cm^3

3. Which sample represents a homogeneous mixture?

   (A) $CH_3OH(l)$
   (B) $CH_3OH(aq)$
   (C) $CH_3OH(g)$
   (D) $CH_3OH(s)$
   (E) None of the above

4. A book is lifted off of the floor and placed on a table that is one meter above the floor. The book has

   (A) gained sound energy
   (B) lost chemical energy
   (C) gained potential energy
   (D) gained kinetic energy
   (E) lost nuclear energy

5. Which statement is incorrect regarding energy?

   (A) Energy can be given off in a reaction.
   (B) Energy can be gained in a reaction.
   (C) Energy cannot be created or destroyed.
   (D) Energy can take various forms.
   (E) Energy has mass and takes up space.

6. What is the mass of an object that has a density of 13 g/mL and a volume of 10 mL?

   (A) 1.3 g/mL
   (B) 0.77 g/mL
   (C) 1.3 g/L
   (D) 130 g
   (E) 130 g/L

7. Which sentence below is incorrect?

   (A) Salads are heterogeneous mixtures.
   (B) NaCl(aq) is a homogeneous mixture.
   (C) Milk is a homogeneous mixture.
   (D) Sand and water make a heterogeneous mixture.
   (E) Pure iron is a heterogeneous mixture.

8. Which type of change is different from the other four?

   (A) Baking a potato
   (B) Rusting of an iron nail
   (C) Burning a piece of paper
   (D) Melting an ice cube
   (E) Ignition of propane

9. Which of the following is not a physical property?

   (A) Color
   (B) Phase
   (C) Odor
   (D) Boiling point
   (E) Reactivity with oxygen

10. Which substance cannot be decomposed chemically?

    (A) Ammonia
    (B) Tellurium
    (C) Methane
    (D) Water
    (E) Lunch

11. The study of matter is called

    (A) Chemistry
    (B) Biology
    (C) Geology
    (D) Physics
    (E) Psychology

12. Refer to the following choices:

    I. solid to liquid
    II. liquid to gas
    III. solid to gas

Which phase change above is endothermic?

    (A) I only
    (B) II only
    (C) III only
    (D) II and III only
    (E) I, II, and III

13. The energy needed to start a reaction is called

    (A) potential energy of the reactants
    (B) potential energy of the products
    (C) activation energy
    (D) heat of reaction
    (E) sound energy

Directions: The following question consists of two statements. Determine whether statement I in the leftmost column is true (T) or false (F) and whether statement II in the rightmost column is true (T) or false (F).

I		II

14. An exothermic reaction releases   BECAUSE   the potential energy of the products is greater than
    more heat than it absorbs   that of the reactants.

ANSWERS:

1. (A)	2. (D)	3. (B)
4. (C)	5. (E)	6. (D)
7. (E)	8. (D)	9. (E)
10. (B)	11. (A)	12. (E)
13. (C)	14. (T, F)	

CHAPTER 2
PHASES OF MATTER

IN THIS CHAPTER YOU WILL LEARN ABOUT . . .

Properties of Liquids, Gases, and Solids
Behavior of Gases
Boyle's Law
Charles' Law
Combined Gas Law
Dalton's Law
Avogadro's Law
Graham's Law
Ideal Gas Law
Phase Changes

Every day you encounter substances that exist in different phases—the air you breathe, the water you drink, and your mom's wooden cooking spoon. This chapter will take a closer look at the phases of matter and the changes in phase that matter can undergo.

GASES

The basic properties of *gases* are these:

- Gases have no definite volume and can be compressed.
- Gases have no definite shape and take the shape of their container.
- Gas molecules are spread far apart.

In addition to these basic properties, there are a number of theories and laws that tell more about how gases behave. These are the major focus of this chapter.

KINETIC MOLECULAR THEORY

The behavior of gases can further be explained with *Kinetic Molecular Theory* or KMT. KMT tells the following:

- Gas molecules are individual particles that travel in a straight-line random motion. This will continue until they collide or are acted upon by another force.

- Gas molecules continuously collide and transfer energy during these collisions. In an isolated sample of gas the net energy is conserved.
- The volume of the individual gas molecules is negligible compared to the volume they occupy.
- No forces of attraction are considered to exist between the gas molecules.

PRESSURE

Gases exert a *pressure* on other objects as they collide. This pressure exerted by a gas can be defined as the amount of force exerted on an area. Anyone who has watched or listened to a weather report can recall hearing about the barometric pressure or atmospheric pressure. These pressures can differ as high and low pressure systems move across a particular region. There are two devices used to measure pressure exerted by gases, the mercury barometer and the manometer. Both devices can be useful depending upon the situation.

The mercury barometer (see Figure 2.1) uses a glass column filled with mercury that is inverted into a container filled with mercury. As you move to higher altitudes the level of the mercury in the glass column drops because there is less air present at the higher altitude. On average, the height of the mercury in the column is 760 millimeters higher than the level of mercury in the container. This average pressure is called *standard pressure* or normal barometric pressure.

There are a number of units that can be used to measure barometric pressure. For example, millimeters can be converted into inches; thus, a standard pressure of 760 mm Hg can also be recorded as 30.0 inches of mercury. Inches of mercury are the units used for weather reports in the United States. Three other very common units that correspond to 760 mm Hg are:

- 1.0 atmospheres (atm)
- 760 torr (after Evangelista Torricelli)
- 101.3 kilopascals (kPa)

One final unit encountered in measuring pressure is pounds per square inch (psi). This unit is usually reserved for measuring the air pressure in a car's tires at gas stations in the United States.

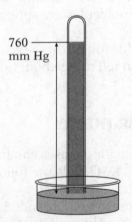

760 mm Hg

Figure 2.1 A Mercury Barometer

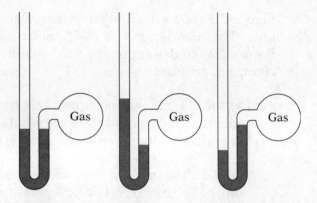

Figure 2.2 Three Manometers

A manometer is used to determine the pressure of a gas that is confined in a vessel. The shape and opening at the top of the manometer allow a certain amount of mercury to be shifted depending upon the atmospheric pressure and the pressure of the gas in the vessel. There are three situations to consider when using a manometer: gas pressure being equal to the atmospheric pressure, gas pressure being greater than the atmospheric pressure, and gas pressure being less than the atmospheric pressure. These situations can be seen in Figure 2.2.

BOYLE'S LAW

Robert Boyle performed experiments to see how varying the pressure on a gas would affect the volume of the gas. His experiments showed that as the pressure on a gas was increased, the volume that the gas occupied decreased. In Figure 2.3, the diagram on the left shows a piston that is not putting much pressure on a sample of a gas. The diagram in the center shows that the piston has been pressed downward, increasing the pressure on the gas. In response to the increase in pressure, the volume of the gas has decreased.

Boyle was able to determine that, at a constant temperature, there was an inverse relationship between pressure and volume. That is, as the value of one factor increased, the value of the other factor decreased. The graph in

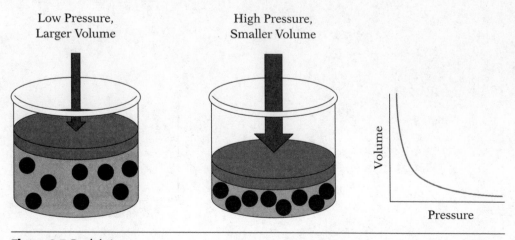

Figure 2.3 Boyle's Law

Figure 2.3 shows the mathematical relationship between pressure and volume. The equation $P_1V_1 = P_2V_2$ can be used to perform calculations using *Boyle's Law* to determine the final volume of a gas after it has undergone a change in pressure.

PROBLEM: A gas that occupies 22.4 liters is subjected to an increase in pressure from 760 torr to 1800 torr. What is the new volume occupied by this gas? Why does your answer make sense?

> *Solution:* The initial volume of the gas is 22.4 liters (V_1) at an initial pressure of 760 torr (P_1). The new pressure is 1800 torr (P_2). Using the equation $P_1V_1 = P_2V_2$, solve for the new volume so that the equation looks like $P_1V_1/P_2 = V_2$. Substituting gives:

$$\frac{(22.4 \text{ liters})(760 \text{ torr})}{(1800 \text{ torr})}$$ The value of the new volume is 9.46 liters.

Does the answer make sense? Yes. The pressure was increased, therefore the volume must decrease. The answer clearly shows a volume that is less than that of the original 22.4 liters.

Think about this: *Does volume decrease as pressure increases? A plastic soda bottle can hold a volume of 20 ounces. If a bottle was emptied of its contents and then flattened by having someone jump on it, how many ounces of liquid would the soda bottle hold now?*

CHARLES' LAW

Jacques Charles did quantitative experiments involving the effect of *temperature* changes on the volume of gases. Through various experiments, he was able to quantify the relationship between temperature and volume. This relationship, unlike the relationship expressed by Boyle's Law, was found to be a direct relationship. That is, as the temperature of a gas increased, so did the volume of the gas. The graph in Figure 2.4 demonstrates a direct relationship:

It is important to examine the topic of temperature before attempting to calculate using Charles' Law:

$$\frac{V_1}{T_1} = \frac{V_2}{T_2}$$

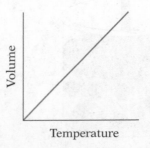

Figure 2.4 Charles' Law

Think about this: *What would happen to a balloon if it were placed in a freezer for 10 minutes?*

TEMPERATURE

Temperature is defined as the average kinetic energy of a sample. What does temperature really measure? Many people confuse temperature with heat. Temperature measures motion, hence the definition "average kinetic energy." There are two important temperature scales to know in chemistry, *Celsius* and *Kelvin*. The Celsius scale is based upon the freezing points and boiling points of water, 0°C and 100°C respectively. The Kelvin scale is based upon the lowest temperature that can be achieved, 0 K or *absolute zero*. It is believed that, at absolute zero, all motion of molecules stops. The relationship between Celsius and Kelvin is K = C + 273. The Kelvin scale is used when solving problems because, unlike the Celsius scale, the Kelvin scale does not result in a negative number or a zero to multiply or divide by.

PROBLEM: What are the boiling and freezing points of water on the Kelvin scale?

Solution: The boiling point of water is 100°C. Because K = C + 273, 100 + 273 = 373 K. Using the same equation, the freezing point of water will be 273 K.

STANDARD TEMPERATURE AND PRESSURE

Because experiments can involve a wide range of conditions that can affect the experiment's outcome, a standard needs to be set to create a common set of conditions. *Standard temperature and pressure,* or STP, is defined as 0°C and 1 atm (273 K and 760 torr).

PROBLEM: A sample of a gas at STP occupies 11.2 liters. The temperature of the gas is changed to 15°C. What is the new volume of the gas? Does your answer make sense?

Solution: First convert the temperatures to Kelvin. The initial temperature of the sample (T_1) is 273 K. The new temperature is (T_2) K = 15 + 273 = 288 K. The initial volume (V_1) is 11.2 liters. Using Charles' Law and substituting gives:

$$\frac{V_1}{T_1} = \frac{V_2}{T_2} \text{ and } \frac{(11.2\,L)}{(273\,K)} = \frac{(V_2)}{(288\,K)} \text{ rearranging gives } \frac{(11.2\,L)(288\,K)}{(273\,K)} = (V_2)$$

Solving gives an answer of 11.8 liters. Is this right? Yes, an increase in the temperature resulted in an increase in the volume of the gas.

AVERAGE KINETIC ENERGY OF GASES

Even though an increase in temperature results in an increase in the average kinetic energy, not all of the gas molecules will have the same amount of kinetic energy. This is why the term "average kinetic energy" is used. The two curves in Figure 2.5 show the distribution of molecular speeds of a gas at two

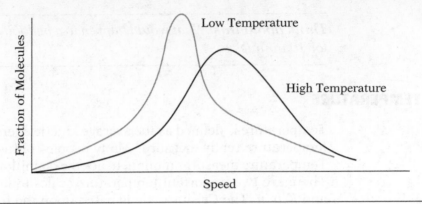

Figure 2.5 Kinetic Energy

different temperatures. Notice that at a higher temperature the gas molecules are moving with more kinetic energy.

COMBINED GAS LAW

Charles' and Boyle's Laws are used when the pressure and temperature of a system are held constant, respectively. What happens if pressure, temperature, and volume are all changed in a problem? The *Combined Gas Law*, $\frac{P_1 V_1}{T_1} = \frac{P_2 V_2}{T_2}$, combines the laws of Charles and Boyle. Notice that if you cover the T's with your finger, Boyle's equation remains. Likewise, if you cover the P's, Charles' equation remains. Rather than memorizing both equations for Charles' and Boyle's Laws, it is much easier to just remember the Combined Gas Law and "cover" the variables that are being held constant.

PROBLEM: A 5.6 liter sample of neon gas is at STP. If the temperature is increased to 298 K and the pressure changed to 600 torr, what will the new volume of the gas sample be?

Solution: The problem starts at STP, meaning that $P_1 = 760$ torr and $T_1 = 273$ K. The initial volume is 5.6 liters, so $V_1 = 5.6$ L. The final pressure P_2 is 600 torr and the final temperature is 298 K.

Next, use the Combined Gas Law: $\frac{P_1 V_1}{T_1} = \frac{P_2 V_2}{T_2}$

Substitution gives: $\frac{(760 \text{ torr})(5.6 \text{ L})}{(273 \text{ K})} = \frac{(600 \text{ torr})(V_2)}{(298 \text{ K})}$

Cross-multiply and divide to solve for the new volume: $\frac{(760)(5.6)(298)}{(273)(600)} = 7.7$ liters. Ask yourself, Does the answer make sense? The temperature was increased and the pressure was decreased, therefore the volume should be greater than the original 5.6 liters.

DALTON'S LAW OF PARTIAL PRESSURES

Dalton discovered that when gases are mixed in the same container and they have the same temperature, the total pressure exerted by the gases is equal to the sum of the pressures exerted by the individual gases. This can be seen in the equation: $P_{total} = P_{gas1} + P_{gas2} + P_{gas3}$. . . . Consider a container that has a mixture of nitrogen and oxygen gas. If the pressure of the oxygen gas is 400 torr and the pressure of the nitrogen gas is 360 torr, what is the total pressure? Using *Dalton's Law of Partial Pressures*, the total pressure is the sum of these individual pressures, 760 torr.

PROBLEM: Refer to the example given above. Write an equation for the partial pressures of nitrogen and oxygen. If the total pressure in the container is 900 torr and the partial pressure of the oxygen gas is 560 torr, what is the partial pressure of the nitrogen gas?

Solution: The equation is $P_{total} = P_{O_2} + P_{N_2}$. The total pressure of 900 torr minus the partial pressure of the oxygen gas, 560 torr, gives the pressure of the nitrogen gas, 340 torr.

AVOGADRO'S LAW

Avogadro's work will be examined in greater detail in Chapter 6. For now just learn this rule: Equal volumes of gases will contribute to the total pressure equally. For example, suppose a container is filled with 50% neon gas and 50% argon gas and the total pressure of the gases is 760 torr. The pressure of each gas would be 380 torr. If the volumes of the gases are not equal, the percentage of the volume that each gas occupies will contribute an equal percentage to the total volume.

PROBLEM: A container has three gases mixed together at STP. The container has (by volume) 10% He, 40% neon, and 50% argon. What is the partial pressure of the three gases?

Solution: Because this mixture of gases is at STP, the total pressure will be standard pressure, 760 torr. Because the percentages of the gases have been given, you can set up the following:

He 10% $(0.10)(760) = 76$ torr

Ne 40% $(0.40)(760) = 304$ torr

Ar 50% $(0.50)(760) = 380$ torr

Check your work. Does the sum of the partial pressures equal the total pressure of 760 torr?

GRAHAM'S LAW OF DIFFUSION/EFFUSION AND DENSITY OF GASES

Have you ever walked into a room and noticed the odor of an open bottle of perfume? If the open bottle of perfume is on one side of the room, why can you smell it on the other side of the room? Gases travel with great speeds and they can spread out or diffuse. According to *Graham's Law*, at the same temperature and pressure, gases diffuse at a rate inversely proportional to the square roots of their molecular masses. This can be seen in the equation:

$$\frac{r_1}{r_2} = \sqrt{\frac{M_2}{M_1}}$$

Even though this may seem a bit difficult to digest, it can be simply put as "lighter, less dense gases will travel at greater speeds."

Because gases are light, you need a larger sample of gas to measure the mass of a given sample. For convenience, you can assume a 22.4-liter sample (for reasons to be discussed later) and use the equation: $D_{gas} = \dfrac{\text{molar mass}}{22.4 \text{ liters}}$

PROBLEM: Calculate the effusion rate of nitrogen gas to oxygen gas. How does this compare to the density of the gases?

Solution:

$$\frac{r_1}{r_2} = \sqrt{\frac{M_2}{M_1}} \text{ and } \frac{r_{N_2}}{r_{O_2}} = \sqrt{\frac{32}{28}} = 1.07$$

Because the value of the ratio of the rates is greater than 1.00, the numerator, the rate for N_2, had a greater value. Because the nitrogen gas is lighter in mass, it should travel at faster speeds. Calculating the densities of these gases,

$$D_{N_2} = \frac{28 \text{ grams}}{22.4 \text{ liters}} = 1.25 \text{ g/L} \qquad D_{O_2} = \frac{32 \text{ grams}}{22.4 \text{ liters}} = 1.43 \text{ g/L}$$

you can see that density and rate of effusion are inversely proportional, N_2 has a greater rate of effusion and a lower density.

Think about this: *The mass of a baseball is 145 grams and the mass of a bowling ball can be more than 22,000 grams. Which one would you be able to throw through the air with a greater speed?*

IDEAL GAS LAW—RELATING PRESSURE, VOLUME, TEMPERATURE, AND MOLES

The *Ideal Gas Law* is derived from Kinetic Molecular Theory. Now that you have examined some gas laws, you can rephrase KMT and realize that ideally, gas molecules:

- Should be as far apart as possible (low pressures and high temperatures are best for this condition)
- Should have as little mass as possible (like H_2 or He)
- Should have no attraction for each other (we will discuss nonpolar molecules and forces of attraction later in the book)

It should be noted that gases do deviate from ideality, and there are equations that can adjust calculations to compensate for nonideal situations. These equations, however, are complex and are beyond the focus of this review.

By definition, an ideal gas obeys the equation PV = nRT. You will notice two new variables in this equation, n and R. The variable n stands for the number of *moles* of the gas and R is a constant. Before you can use this equation you must first become familiar with the mole and what it stands for.

If you buy a dozen eggs you expect to open the carton and find 12 eggs inside. The term "dozen" is used as a substitute for the word "twelve." The same idea applies when using the term "mole." A mole of "something" is equal to 6.02×10^{23} of those "things." For example a mole of carbon atoms is 6.02×10^{23} carbon atoms. The number 6.02×10^{23} is also known as *Avogadro's Number*.

As for the constant, R, the value of this constant depends upon the units being used for pressure, temperature, and volume. When you use the conventional units of liters, atmospheres, and Kelvin, the value and units are 0.0820 (L•atm)/(mol•K).

PROBLEM: How many molecules of H_2 gas are present in a 11.2 liter sample at 273 K and a pressure of 760 mm Hg?

Solution: The pressure is 760 mm Hg, this is equal to 1.0 atm. The volume of the gas is 11.2 liters and the temperature is 273 K. From the above you know that $R = 0.0820$ (L•atm)/(mol•K). The Ideal Gas Law is PV = nRT. Substituting gives: (1.0 atm)(11.2 L) = (n)(0.0820L•atm)/(mol•K)(273 K).

$$n = \frac{(1.0 \text{ atm})(11.2 \text{ L})}{(0.0820 \text{ L} \cdot \text{atm}/\text{mol} \cdot \text{K})(273 \text{ K})}$$

Solving gives an answer of 0.50 moles of H_2 gas. Because one mole of a gas contains 6.02×10^{23} molecules of the gas, 0.50 moles of the gas will contain half of Avogadro's number or 3.01×10^{23} molecules of H_2 gas.

LIQUIDS

Liquids are characterized by their definite volume. Unlike gases, liquids (for the most part) cannot be compressed. Liquids, like gases, do not have a definite shape and will take the shape of the container they are placed in. The molecules of a liquid are constantly touching one another because of the forces that exist between them and hold them together. These forces are not strong enough to hold the molecules in a fixed position as is the case for solids.

Liquids are constantly evaporating at their surface. That is, the molecules at the surface of the liquid can achieve enough kinetic energy to overcome the forces between them and they can move into the gas phase. This process is called *vaporization* or evaporation. As the molecules of the liquid enter the gas phase, they leave the liquid phase with a certain amount of force. This amount of force is called the *vapor pressure*. Vapor pressure depends upon the temperature of the liquid. Think about a pot of water that is being heated in preparation for dinner. The water starts out cold and you do not see any steam. As the temperature of the water increases you begin to see more steam. As the temperature of the water molecules increases, the molecules have more kinetic energy, which allows them to leave the liquid phase with more force and pressure. You can then conclude that as the temperature of a liquid increases, the vapor pressure increases as well. This is a direct relationship.

BOILING POINT

As mentioned earlier, the *boiling point* of water is 100°C or 373 K. Why does this phenomenon occur at this temperature? Does it always occur at this temperature? Is it true that water will boil at 100°C or 373 K provided that the atmospheric pressure is 760 torr? A temperature of 100°C or 373 K is what is referred to as the normal boiling point of water, the temperature at which water will boil when the atmospheric pressure is 760 torr. But why does water boil at this temperature? When water is heated to 373 K, the vapor pressure of the water molecules is 760 torr, a vapor pressure that is exactly the same as the atmospheric pressure! That answers the question: A liquid boils when the atmospheric pressure is equal to the vapor pressure of the liquid.

Does water always boil at 100°C? No, it does not, because atmospheric pressure can change. If a low pressure system is over the region where you live, atmospheric pressure is lower and water will boil at a temperature below 100°C because a lower temperature is needed to achieve the lower required vapor pressure.

Think about this: *Have you ever heard the expression, "The watched pot never boils"? Maybe the person who is watching and waiting for the pot to boil is in an area that has a high pressure system over it. In this situation, a higher temperature would be required to make the water boil.*

SOLIDS

Solids are characterized by their definite shape and volume. The atoms in a solid have a rigid, fixed, regular geometric pattern. These properties result from the fact that atoms in a solid constantly vibrate, but vibrate in place. When solids are heated to high enough temperatures, they have enough kinetic energy to undergo the process of *melting* and turning into a liquid. The *melting point* of a solid is the temperature at which this occurs. When a liquid changes into a solid the process is called *freezing*. It should be pointed out that the melting and freezing points of a solid are the exact same temperature. For example, ice melts at 273 K and water freezes at 273 K. This is why solids and liquids are listed together in the phase change diagrams that follow.

SUBLIMATION AND DEPOSITION

Why is it that when you buy ice cream from a vendor in a park on a hot summer day, the ice cream wrapper is not soaked with water from the "ice" that is used to keep it cold? Is the vendor using ice at all? "Dry ice" is the term for solid carbon dioxide. Dry ice can change from the solid phase right to a gas phase, without any apparent liquid phase in between. This process is called *sublimation*. Some other substances that can sublime are mothballs (naphthalene) and solid iodine. *Deposition* can be thought of being "the opposite of" sublimation. In this process a gas will form a solid, again without any apparent liquid phase in between.

PHASE CHANGES

Matter can exist in three phases: solid, liquid, and gas. Figure 2.6 summarizes the names of the changes that phases can undergo.

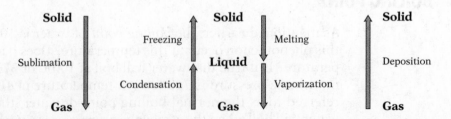

Figure 2.6 Phase Change Comparison

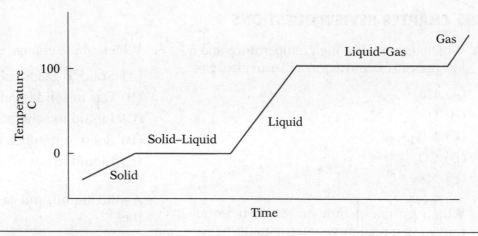

Figure 2.7 Heating Curve for Water

A heating curve is also helpful in looking at phase changes over time. Figure 2.7 is an example of a heating curve, specifically a heating curve for water.

In this diagram, heat is being applied to a sample of ice. The temperature of the ice will increase until it reaches 0°C. At this point the solid and liquid phases exist at the same time. Notice that there is no change in the average kinetic energy during the phase change. If heat is being added and there is no change in the average kinetic energy, then the sample must be gaining potential energy. After all of the ice has become water, the temperature (average kinetic energy) will rise again until it reaches the boiling point. Again, there is a phase change and the sample is gaining potential energy while the phase change occurs.

A phase diagram is helpful when looking at the effects of temperature and pressure on a substance. Because pressure can have an effect on the volume of gases and the boiling points of liquids, it is important not to overlook pressure when examining changes in the phase. A general phase diagram is shown in Figure 2.8.

The point where the solid, liquid, and gas phases can all exist at the same time, given a specific temperature and pressure, is called the *triple point*. For H_2O, the triple point exists when the pressure is 4.57 atm and the temperature is 0.01°C.

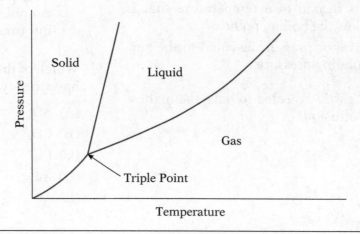

Figure 2.8 Phase Diagram

CHAPTER REVIEW QUESTIONS

1. Which gas under a high temperature and a low pressure behaves most like an ideal gas?

   (A) He
   (B) O_2
   (C) NH_3
   (D) CO_2
   (E) Ne

2. Which sample demonstrates particles arranged in a regular geometric pattern?

   (A) $CO_2(g)$
   (B) $CO_2(s)$
   (C) $CO_2(l)$
   (D) $CO_2(aq)$
   (E) None of the above

3. At which temperature does a water sample have the highest average kinetic energy?

   (A) 0 degrees Celsius
   (B) 100 degrees Celsius
   (C) 0 K
   (D) 100 K
   (E) 273 K

4. A liquid will boil when

   (A) its freezing point is equal to its melting point
   (B) a salt has been added to the liquid
   (C) its vapor pressure is equal to the melting point
   (D) it is heated to a temperature that is below the boiling point
   (E) its vapor pressure is equal to the surrounding pressure

5. Which gas is expected to have the highest rate of effusion?

   (A) O_2
   (B) F_2
   (C) H_2O
   (D) He
   (E) CH_4

6. Which phase change is described correctly?

   (A) Solid to gas is called deposition.
   (B) Gas to solid is called sublimation.
   (C) Liquid to solid is called freezing.
   (D) Solid to liquid is called vaporization.
   (E) Liquid to gas is called condensation.

7. A solid, liquid, and gas can exist together at the

   (A) sublimation point
   (B) triple point
   (C) boiling point
   (D) freezing point
   (E) melting point

8. A mixture of gases exists in a sealed container with the following percentages:

   helium 40%, neon 50%, and argon 10%

   If the total pressure of the gases is 1100 torr, then which of the following is true about these gases?

   (A) Volume and temperature have an inversely proportional relationship.
   (B) Volume and pressure have a direct relationship.
   (C) The partial pressure of the neon gas is 550 torr.
   (D) The partial pressure of the argon gas is 100 torr.
   (E) The partial pressures of the gases cannot be calculated with the given information.

9. Which of the following gases is expected to have the lowest density at STP?

   (A) SO_2
   (B) CO_2
   (C) Cl_2
   (D) Xe
   (E) Ar

10. An ideal gas at STP occupies 22.4 liters. If the pressure on the gas is increased to 1000 torr and the temperature of the gas is reduced to 250 K, what can be said about the gas?

    (A) The number of moles of the gas has changed.
    (B) The volume of the gas has increased.
    (C) The volume of the gas has decreased.
    (D) The pressure and the temperature have an inversely proportional relationship.
    (E) None of the above.

11. Which is inconsistent with the Kinetic Molecular Theory?

    (A) Gas molecules have forces of attraction for each other.
    (B) Gas molecules move in a random, straight-line motion.
    (C) Gas molecules have a negligible volume compared to the volume they occupy.
    (D) Collisions between gas molecules lead to a transfer of energy that is conserved.
    (E) All of the above statements are correct.

12. Refer to the following choices:

    I. Boyle's Law
    II. Charles' Law
    III. Combined Gas Law

    Which of the above can be used to calculate changes in volume with changes in pressure at constant temperature?

    (A) I only
    (B) II only
    (C) II and III only
    (D) I and III only
    (E) I and II only

13. Referring to the heating curve in Figure 2.7, which of the following is not associated with the heating curve for water?

    (A) Addition of heat energy
    (B) Melting
    (C) Boiling
    (D) Increase in kinetic energy
    (E) Sublimation

Directions: The following question consists of two statements. Determine whether statement I in the leftmost column is true (T) or false (F) and whether statement II in the rightmost column is true (T) or false (F).

I

II

14. Increasing the pressure on a gas forces the gas to occupy a smaller volume   BECAUSE   gases are compressible.

ANSWERS:

1. (A)	2. (B)	3. (B)
4. (E)	5. (D)	6. (C)
7. (B)	8. (C)	9. (E)
10. (C)	11. (A)	12. (D)
13. (E)	14. (T, T, CE)	

CHAPTER 3
ATOMIC STRUCTURE

IN THIS CHAPTER YOU WILL LEARN ABOUT...

The History of the Atom
Dalton's Atomic Theory
Rutherford's Gold Foil Experiment
Subatomic Particles
Isotopes
The Bohr Model of the Atom
Wave-Mechanical Model
Valence Electrons and Dot Diagrams
Hybridization
Effective Nuclear Charge
Quantum Numbers
Ions and the Octet Rule
Isoelectronic Series

A BRIEF HISTORY OF THE ATOM

Our idea of the atom has come a long way over the last several thousand years. The early Greek idea of the substances was that everything was made up of earth, wind, fire, or water. The idea of an indivisible particle arrived later on, and the word "atom" was derived from the Greek word "atomos," which means "indivisible."

In the early 1800s John Dalton formulated his *Atomic Theory*, which can be summarized as follows:

- All matter is composed of atoms.
- All atoms of a given element are alike (however, in Chapter 3 this will be disputed).
- Compounds are made up of atoms combining in fixed proportions.
- A chemical reaction involves the rearrangement of atoms. The atoms are not created or destroyed in a chemical reaction.

In the early 1900s a scientist named Robert Millikan performed an experiment using drops of oil that carried a charge. Although the details of the experiment can be quite extensive, essentially the drops were allowed to fall through an electric field at a certain rate. From this experiment Millikan was able to determine the mass and charge of the electrons that were on the oil drops.

Also in the early 1900s Ernest Rutherford performed his famous Gold Foil Experiment. Rutherford set up an experiment in which a radioactive substance released alpha particles. These particles were aimed at a thin sheet of gold foil. A screen coated with zinc sulfide was set up around the gold foil to detect the alpha particles when they hit the screen. Rutherford's experiment

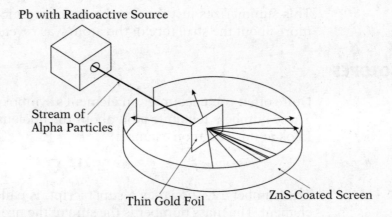

Figure 3.1 Rutherford's Gold Foil Experiment

showed that while some of the alpha particles were deflected and bounced back, the majority of the alpha particles were able to go through the gold foil. This is demonstrated in Figure 3.1.

The next question to answer is: Why can this take place? How can an alpha particle, a particle that has mass, go through the gold foil, another substance with mass? Because the gold was hammered into a thin sheet (some say about 100 atoms thick), many alpha particles were able to pass through the gold foil. Not every alpha particle was able to pass through, however, and some were bounced back at the source of radiation. Rutherford drew the following conclusions from this experiment:

- The atom is mainly empty space. (This conclusion is often referred to as the "empty space concept.")
- The size of the nucleus of the atom is small compared to the size of the entire atom.
- The mass of the atom is concentrated in the nucleus of the atom.
- The nucleus has a positive charge.

SUBATOMIC PARTICLES

A better understanding of the structure of the atom came about through additional experiments in the early 1900s. The discovery of the subatomic particles was a major breakthrough in atomic structure. These particles were classified as *electrons* and *nucleons*. The nucleons were later found to be *neutrons* and *protons*. The properties of these particles can be compared side by side:

	Protons	**Neutrons**	**Electrons**
Discovered by	Goldstein	Chadwick	Thomson
Mass	1 atomic mass unit (AMU)	1 AMU	1/1836 AMU (discovered by Millikan in his oil drop experiment)
Charge	1+	No charge	1– (discovered by Millikan in his oil drop experiment)
Location	Nucleus	Nucleus	Principal energy levels—orbiting around the nucleus

This summarizes just the very basics of atomic structure. As time went on, more about the structure of the atom was revealed.

ISOTOPES

Quite often you encounter an element's symbol written with a few numbers. These numbers tell quite a bit about the nucleus of the element in question. Look at this common example:

$$^{12}_{6}C$$

The number 12, shown as a superscript, is called the *mass number* for the element. The mass number is the sum of the protons and neutrons in the nucleus of the atom. Recall from Rutherford's experiment that the nucleus contains the atom's mass. Because protons and neutrons are the particles in the nucleus of the atom, they make up the mass of the atom because the masses of the electrons are minimal in comparison. The number 6, shown as a subscript, is called the *atomic number*. This can be defined as the number of protons in the nucleus, the nuclear charge (protons are the only nucleons with a charge), or the number of electrons in a neutral atom. How many neutrons are in carbon-12? To find the number of neutrons in the atom, subtract the atomic number from the mass number. In this case there are 6 neutrons in this atom.

There exist other carbon atoms with different mass numbers: carbon-13 and carbon-14. Here is a comparison of these carbon atoms:

	Carbon-12	Carbon-13	Carbon-14
Number of Protons / Atomic Number	6	6	6
Number of Neutrons	6	7	8
Mass Number	12	13	14

Notice that there is something different about the nucleus of these carbon atoms. These atoms are *isotopes* of one another. The similarities and differences are shown below:

Isotopes have the same: Isotopes have a different:

- Atomic number   - Number of neutrons
- Number of protons   - Mass number
- Name of the element

Isotopes can help determine the *atomic mass* (not mass number) of an element. The atomic mass for an element can be found on the periodic table. The atomic mass is a number that contains decimal places. Why? Can there be a fraction of a proton or neutron? The atomic mass is not a whole number because the atomic mass takes into account all of the masses of the isotopes of an atom and their relative abundance. For example, bromine has two isotopes, bromine-79 and bromine-81. It has been discovered that 50% of all bromine atoms are bromine-79 and 50% are bromine-81. From this you can calculate why the atomic mass of bromine is 80:

$$(79)\ (0.50) = 39.5$$
$$\underline{(81)\ (0.50) = 40.5}$$
$$= 80.0 \text{ as the atomic mass}$$

PROBLEM: Chlorine has two isotopes: 75% of all chlorine atoms are chlorine-35 and 25% are chlorine-37. How many neutrons are in these isotopes? What is the atomic mass of chlorine?

Solution: Both isotopes of chlorine have 17 protons. Chlorine-35 will have 18 neutrons to make the total mass number of 35. Chlorine-37 will have 20 neutrons to make a total mass number of 37. Multiply the masses of the two isotopes by their relative abundances:

$$(35)\ (0.75) = 26.25$$
$$\underline{(37)\ (0.25) = 9.25}$$
$$= 35.50$$

Checking the periodic table in Appendix 3 shows that this is close to the mass listed for chlorine, 35.45.

THE BOHR MODEL OF THE ATOM

Niels Bohr proposed a model of the atom in which the electrons move around the nucleus in fixed orbits. In this model, each orbit is at a fixed distance from the nucleus and the electrons in these orbits each have a certain amount of energy. The electrons have to be in one of the orbits; they cannot be in-between. Low-energy electrons orbit closer to the nucleus while higher-energy electrons orbit farther from the nucleus. When the electrons are in their lowest energy state they are said to be in the *ground state*. In this state they are orbiting as close to the nucleus as possible. When energy is added to atoms, for example in the form of heat or electricity, the electrons move to a higher energy level called the *excited state*. Because nature prefers a lower energy state, the electrons give off their added energy in the form of light. This allows the electrons to return to their ground state as diagrammed in Figure 3.2.

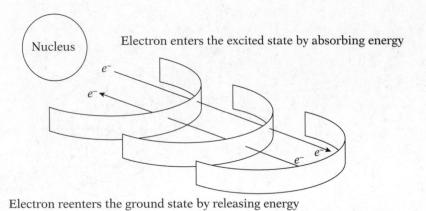

Figure 3.2 Ground and Excited States of the Atom

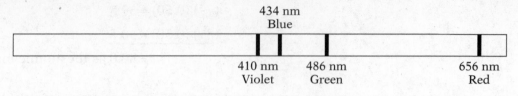

Figure 3.3 Line Spectrum for Hydrogen

The light given off from an excited element can be passed through a prism or diffraction gradient to determine the exact wavelengths of light being given off by the excited atom. These exact wavelengths of light are what are called an element's *line spectrum*. The line spectrum for hydrogen is shown in Figure 3.3.

Each element has its own line spectrum. This is why the line spectrum for an element is also considered to be a "fingerprint" for that particular element. Because the amounts of light given off by the excited atoms were in fixed amounts, Bohr termed them "quantized" amounts of light. These fixed amounts of energy proved that the electrons could only make certain jumps between the orbits that were at fixed distances in the atom. Because of these fixed, circular orbits, Bohr's model of the atom is often referred to as the "solar system model" of the atom (see Figure 3.4).

WAVE-MECHANICAL MODEL

The wave-mechanical model of the atom shows a more complex structure of the atom and the way electrons configure themselves in the principal energy levels. Principal energy levels are divided into sublevels, each with its own distinct set of *orbitals*. This more complex structure is outlined with the help of this diagram. The principal energy levels in the atom are numbered 1 through 7.

PEL#   **1**   **2**   **3**   **4**

The first principal energy level has just one subshell, the second principal energy level has two, the third has three, etc. Now the outline can be enhanced:

PEL#   **1**   **2**   **3**   **4**

Subshell   )   ))   )))   ))))

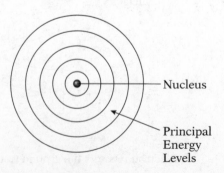

Figure 3.4 Solar System Model of the Atom

The first subshell in each principal energy level is given the letter "s." The second sublevel is given the letter "p," the third "d," and the fourth "f." Now the model is labeled even further:

PEL#	1		2	3	4
Subshell	)		))	)))	))))
	s		s p	s p d	s p d f

The s subshell contains one orbital or region where an electron can be found. The p sublevel has three orbitals, while the d and f levels have five and seven orbitals, respectively. The shapes of the s and p orbitals are shown in Figure 3.5.

Each orbital is allowed to carry a maximum of two electrons and, according to the *Pauli Exclusion Principle*, the electrons must have opposite spins.

How do electrons fill these principal energy levels? Consider the simplest example, hydrogen and its one electron. Because electrons want to be in the lowest energy state possible, the one electron will be located in the 1s orbital. This can be diagrammed by simply writing $1s^1$ or by drawing a diagram as follows: $\frac{\uparrow}{1s}$.

Because the 1s orbital can hold up to two electrons, the electron configuration for helium would look like this: $1s^2$ or $\frac{\uparrow\downarrow}{1s}$. Notice that the arrows are pointing in different directions. This shows the opposite spins of the electrons. The next example will show how *Hund's rule* works. Consider the electron configuration of nitrogen in its ground state. Nitrogen has seven electrons total. The first two electrons will fill the 1s orbital. Looking at the diagrams above, after the 1s orbital is full, the 2s orbital then takes in electrons. The remaining three electrons will go in the 2p orbital. The electron configuration looks like this: $1s^2 2s^2 2p^3$. What is of concern here is how the electrons orient themselves in the 2p orbitals. There are two possibilities:

$$\frac{\uparrow\downarrow}{1s}\ \frac{\uparrow\downarrow}{2s}\ \frac{\uparrow\downarrow\ \uparrow}{2p}\quad \text{or}\quad \frac{\uparrow\downarrow}{1s}\ \frac{\uparrow\downarrow}{2s}\ \frac{\uparrow\ \uparrow\ \uparrow}{2p}$$

According to Hund's rule, the electrons fill their orbitals singly and then they begin to pair up. This means that the second situation shown above is the correct filling order for the electrons.

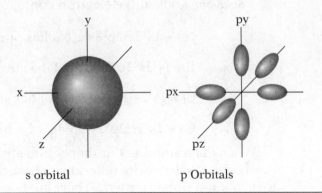

Figure 3.5 Shapes of s and p Orbitals

PROBLEM: Give the electron configurations for S and Ca.

Solution: Sulfur has 16 electrons. The shorthand notation is $1s^2 2s^2 2p^6 3s^2 3p^4$ or

$$\underline{\uparrow\downarrow} \;\; \underline{\uparrow\downarrow} \;\; \underline{\uparrow\downarrow} \; \underline{\uparrow\downarrow} \; \underline{\uparrow\downarrow} \;\; \underline{\uparrow\downarrow} \; \underline{\uparrow} \; \underline{\uparrow}$$

$$1s \;\; 2s \;\;\;\; 2p \;\;\;\;\;\;\;\; 3s \;\; 3p$$

Ca has 20 electrons and has the configuration that looks like this: $1s^2 2s^2 2p^6 3s^2 3p^6 4s^2$.

$$\underline{\uparrow\downarrow} \;\; \underline{\uparrow\downarrow} \;\; \underline{\uparrow\downarrow} \; \underline{\uparrow\downarrow} \; \underline{\uparrow\downarrow} \;\; \underline{\uparrow\downarrow} \; \underline{\uparrow\downarrow} \; \underline{\uparrow\downarrow} \; \underline{\uparrow\downarrow} \;\; \underline{\uparrow\downarrow}$$

$$1s \;\; 2s \;\; 2p \;\;\;\;\;\; 3s \;\; 3p \;\;\;\;\;\; 4s$$

Notice that in the previous problem the 3d orbital has been skipped over. The 3d orbital fills up after the 4s orbital has filled up. This is because the 4s orbital actually has less energy than the 3d orbital (unusual but true). This means that the filling order through element number 36, Kr, is $1s^2 2s^2 2p^6 3s^2 3p^6 4s^2 3d^{10} 4p^6$. Also note the maximum number of electrons that can be held in a principal energy level. The first principal energy level can hold up to 2 electrons ($1s^2$), the second up to 8 ($2s^2 2p^6$), and the third up to 18 ($3s^2 3p^6 3d^{10}$). This pattern follows the equation $2n^2$, where n is the number for the principal energy level. For example, to find the maximum number of electrons that can be held in the fourth principal energy level you can say that $2(4)^2 = 32$ electrons maximum.

VALENCE ELECTRONS AND DOT DIAGRAMS

Valence electrons play a huge role in bonding, as will be shown later. Valence electrons are the electrons that are in the outermost principal energy level (not to be confused with the outermost subshell). These electrons are important because they are the electrons that are lost, gained, or shared when forming chemical bonds. The valence electrons of an atom are the electrons that interact with the valence electrons of another atom to form these bonds.

PROBLEM: How many valence electrons are in the elements Na, P, Cl, and Ca?

Solution: Look at the electron configurations first.

Na = $1s^2 2s^2 2p^6 \mathbf{3s^1}$. Na has one valence electron.

P = $1s^2 2s^2 2p^6 \mathbf{3s^2 3p^3}$. P has five valence electrons.

Cl = $1s^2 2s^2 2p^6 \mathbf{3s^2 3p^5}$. Cl has seven valence electrons.

Ca = $1s^2 2s^2 2p^6 3s^2 3p^6 \mathbf{4s^2}$. Ca has two valence electrons.

When diagramming the shapes and structures of atoms it is important to have correct valence dot diagrams. These diagrams are easy to draw because they follow a simple pattern. There are two rules you need to remember when drawing valence dot diagrams (see Figure 3.6):

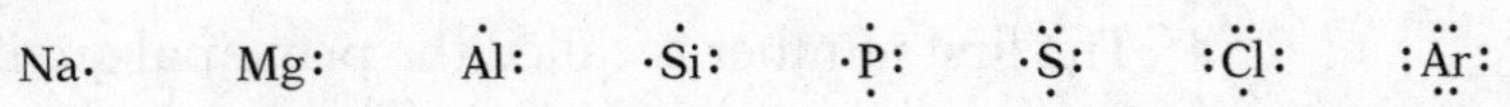

Na·   Mg: Ȧl: ·Ṡi: ·P̈:   ·S̈:   :C̈l: :Ä̇r:

Figure 3.6 Valence Dot Diagrams

- The first two electrons go together on the same side of the symbol, as if they were filling the s orbital.
- The next six electrons fill singly then pair up according to Hund's rule, as if they were filling the p orbital. A good way to remember this rule is to say to yourself, "Single, single, single, then paired, paired, paired."

PROBLEM: Give the valence dot diagram for N, Ne, O, and Ca.

Solution: Determine the electron configurations and find the valence electrons:

N—$1s^2 2s^2 2p^3$ Ne—$1s^2 2s^2 2p^6$ O—$1s^2 2s^2 2p^4$ Ca—$1s^2 2s^2 2p^6 3s^2 3p^6 4s^2$

Then draw the diagrams as in Figure 3.7.

EFFECTIVE NUCLEAR CHARGE—Z_{EFF}

All electrons in an atom experience an attraction to the positively charged nucleus. However, at the same time, there is a repulsion felt between the electrons because of their negative charge. The effective nuclear charge acting upon an electron or set of electrons can be calculated using the following:

$$Z_{eff} = Z - S$$

where Z is the number of protons in the nucleus and S is the number of electrons between the valence electrons and the nucleus. If we were to calculate the effective nuclear charge for the valence electron in a sodium atom we would find that value would be 1+. Looking at the electron configuration of sodium we find that there are 10 electrons between the nucleus and the valence electron:
$1s^2 2s^2 2p^6 3s^1$.
Z in this case is 11 and S is 10, giving Z_{eff} = 1+.

PROBLEM: What is the effective nuclear charge on the valance electrons of Br? Assume complete shielding.

Solution: Br has 35 protons and 28 core electrons: $1s^2 2s^2 2p^6 3s^2 3p^6 4s^2 3d^{10} 4p^5$
The effective nuclear charge will be 7+.

QUANTUM NUMBERS

To help keep track of each electron present in an atom, four quantum numbers are assigned to each electron. The rules for assigning quantum numbers are as follows:

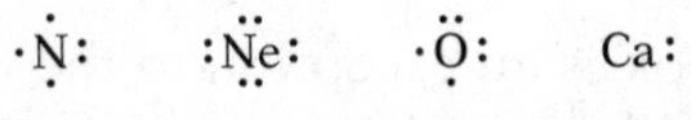

·N̈:   :N̈e: ·Ö: Ca:

Figure 3.7 Valence Dot Solutions

- The first number is called the principal quantum number, n, and it can be any whole number integer. The first quantum number, n, represents the principal energy level that the electron in question is in. For example the one electron in H is $1s^1$ and n will equal 1. The valence electron of Li, $1s^22s^1$, will have n equal to 2.
- The second number is called the angular momentum quantum number, l, and it can be an integer from 0 to $n - 1$. The second quantum number represents the sublevel that the electron is in. If the electron is in the s orbital then $l = 0$. An electron in the p orbital will have $l = 1$, and so on.
- The third number is called the magnetic quantum number, m_l, and it can be an integer that ranges from $-l$ to $+l$. The third quantum number helps us identify in which region of each sublevel the electron in question is located. These regions are specific orbitals.
- The fourth quantum number is m_s and it can only be a value of +1/2 or -1/2. The fourth quantum number represents the direction of the electron's spin. The opposite signs represent the opposite spins.

Assigning quantum numbers to the electrons of a few simple atoms will help us learn how to assign these numbers.

PROBLEM: An atom of hydrogen has one electron and it is labeled $1s^1$. What will the quantum number be?

Solution: In this case n will be equal to 1 because the electron is in principal energy level 1.

l must be 0 because l is a number from 0 to $n - 1$.
m_l is equal to 0 because the value of l is 0.
m_s is equal to +1/2.
The quantum numbers are 1, 0, 0, +1/2.

PROBLEM: Assign quantum numbers to the valence electron of a lithium atom.

Solution: The sole valence electron for lithium has an electron configuration of [He]$2s^1$. From this we can see that:

$n = 2$ because the valence electron is in the second principal energy level.
l can be either 0 or 1 ($2 - 1 = 1$). We assign the 0 first as $l = 1$ because it will not be used until we start to fill the p orbitals.
m_l is equal to 0 because the value of l is 0.
m_s is equal to +1/2.
The quantum numbers are 2, 0, 0, +1/2.

PROBLEM: Assign quantum numbers for the tenth electron to fill a neon atom.

Solution: $n = 2, l = 1, m_l = 1$ and $m_s = -1/2$.

HYBRIDIZATION

Carbon is an exception to the rules for writing valence dot diagrams as described above. A close examination of carbon and its valence electrons shows a different story, the proof being that carbon can make four bonds. Carbon has six electrons and its electron configuration is $1s^2 2s^2 2p^2$. The valence dot diagram according to this arrangement should look like:

$$\cdot \ddot{C} \colon$$

According to this dot diagram, carbon should make only two bonds. However, this is not the case as all. Instead, carbon has an arrangement that is $1s^2 2s^1 2p^3$ or

$$\frac{\uparrow\downarrow}{1s} \quad \frac{\uparrow}{2s} \quad \frac{\uparrow}{}\frac{\uparrow}{2p}\frac{\uparrow}{}$$

This is called the sp^3 *hybridization* of the carbon atom. One of the electrons has moved from the 2s orbital to the 2p orbital. This now gives carbon four single electrons rather than two paired electron and two unpaired electrons. It is the single electrons that bond in an effort to become paired up. Because there are four single electrons in the sp^3 hybridized carbon atom, carbon can bond four different atoms to its four single electrons. In order to have the four electrons as unpaired electrons, the sp^3 hybridization also changes the shapes of the s and p orbitals that are involved in the bonding. This change is shown in Figure 3.8.

The new arrangement is called a tetrahedral shape because the atoms are at 109.5-degree angles to each other. This can be seen in the three-dimensional shape of methane, a molecule with a tetrahedral molecular geometry. Should only three atoms bond to the sp^3 hybridized atom, then the molecular geometry is said to be trigonal pyramidal. Ammonia is an example of this. Finally, in water the oxygen is sp^3 hybridized but because only two atoms have bonded to the oxygen, the molecular geometry is said to be bent (see Figure 3.9).

In an sp^2 hybridization one 2s and two 2p orbitals combine to form a new shape as shown in Figure 3.10.

In this hybridized state the carbon will make two single bonds and one double bond. This will allow the carbon atom to bond to three different atoms. The orientation of these atoms around the carbon atom will be trigonal planar molecular geometry. The angle of these atoms to one another is 120 degrees. This is demonstrated in the diagram of ethene shown in Figure 3.11.

When an atom is sp hybridized, one s orbital combines with one p orbital. This allows an atom like carbon to make two double bonds or one single and

Figure 3.8 sp^3 Hybridized Carbon Atom

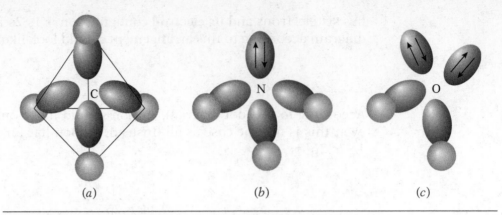

Figure 3.9 Tetrahedral, Trigonal Pyramidal, and Bent Molecular Geometries

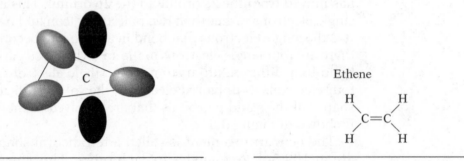

Figure 3.10 The sp² Hybridized Carbon Atom

Ethene

Figure 3.11 Molecular Geometry of Ethene

one triple bond. The two atoms that bond to the carbon atom in this case will orient themselves 180 degrees apart. This is called a linear molecular geometry (see Figure 3.12).

Here is a side-by-side comparison of the three states of hybridization:

	sp³ Hybridization	sp² Hybridization	sp Hybridization
Number of Atoms Bonded to the Central Atom	4	3	2
Angle between Atoms Bonded to Central Atom	109.5°	120°	180°
Molecular Geometry	Tetrahedral with four atoms bonded. Trigonal pyramidal with three atoms bonded. Bent with two atoms bonded.	Trigonal planar with three atoms bonded.	Linear with two atoms bonded.
Types of Bonds Found	Four single bonds.	One double bond and two single bonds.	One single and one triple bond. (Or) Two double bonds.

An sp Hybridized
Carbon Atom

H—C≡N and O=C=O

Examples of sp Hybridized
Carbon Atoms

Figure 3.12 sp Hybridized Carbon Atoms

PROBLEMS: What is the hybridization of carbon in CCl_4, in HCN, and in CH_2O?

Solutions: In CCl_4 the carbon atom forms all single bonds. The carbon is sp^3 hybridized and the chlorine atoms will be 109.5 degrees apart. The carbon in HCN has a single bond with the hydrogen atom and a triple bond with the nitrogen atom. This carbon atom is sp hybridized. The carbon atom in CH_2O has a double bond with the oxygen atom and single bonds with the hydrogen atoms. This carbon atom is sp^2 hybridized.

▆▆ IONS AND THE OCTET RULE

The number of electrons in an atom can be found by looking at the atomic number, provided the atom is neutral. In a neutral atom the number of protons is equal to the number of electrons. The balancing of the positive and negative charge is what makes the atom neutral. But what happens if an atom gains or loses electrons? When an atom loses or gains electrons it becomes an *ion*. The atom now has an unequal number of protons and electrons (the number of protons does not change). When an atom loses electrons, it loses negative charges. In this case, the protons outnumber the electrons and the ion is positively charged. An ion with a positive charge is called a *cation*. When an atom gains electrons it gains negative charges. In this case, the electrons outnumber the protons and the ion is negatively charged. An ion with a negative charge is called an *anion*.

Think about this: *If you lost all the negative feelings in your life, how would you end up feeling? Positive, of course!*

Just how many electrons will an atom gain or lose? How much of a charge will an ion take on? There is a simple pattern to follow but in the end you will see that it all depends upon the number of valence electrons that an atom has. There is one thing that all ions have in common: They form so that they have eight valence electrons in the outermost principal energy level. This is called the *octet rule*. Having eight electrons in the valence shell gives the atom a stable electron configuration. Once eight valence electrons have been achieved, it is very difficult to modify this stable configuration.

To determine the charge an atom will take on as an ion, first look at the electron configuration of the atom. Sodium, for example, has 11 electrons and the electron configuration of $1s^2 2s^2 2p^6 3s^1$. In order to achieve an octet, sodium has two options—gain seven electrons or lose just one. Losing one electron is a much easier task than gaining seven electrons. When the sodium atom loses one electron (loses a negative charge), the sodium ion has a charge of 1+ and the atom is written as Na^{1+}. After the one electron has been lost, the new electron configuration of sodium is $1s^2\mathbf{2s^22p^6}$. The eight valence electrons are shown in bold to emphasize the octet rule.

Another example would be the ion formed for sulfur. Sulfur has the electron configuration of $1s^22s^22p^6\mathbf{3s^23p^4}$. Sulfur has six valence electrons as shown in bold. For sulfur to achieve a stable octet it could either lose six electrons or gain two electrons. Gaining two electrons is a more feasible task, and the gain of two electrons will give sulfur ion a 2– charge. The new electron configuration for sulfur is $1s^22s^22p^6\mathbf{3s^23p^6}$. Again, the eight valence electrons have been emphasized.

PROBLEM: What will be the charges of the following elements when they form an ion: Ca, Al, F, N, Ne?

Solution: Start with the electron configurations for each atom:

$$Ca—1s^22s^22p^63s^23p^64s^2$$

$$Al—1s^22s^22p^63s^23p^1$$

$$F—1s^22s^22p^5$$

$$N—1s^22s^22p^3$$

$$Ne—1s^22s^22p^6$$

Now look at the configurations and decide which is the easiest way to obtain a stable octet. Calcium will lose two electrons and become Ca^{2+}. Aluminum will lose three electrons and become Al^{3+}. Fluorine will gain one electron and become F^{1-}. Nitrogen will gain three electrons and become N^{3-}. Neon will not lose or gain any electrons because of its stable octet.

EXCEPTIONS TO THE OCTET RULE

There are exceptions to the octet rule. Helium, for example, is incredibly stable with just two valence electrons in its outermost principal energy level. The same holds true for lithium and beryllium ions as well. This indicates that it isn't so much having an octet that stabilizes the atom, as it is the issue of having a full outermost principal energy level.

One last exception to the octet rule lies in the bonding of the atom boron. Boron prefers six electrons in its outermost principal energy level. This allows compounds containing boron to make three bonds in a trigonal planar arrangement. Two examples are BH_3 and BF_3 as shown in Figure 3.13.

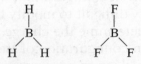

Figure 3.13 The Trigonal Planar Geometry of Boron

ISOELECTRONIC SERIES

Atoms will gain, lose, or share valence electrons as to achieve an octet (in most cases) or electron configuration that is similar to that of a noble gas. For example, all of the following atoms and ions have 18 electrons and an electron configuration similar to that of a noble gas, argon: S^{2-}, Cl^{1-}, Ar, K^{1+}, and Ca^{2+}. All of these atoms/ions have an electron configuration of $1s^2 2s^2 2p^6 3s^2 3p^6$.

Knowing the information just presented, we can now predict the relative size of the atoms and ions presented. Because they all have 18 electrons we can look at the number of protons present as well. The ion with the greatest number of protons, Ca, will have the smallest radius because it has the greatest nuclear "pull" on the 18 electrons. Sulfur, with just 16 protons, will have the least nuclear attraction for the 18 electrons that it has. This helps explain why non-metal atoms are smaller than their respective ions. Just the same, it also explains why metal atoms are larger than their respective ions.

■■ CHAPTER REVIEW QUESTIONS

1. Which of the following isotopes has the greatest number of neutrons?

   (A) ^{35}Cl

   (B) ^{31}P

   (C) ^{40}Ar

   (D) ^{41}Ca

   (E) ^{14}C

2. An atom has eight electrons in a 3d subshell. How many orbitals in this subshell have an unpaired electron?

   (A) 1

   (B) 2

   (C) 3

   (D) 4

   (E) 5

3. Which principal energy level has exactly four subshells?

   (A) 1

   (B) 2

   (C) 3

   (D) 4

   (E) 5

4. An atom in the ground state has seven valence electrons. Which electron configuration could represent the valence electron configuration of this atom in the ground state?

   (A) $3s^13p^6$

   (B) $3s^63p^1$

   (C) $3s^13p^43d^2$

   (D) $3s^23p^43d^1$

   (E) $3s^23p^5$

5. How many valence electrons are in an atom with the configuration $1s^22s^22p^63s^23p^2$?

   (A) 6

   (B) 5

   (C) 4

   (D) 3

   (E) 2

6. Which electron configuration demonstrates an atom in the excited state?

   (A) $1s^22s^1$

   (B) $1s^22s^22p^4$

   (C) $1s^22s^2$

   (D) $1s^22s^22p^63s^2$

   (E) $1s^22s^23s^1$

7. Which pair of symbols below show different isotopes of the same element?

   (A) $^{39}_{18}A$ and $^{39}_{19}R$

   (B) $^{60}_{27}X$ and $^{59}_{28}Y$

   (C) $^{12}_{6}L$ and $^{14}_{6}L$

   (D) $^{37}_{17}X$ and $^{37}_{17}X$

   (E) $^{3}_{2}E$ and $^{3}_{1}G$

8. Which of the following is not a conclusion Rutherford made from his experiment with alpha particles being shot at a thin sheet of gold foil?

   (A) An atom has a very small, compact nucleus.

   (B) An atom is mainly empty space.

   (C) An atom's mass is concentrated in the nucleus.

   (D) An atom has a very dense nucleus.

   (E) An atom has a negatively charged nucleus.

9. Which atom is not paired with its correct ion and ionic charge?

    (A) Rb / Rb^{1-}

    (B) Mg / Mg^{2+}

    (C) F / F^{1-}

    (D) Li / Li^{1+}

    (E) Br / Br^{1-}

10. Which of the following statements is false regarding sub-atomic particles?

    (A) The proton has a positive one charge.

    (B) The neutron has no charge.

    (C) The electrons are found in regions of the atom called orbitals.

    (D) The electrons have a greater mass than the protons.

    (E) Protons and neutrons are the nucleons of the atom.

11. Which is inconsistent with the concept of an isotope?

    (A) Same atomic number

    (B) Different number of neutrons

    (C) Same mass number

    (D) Same name of the element

    (E) Same number of protons

12. A mysterious element has the following relative abundances:
    X-34 15%   X-35 20%   X-36 65%
    Which of the following is true?

    (A) The atomic mass of this element is closer to 34.1.

    (B) The atomic mass of this element is closer to 34.9.

    (C) The atomic mass of this element cannot be determined without knowing exactly what X is.

    (D) A mass spectrophotometer would not be helpful in determining the percentages of the isotopes.

    (E) The atomic mass of this element is approximately 35.5.

13. Which of the following ions will be the smallest in the isoelectronic series?

    (A) O^{2-}

    (B) F^{1-}

    (C) Ne

    (D) Na^{1+}

    (E) Mg^{2+}

14. What is the correct set of quantum numbers for the eighth electron that fills the orbitals in an atom of oxygen?

    (A) $n = 2, l = 1, m_l = -1, m_s = -1/2$

    (B) $n = 2, l = 1, m_l = +1, m_s = -1/2$

    (C) $n = 2, l = 1, m_l = +1, m_s = +1/2$

    (D) $n = 2, l = 0, m_l = -1, m_s = +1/2$

    (E) $n = 1, l = 1, m_l = +1, m_s = -1/2$

15. Which of the following is not true about the effective nuclear charge felt by the valance electrons of the following atoms?

    (A) Z_{eff} for the valence electrons of Mg is 2+.

    (B) Z_{eff} for the valence electrons of Na is 2+.

    (C) Z_{eff} for the valence electrons of Be is 2+.

    (D) Z_{eff} for the valence electrons of Ne is 8+.

    (E) Z_{eff} for the valence electrons of Li is 1+.

ANSWERS:

1. (C)	2. (B)	3. (D)
4. (E)	5. (C)	6. (E)
7. (C)	8. (E)	9. (A)
10. (D)	11. (C)	12. (E)
13. (E)	14. (A)	15. (B)

CHAPTER 4

THE PERIODIC TABLE AND PERIODIC TRENDS

IN THIS CHAPTER YOU WILL LEARN ABOUT...

History and Arrangement of the Periodic Table
Metals, Nonmetals, and Semimetals
The Families
Electronegativity
Ionization Energy
Atomic Radius
Ionic Radius

HISTORY OF THE PERIODIC TABLE

The periodic table has been developed and perfected over many years. Although there are many scientists who have contributed to the periodic table, the two scientists who are given the most credit are Dmitry Mendeleyev and Henry Moseley. Mendeleyev, even though his periodic table had elements missing from it, is given the most credit for the periodic table and periodic trends. Later on, Moseley used a technique called x-ray crystallography and discovered the idea of the atomic number. This discovery is the basis for the arrangement of the modern periodic table. There were proposed periodic tables based upon atomic mass, but these arrangements did not suffice because of isotopes that can exist for an element.

ARRANGEMENT OF THE PERIODIC TABLE

The periodic table contains a number of *periods* and *groups*. The periods are the horizontal rows. They are numbered 1 through 7. The groups (or *families*) are the vertical columns. They are numbered 1 through 18. You will be provided with a periodic table when you take the SAT II: Chemistry test. **NOTE: A complete Periodic Table is provided in Appendix 3 at the back of this book.**

METALS, NONMETALS, AND SEMIMETALS

Two categories of elements on the periodic table are the *metals* and the *nonmetals*. Their properties are summarized in the chart below:

Metals	Nonmetals
• Are *ductile* and can be rolled into thin wires	• Are soft and brittle
• Are *malleable* and can be hammered into thin sheets	• Lack luster
• Conduct heat	• Are poor conductors of heat
• Conduct electricity	• Are poor conductors of electricity
• Have a shiny luster	• Tend to gain electrons and form anions
• Tend to lose electrons and become cations	
• Make up two-thirds of the periodic table	

The *semimetals,* or *metalloids,* are known to exhibit some of the properties of metals and some of those of nonmetals. The semimetals are B, Si, Ge, As, Sb, Te, and At. They are highlighted in bold in the partial periodic table in Figure 4.1. The elements located to the left of the semimetals are the metals; those to the right of the semimetals are the nonmetals. Identifying an element as a metal, nonmetal, or semimetal is important in identifying periodic trends and in identifying the types of bonds that atoms will form with each other.

PROBLEM: Identify the following elements as metals, nonmetals, or semimetals: potassium, calcium, bromine, hydrogen, and neon.

PERIODIC TABLE OF THE ELEMENTS

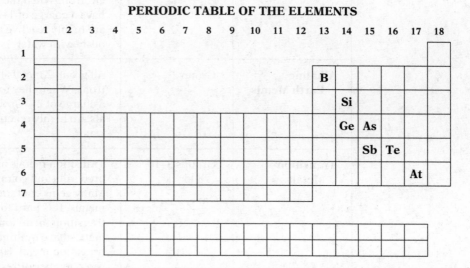

Figure 4.1 The Semimetals

Solution: K and Ca are located on the left side of the periodic table and are metals. Br and Ne are on the right side of the semimetals and are nonmetals. Hydrogen, although on the left side of the periodic table, is a nonmetal. If you're still not convinced about hydrogen, ask yourself about the properties of hydrogen gas and see where those properties fit in the comparison chart on p. 77.

THE FAMILIES

Some groups or families are given special names and have certain properties that should be addressed. But first you must understand why elements are put into the same group. Think about a family you know, not a chemical family, but a human family. Children look like their parents. They learn to do things from their parents and do them in the same way. The same holds true for the elements in the families of the periodic table; they react the same way (for the most part). As you learned in the last chapter, each element has a certain number of valence electrons. As you will learn in the next chapter, it is the number of valence electrons of an atom that determines its chemical reactivity. Because the elements in a family have the same number of valence electrons, they will have a similar chemical reactivity. For example, Na and K can be compared in electron configuration and ions formed:

$$\text{Na}-1s^2 2s^2 2p^6 \mathbf{3s^1} \quad \text{and} \quad \text{K}-1s^2 2s^2 2p^6 3s^2 3p^6 \mathbf{4s^1}$$

Both atoms have 1 valence electron and will lose this one electron to form ions with charges of 1+. This similar charge will mean that both elements have a similar chemical reactivity.

The important families and groups are listed below followed by their important characteristics. These characteristics will become more familiar to you as you study the chapter on bonding.

Name	Group Number	Special Properties
Alkali Metals	Group 1	All group 1 metals have one valence electron. When they form ions, they will have a charge of 1+. Group 1 alkali metals are highly reactive and will react vigorously with water.
Alkaline Earth Metals	Group 2	All group 2 metals have two valence electrons. When they form ions, they will have a charge of 2+. Group 2 alkaline earth metals are highly reactive and will react with water.
Transition Metals	Groups 3–10, d block	Transition metals are famous for the colored salts and colored solutions they form. Many gems contain numerous transition metals. It is hard to predict the charge of a transition metal ion because the transition metals have multiple oxidation states. One transition metal, Hg, exists as a liquid at room temperature.

Halogens	Group 17	Halogens (salt formers) have seven valence electrons and form ions with a charge of 1–. The halogens exist in three phases at room temperature. Fluorine is a pale-yellow gas, chlorine is a green gas, bromine is a brown-orange liquid, and iodine is a purple solid.
Noble (Inert) Gases	Group 18	Noble gases have a full outer shell and will not react to form ions or share electrons.
Lanthanides and Actinides	f Block	These elements have their valence electrons located in the f orbitals and are radioactive in nature.

There are important periodic trends that occur across the periods and up and down the groups. It is best to remember the trends of just a few elements. This will simplify the trends greatly and make the periodic trend questions the easiest to answer on the test.

ELECTRONEGATIVITY

Electronegativity is a measure of an atom's ability to attract electrons. The electronegativities of the elements are given a value of between 0.0 and 4.0. The greatest electronegativity value goes to fluorine, 4.0. So where is the element with the lowest electronegativity? Look furthest from fluorine and across to the bottom left of the periodic table. Francium, Fr, has an electronegativity of 0.7. This should make sense because nonmetals tend to gain electrons and have a higher electronegativity value, whereas metals tend to lose electrons and have a lower electronegativity value. Because they don't react, the noble gases do not have a value for electronegativity.

PROBLEM: Which is expected to have a lower electronegativity, Na or S?

Solution: Na has a lower electronegativity because it is further from fluorine on the periodic table.

IONIZATION ENERGY

Ionization energy, as its name suggests, is the energy needed to remove an electron from an atom and form an ion. This concept should be easy to recognize in the periodic table once you have grasped the idea of electronegativity. It takes a lot of energy to remove electrons from the very stable octets of the noble gases. For example, for helium the first ionization energy is 2372 kJ/mol, whereas neon has a first ionization energy of 2081 kJ/mol. Fluorine, with the highest electronegativity and the ability to "hold onto" electrons, has a first ionization energy of 1681 kJ/mol.

You might have guessed by now that the opposite holds true for the metals as you move further away from fluorine and the noble gases. The proof lies in the first ionization energies for iron (762 kJ/mol) and potassium

(419 kJ/mol). These values are just a fraction of the first ionization energies for certain nonmetals.

PROBLEM: Which is expected to have a greater ionization energy, Ca or Br?

Solution: Br is located closer to F and will have a higher ionization energy.

ATOMIC RADIUS

The *atomic radius* of an atom can be defined as the distance from its nucleus to the outermost electron of that atom. As you go down a group, the radius of the atoms will increase as the atoms fill more principal energy levels with electrons. The proof for this trend can be seen in lithium, which has an atomic radius of 155 picometers (10^{-12} meters), and cesium, which has an atomic radius of 267 picometers. You might expect the same to happen as you examine the elements from left to right across a period. If lithium has fewer electrons than fluorine, then lithium should have a smaller radius than fluorine, right? Wrong! Fluorine has nine electrons and lithium has just three, yet fluorine has an atomic radius of 57 picometers and lithium a radius 155 picometers. Why the difference? Fluorine has more protons and positive charge in its nucleus than does lithium. It turns out that when looking at atomic radii across a period, it is the nuclear charge (and not the number of electrons) that determines the radius of the atom.

IONIC RADIUS

As covered in the previous chapter, atoms can gain or lose electrons. The resulting ions can be expected to be of a different radius than that of the original atom. When a nonmetal gains an electron, the ionic radius of the anion will be bigger than that of the nonmetal atom. This is shown in Figure 4.2.

The opposite holds true for metal atoms and cations. Metals lose electrons and will experience a decrease in their radius as shown in Figure 4.3.

THE s, p, d, AND f BLOCKS

The location of an element on the periodic table can tell a lot about the number of valence electrons the element has and in which subshell these valence electrons can be located. These blocks are outlined in Figure 4.4.

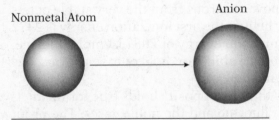

Nonmetal Atom   Anion

Figure 4.2 Relative Sizes of Anions

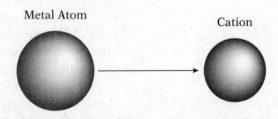

Metal Atom   Cation

Figure 4.3 Relative Sizes of Cations

PERIODIC TABLE OF THE ELEMENTS

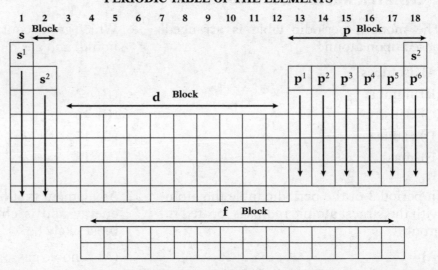

Figure 4.4 s, p, d, and f Blocks of the Periodic Table

The alkali and alkaline earth metals have their valence electrons in the s subshells. Groups 13 through 18 have their valence electrons located in the p subshells. The transition elements have their valence electrons in the d subshells, and finally, the lanthanides and actinides have their valence electrons in the f sublevel.

CHAPTER REVIEW QUESTIONS

1. The modern periodic table is arranged based upon atomic

    (A) isotopes
    (B) number
    (C) density
    (D) radius
    (E) mass

2. In period 3 of the periodic table the atom with the largest atomic radius is located in group

    (A) 1
    (B) 3
    (C) 13
    (D) 17
    (E) 18

3. The elements that display the greatest non-metallic character are located toward which corner of the periodic table?

    (A) Upper left
    (B) Dead center
    (C) Lower right
    (D) Lower left
    (E) Upper right

4. Which two elements will display the most similar chemical properties?

    (A) Aluminum and calcium
    (B) Nickel and phosphorus
    (C) Chlorine and sulfur
    (D) Carbon and sulfur
    (E) Lithium and potassium

5. Assuming the ground state, all of the elements located in group 13 of the periodic table will have the same number of

    (A) nuclear particles
    (B) occupied principal energy levels
    (C) electrons
    (D) valence electrons
    (E) neutrons

6. Which group contains elements in the solid, liquid, and gas phases at 298 K and 1 atm?

    (A) 1
    (B) 2
    (C) 16
    (D) 17
    (E) 18

7. An element that has a high first ionization energy and is chemically inactive would most likely be

    (A) a noble gas
    (B) a transition element
    (C) an alkali metal
    (D) a halogen
    (E) an alkaline earth metal

8. Which salt solution is most likely to be colored?

    (A) $KClO_3$ (aq)
    (B) KNO_3 (aq)
    (C) K_2CrO_4 (aq)
    (D) K_2SO_4 (aq)
    (E) KCl (aq)

9. As the elements of period 2 are considered from left to right, there is generally a decrease in

    (A) ionization energy
    (B) electronegativity
    (C) metallic character
    (D) nonmetallic character
    (E) none of the above

10. Which element is a liquid at room temperature?

    (A) K
    (B) Hg
    (C) I_2
    (D) Mg
    (E) Kr

11. At STP, which element is most expected to exist as a monatomic gas?

   (A) Calcium
   (B) Hydrogen
   (C) Nitrogen
   (D) Neon
   (E) Bromine

12. Nonmetals are poor conductors of heat and they also tend to

   (A) be brittle
   (B) conduct an electrical current
   (C) have a shiny luster
   (D) be malleable
   (E) lose electrons

13. Which statement does not explain why elements in a group are placed together?

   (A) They tend to have the same number of valence electrons.
   (B) They tend to have a similar oxidation number.
   (C) They tend to have the same electronegativities.
   (D) They tend to have the same chemical reactivity.
   (E) They tend to have the same charge when they form ions.

14. Refer to the following:

   I. F, C
   II. Na, Mg
   III. Fe, Co

   Which of the above have multiple oxidation states, colored salts, and valence electrons in the d orbitals?

   (A) I only
   (B) II only
   (C) III only
   (D) II and III only
   (E) I, II, and III

15. Which metal is not correctly paired with its color when put into a flame?

   (A) Lithium—Red
   (B) Potassium—Lilac
   (C) Sodium—Yellow
   (D) Copper—Orange
   (E) Magnesium—White

Directions: The following question consists of two statements. Determine whether statement I in the leftmost column is true (T) or false (F) and whether statement II in the rightmost column is true (T) or false (F).

I		II
16. A potassium atom is larger than a potassium ion	BECAUSE	metals gain electrons and experience an increase in their radius.
17. Oxygen and nitrogen are in the same family	BECAUSE	elements in the same family have the same number of valence electrons leading to ions that form similar chemical bonds.

ANSWERS:

1. (B)	2. (A)	3. (E)	4. (E)	5. (D)	6. (D)
7. (A)	8. (C)	9. (C)	10. (B)	11. (D)	12. (A)
13. (C)	14. (C)	15. (D)	16. (T, F)	17. (F, T)	

CHAPTER 5
BONDING

IN THIS CHAPTER YOU WILL LEARN ABOUT . . .

Intramolecular and Intermolecular Bonds

Ionic Bonding

Covalent Bonding

Sigma and Pi Bonds

Network, Coordinate Covalent, and Metallic Bonds

Dipole Forces and Polarities of Molecules

Hydrogen Bonding

Van der Waals Forces

Molecule-Ion Attraction

Naming Compounds

Determining Chemical Formulas

Stock Method

INTRAMOLECULAR BOND VERSUS INTERMOLECULAR BONDS

This chapter examines what it is that holds matter together. The "glues" that are responsible for holding atoms together with other atoms and molecules with other molecules are called bonds. *Intramolecular bonds* are the bonds that are found within molecules. In other words, intramolecular bonds hold atoms to other atoms. These bonds vary depending on the types of elements involved in the bonding process. *Intermolecular bonds* are the bonds between molecules. These bonds are what give substances their varying melting points, boiling points, and vapor pressures.

The rules that govern bonds between atoms and molecules can be quite tricky. It may be worthwhile to review the chapters involving atomic structure and the periodic table/trends before you tackle the material presented here.

IONIC BONDING

Ionic bonds are very strong bonds that are formed between a cation and an anion. The ionic bond is formed when a metal loses or transfers an electron (or electrons) to a nonmetal so that the metal and nonmetal form ions that have a full outermost principal energy level. The cations and anions thus formed then attract each other's opposite charges. The attraction between oppositely charged particles is called an electrostatic force.

The reaction between Na and Cl to form NaCl gives a good picture of how this works. Sodium has an electron configuration of $1s^22s^22p^63s^1$, whereas chlorine has a configuration of $1s^2 2s^2 2p^6 3s^2 3p^5$. Sodium has one valence electron that needs to be given away to achieve an octet; chlorine has seven valence electrons and needs just one more to complete its outermost principal energy level. The one valence electron in sodium is transferred to chlorine as shown in Figure 5.1.

$$Na \cdot + \: \ddot{\underset{\cdot\cdot}{Cl}} : \quad \longrightarrow \quad Na^{1+} \: [: \ddot{\underset{\cdot\cdot}{Cl}} :]^{1-}$$

Figure 5.1 Electron Transfer in Sodium Chloride

The sodium and chlorine ions attracted by the opposite charges form a *lattice* in which each sodium ion is surrounded by six chlorine ions and six chlorine ions are surrounded by six sodium ions. The lattice demonstrates why ionic compounds do not form molecules. Instead there is continuous pattern of chlorine and sodium ions packed together as shown in Figure 5.2.

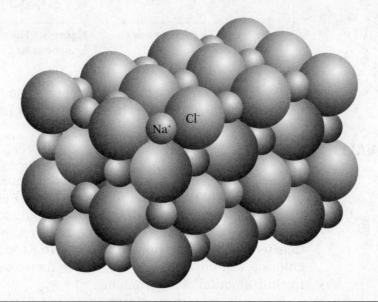

Figure 5.2 Example of a Crystal Lattice

Sodium oxide illustrates a slightly different situation. Here sodium has one valence electron and oxygen has six valence electrons. In this case it will take two sodium atoms to give up their one valence electron each to oxygen. This completes the octets for all three atoms as shown in Figure 5.3.

$$Na \cdot + \cdot \ddot{\underset{\cdot\cdot}{O}} : \quad \longrightarrow \quad 2Na^{1+} \: [: \ddot{\underset{\cdot\cdot}{O}} :]^{2-}$$
$$Na \cdot$$

Figure 5.3 Electron Transfer in Sodium Oxide

On the basis of the diagram in Figure 5.3, you can see that the formula for sodium oxide is Na_2O. There is no need to worry about predicting chemical formulas at this point. For now you just need to know where the two valence electrons came from to give oxygen a full octet.

PROBLEM: Diagram the reaction that takes place between calcium and oxygen to form calcium oxide.

Solution: Recognizing that calcium is a metal and oxygen is a nonmetal signals that the reaction will transfer electrons and the compound formed will be ionic. Because calcium is a metal, it will lose electrons to the nonmetal, oxygen. Calcium has two valence electrons and oxygen has six. Calcium will lose both electrons to the oxygen as shown in Figure 5.4.

The electron configurations for both atoms also help clarify the reaction that takes place (see Figure 5.5).

As Atoms:

$Ca - 1s^2 2s^2 2p^6 3s^2 3p^6 4s^2$ $O - 1s^2 2s^2 2p^4$

As Ions:

$Ca^{2+} - 1s^2 2s^2 2p^6 3s^2 3p^6$ $O^{2-} - 1s^2 2s^2 2p^6$

$Ca :{\cdot}+{\cdot} \ddot{O}: \longrightarrow Ca^{2+} [:\ddot{O}:]^{2-}$

Figure 5.4 Electron Transfer in Calcium Oxide

Figure 5.5 Electron Transfer and Electron Configurations for Calcium Oxide

Because it takes just one calcium atom and one oxygen atom to satisfy the octets for both atoms, the chemical formula for calcium oxide is CaO.

COVALENT BONDING

Covalent bonds are formed when two nonmetal atoms share electrons in order to satisfy their need to have a full outermost principal energy level. Covalent bonds are not as strong as the bonds formed between ions. For example, it would take a high flame and a temperature of almost 800 degrees Celsius to break the bonds between the sodium and chlorine in sodium chloride. The covalent bonds found in methane can be broken instantly with the introduction of a lit match.

It is not enough to simply say that a compound has covalent bonds because there are different types of covalent bonds. One type of covalent bond is called the *nonpolar covalent bond*. In this case the sharing of electrons is equal between the atoms. This occurs because the electronegativities of the atoms involved are (almost) the same. For example, hydrogen gas has an equal sharing of electrons between its two atoms:

$$H . + . H \rightarrow H:H$$

This diagram showing how the valence electrons interact is called a *Lewis structure*. In this case both hydrogen atoms have satisfied their need to have a full outermost principal energy level. Because both hydrogen atoms have the same electronegativity, the atoms will share the electrons equally. This will be the case with any diatomic molecule, such as chlorine gas (see Figure 5.6).

$$:\ddot{Cl}. + .\ddot{Cl}: \longrightarrow :\ddot{Cl}:\ddot{Cl}:$$

Figure 5.6 Electrons Shared in a Covalent Compound

Notice how each chlorine atom in Figure 5.6 has eight electrons around it. Also, notice the shared pair of electrons between the two atoms.

Because the hydrogen and chlorine atoms share a common pair of electrons, the two "dots" can be replaced with a "dash" to represent that a bond has been made. The valence dot diagrams can be rewritten as: H—H and Cl—Cl. The bond that is represented by the "dash" is called a single bond because there is one pair of electrons being shared between the two atoms.

The structure of diatomic nitrogen tells a different story. When you put two atoms of nitrogen next to each other, you see that each atom has three single electrons that want to pair up as shown in Figure 5.7.

The two nitrogen atoms share six electrons or three pairs of electrons. This means that there are three bonds between the two nitrogen atoms, N ≡ N. This is called a *triple bond*.

The other case to be examined is one that involves *double bonds*. Carbon dioxide has two double bonds that form as shown in Figure 5.8.

The bonds found between carbon and oxygen raise a new issue regarding bonding. Because the electronegativities for carbon and oxygen are different (they differ by 0.5 to 1.7), the bond is called a *polar covalent bond*. The polar covalent bond is characterized by the atoms having an unequal sharing of electrons. Because the negatively charged electrons spend more time with the more-electronegative element, the more-electronegative element will experience a negative charge, hence the reason it is called electronegativity. Hydrogen chloride has a polar covalent bond between the hydrogen and chlorine atoms. The buildup of negative charge on the more-electronegative chlorine can be shown with the use of a dipole arrow as in Figure 5.9.

Figure 5.7 Electrons Shared to Make a Triple Bond

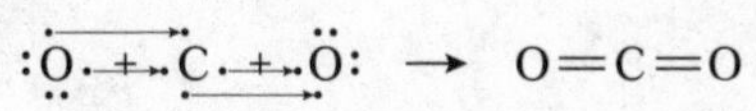

Figure 5.8 Electrons Shared to Make a Double Bond

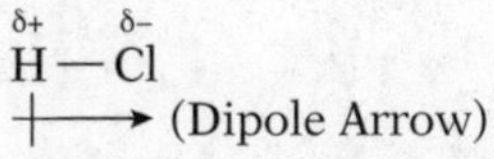

Figure 5.9 The Polar Bond of HCl

Because the hydrogen atom sees its electrons being attracted to the chlorine atom, the hydrogen atom experiences a positive charge as its one negative charge spends more time with chlorine.

PROBLEM: Draw the Lewis structure for formaldehyde, CH_2O. Which bonds are going to be polar covalent? Nonpolar covalent?

> *Solution:* Because carbon has the greatest number of single valence electrons it will be the atom that is placed in the middle of the molecule. The two hydrogen atoms will make single bonds with the carbon atom as shown in Figure 5.10.

The remaining two unpaired electrons on the carbon atom will bond with the two unpaired electrons found around the oxygen atom as shown in Figure 5.11.

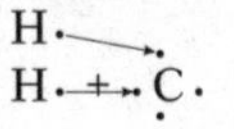

Figure 5.10 Hydrogen Bonds to Carbon

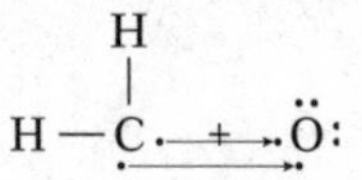

Figure 5.11 Carbon and Oxygen Form a Double Bond

The final structure is shown in Figure 5.12. Because there is little difference between the electronegativities between hydrogen and carbon, the bond between the two is nonpolar covalent. Because the difference in electronegativities between oxygen and carbon is greater, the bond between the two atoms will be polar covalent. In general, a bond is nonpolar covalent if the electronegativity difference between the atoms is 0 to 0.4. If the difference is 0.5 to 1.7, then the bond is polar covalent.

$$
\begin{array}{c}
H \\
\diagdown \\
C = O \\
\diagup \\
H
\end{array}
$$

Figure 5.12 Formaldehyde

SIGMA AND PI BONDS

The formation of single, double, and triple bonds in a molecule depends upon the types of hybridized orbitals that are sharing electrons. For example, when two hydrogen atoms bond to form $H_2(g)$, there is an overlap of s orbitals as shown in Figure 5.13.

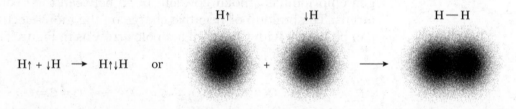

$$H\uparrow + \downarrow H \longrightarrow H\uparrow\downarrow H \quad \text{or}$$

Figure 5.13 s Orbital Overlap

This overlap is what allows the hydrogen atoms to form a single bond. The first bond that forms between two atoms is called a *sigma bond* (σ). The sigma bond arises from the overlap of two s orbitals or from the overlap of one s and one p orbital, or from the overlap of two p orbitals. The bonds in a molecule of methane (Figure 5.14) are an example of a situation in which hybridized p orbitals overlap with an s orbital.

A *pi bond* is the second bond that is formed when two sp^2 hybridized atoms have orbitals that overlap. The first bond that is made is from the joining of

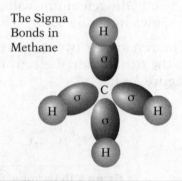

The Sigma Bonds in Methane

Figure 5.14 The Formation of Sigma Bonds in Methane

two sp² hybridized orbitals. The second bond that is formed, the pi bond, is the result of the p orbitals in the *y* axis overlapping as the atoms get close enough to do so. Because the p orbitals that lie in the *y* axis need to be close enough to bond, a double bond is shorter than a single bond. However, the double bond is stronger than a single bond. The overlap is shown in Figure 5.15.

Finally, there is the case for sp³ hybridization and the formation of a second pi bond. The second pi bond is the result of the overlap of the p orbitals in the *z* axis. Because a sigma bond is formed in the *x* axis and two pi bonds are formed in the *y* and *z* axes, a triple bond is formed. The triple bond will be shorter than the double bond and the triple bond will be stronger than the double bond as well. (See Figure 5.16.)

Overlapping Orbitals of Two sp² Hybridized Carbon Atoms

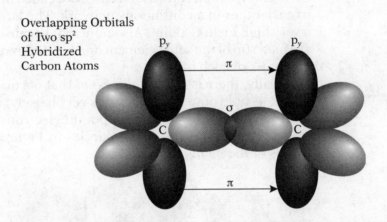

Figure 5.15 The Formation of a Pi Bond

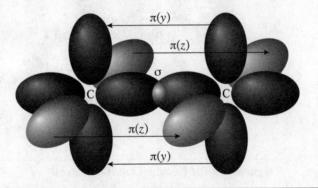

Figure 5.16 Two Pi Bonds Form to Make a Triple Bond

NETWORK, COORDINATE COVALENT, AND METALLIC BONDS

There is another important type of covalent bonding besides the nonpolar and polar covalent bonds just discussed. An example is shown in Figure 5.17.

In this example the free pair of electrons that is located on the nitrogen atom donates two electrons toward the bond that is formed with the hydrogen ion. Normally when a covalent bond is formed, one electron comes from each of the atoms that are bonding. In this case, the hydrogen ion did not donate any electrons toward this bond. When one atom donates both electrons in the covalent bond the bond is called a *coordinate covalent bond*.

Water provides another example of a coordinate covalent bond as shown in Figure 5.18.

$$H-\overset{\overset{\displaystyle H}{|}}{\underset{\underset{\displaystyle H}{|}}{\ddot{N}}}-H + H^{1+} \quad \rightarrow \quad H-\overset{\overset{\displaystyle H^{1+}}{|}}{\underset{\underset{\displaystyle H}{|}}{N}}-H \qquad\qquad H-\overset{\displaystyle \ddot{O}:}{\underset{\displaystyle H}{|}} + H^{1+} \quad \rightarrow \quad H-\overset{\displaystyle \ddot{O}:H^{1+}}{\underset{\displaystyle H}{|}}$$

Figure 5.17 The Formation of a
Coordinate Covalent Bond

Figure 5.18 The Formation of a
Coordinate Covalent Bond

In every example seen so far the covalent bonds have held atoms together in order to make molecules. However, there exist substances such as diamond and graphite where the carbon atoms are covalently bonded but do not bond to form molecules. Such cases are called *network solids*; the atoms bond to each other in a continuous network. The large network gives these solids a very high melting point. Also note that because both diamond and graphite are made up of the same element and are different substances, they are labeled *allotropes* of each other.

Finally, there is the *metallic bond* that occurs between metals. The atoms of metals hold onto their electrons very loosely, which is why metals conduct electricity so well. The loosely bound electrons are often referred to as the "sea of electrons." The darker circles in Figure 5.19 represent the electron clouds of the metal atom.

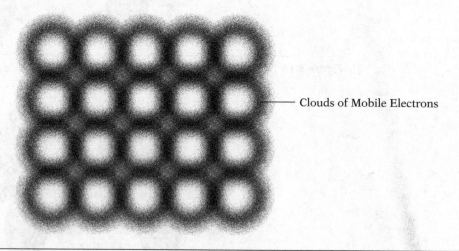

Clouds of Mobile Electrons

Figure 5.19 The Electron Clouds of a Metal

The chart below summarizes the intramolecular bonds just studied.

Type of Intramolecular Bond	Types of Elements Involved	Electron Movement	Difference in Electronegativity	Molecules Present?	Strength of Bond
Ionic	Metals and nonmetals	Electrons are transferred.	Electronegativity differences are 1.8 and higher.	No	Strong. Have a high melting point. Ex. NaCl
Nonpolar Covalent	Nonmetals and non-metals	Electrons are shared evenly.	Electronegativity differences are 0.4 and lower.	Yes	Weaker. Have low melting points. Ex. wax

(continued)

Type of Intramolecular Bond	Types of Elements Involved	Electron Movement	Difference in Electronegativity	Molecules Present?	Strength of Bond
Polar Covalent	Nonmetals and non-metals	Electrons are not shared evenly. Electrons spend more time with more electronega-tive element.	Electronegativity differences are 0.5 to 1.7.	Yes	Weaker. Have low melting points. Ex. water, ice
Coordinate Covalent	Nonmetals and a cation	Both electrons in the bond came from one element.		Yes	Weaker. Can be removed in acid-base reactions. Ex. H_3O^{1+}
Network	Nonmetals and non-metals	Electrons are shared.		No	Strong. Have a high melting point. Ex. diamond
Metallic	Metals and metals	Electrons are loosely bound by the metal atoms.		No	Strong. Have a high melting point. Ex. iron

PROBLEM: Name the type of bonding found between the atoms in KCl, H_3O^{1+}, CCl_4, SiO_2, Fe(s), F_2, and HBr.

Solution: KCl has a metal and a nonmetal ion attracted to one another, and it will be ionic. H_3O^{1+} has polar covalent bonds and one coordinate covalent bond. The bond between C and Cl will be polar covalent because of the difference in electronegativities. SiO_2 is sand and is a network solid. A sample of iron will have metallic bonds because only metal atoms are present. Fluorine is diatomic and will have nonpolar covalent bonds. HBr will have a polar covalent bond because of the great difference in electronegativity between these two nonmetals.

▓ DIPOLE FORCES AND POLARITIES OF MOLECULES

Because bonds can be polar and molecules can have certain shapes, electrons can "build up" on one side of a molecule and make one end carry a slight negative charge. When a molecule has this type of "buildup" of negative charge on one side and a positive charge on another side, the molecule is said to be a *dipole*. This is the case with HCl as shown in Figure 5.20.

$$\overset{\delta^+}{H} - \overset{\delta^-}{Cl}$$

Figure 5.20 The Polar Bonds of HCl

In a molecule of HCl, not only are the bonds polar covalent, but because the electrons spend more time with chlorine than hydrogen, the chlorine end of the molecule has a negative charge on it. HCl is a dipole, or polar, molecule because the differences in electronegativity have created the "two poles." Referring to the dipole arrow, there is no counterbalance of charges in this molecule and it is classified as polar.

A similar situation exists with water. In a water molecule the bonds are polar covalent because of the big difference in electronegativity between hydrogen and oxygen. Because of the bent molecular geometry of a water molecule, the dipole arrows cannot counterbalance. There is an overall dipole moment in the molecule as shown in Figure 5.21.

Reexamining the shape and polar bonds found in ammonia shows a situation similar to that of water. Because there is no symmetry in the molecule of ammonia, there is no counterbalance of forces and there is an overall dipole moment in a molecule of ammonia as shown in Figure 5.22.

Carbon dioxide and carbon tetrachloride tell a different story about polar bonds and overall dipole moment. Both carbon dioxide and carbon tetrachloride have polar bonds, as diagrammed by the dipole arrows shown in Figure 5.23.

In these two cases the dipole arrows cancel each other out because of the shape of the molecules. The linear shape of the molecule of carbon dioxide puts the dipole arrows in opposite directions to counterbalance each other. The same holds true for the tetrahedral molecular geometry found in carbon tetrachloride. Despite having polar bonds, these two molecules are nonpolar. There is no overall dipole moment in these molecules because the dipole arrows are of the same magnitude but lie in opposite directions in the molecule. This counterbalance causes the molecule to be nonpolar.

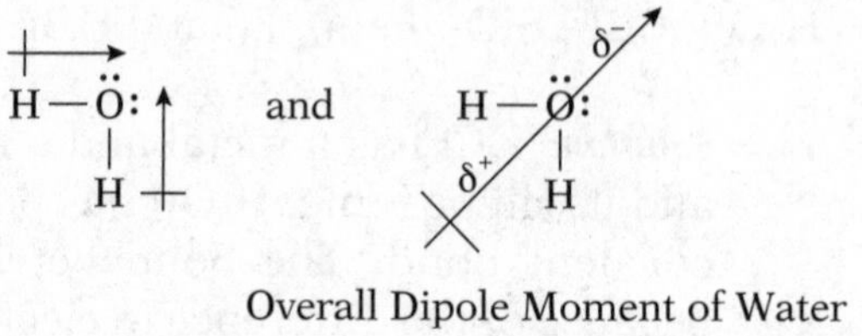

Overall Dipole Moment of Water

Figure 5.21 The Dipole Moment of Water

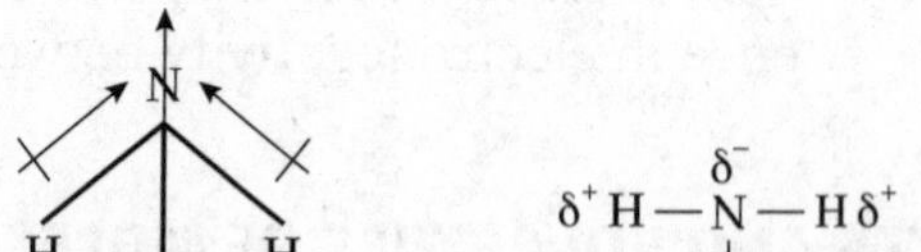

Figure 5.22 The Dipole Moment of Ammonia

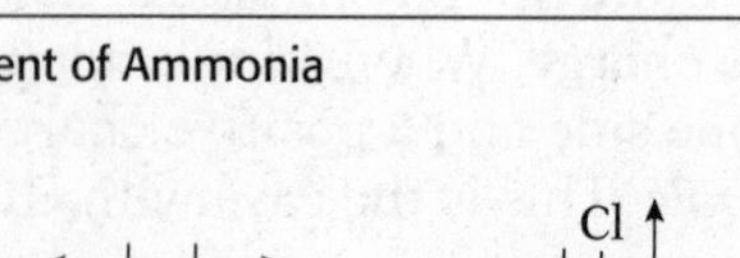

Figure 5.23 Nonpolar Molecules

HYDROGEN BONDING

Hydrogen bonding is a weak force that comes about when hydrogen is bonded to fluorine, oxygen, or nitrogen. A good mnemonic device to use is "We heard about hydrogen bonding on the FON (phone)." When hydrogen is bonded to fluorine, oxygen, or nitrogen, the hydrogen will form a weak hydrogen bond with a neighboring fluorine, oxygen, or nitrogen atom. The dashed line in Figure 5.24 shows the hydrogen bonds being formed between the hydrogen and oxygen atoms in water.

One important application of hydrogen bonding lies in our genetic code— DNA. In a molecule of DNA, two strands are held side by side at the nitrogen bases. It is hydrogen bonding that holds the nitrogen base of one strand to a nitrogen base of the second strand. Hydrogen bonds are strong enough in their greater numbers to hold the strands side by side and help DNA create its double helix structure. However, when DNA replicates, the hydrogen bonds are weak enough to be broken so that each strand can be replicated individually. (See Figure 5.25.)

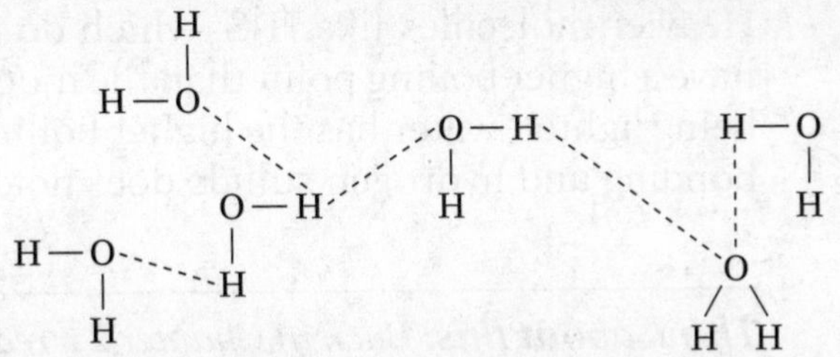

Figure 5.24 Hydrogen Bonding in Water

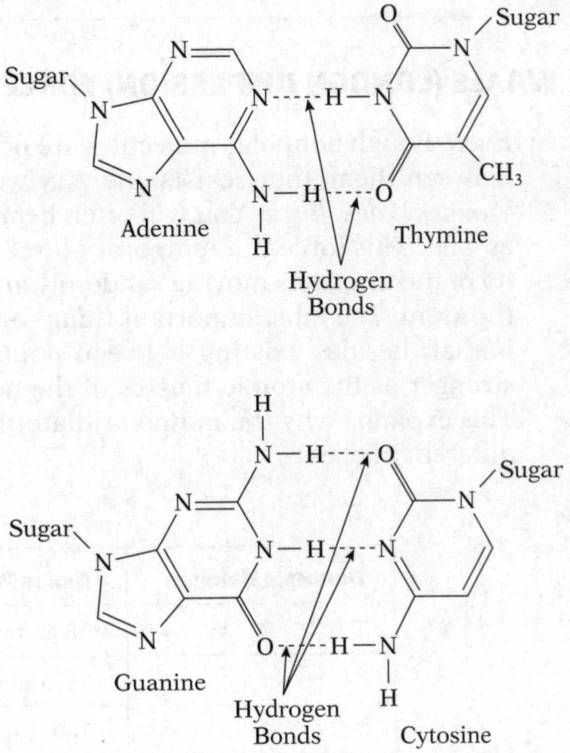

Figure 5.25 Hydrogen Bonding in DNA

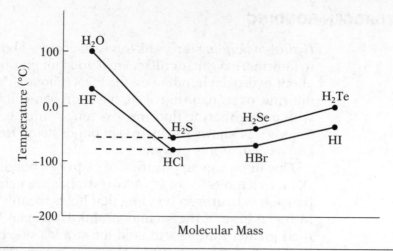

Figure 5.26 Difference in Boiling Points of H_2O and H_2S

Hydrogen bonds are also what give water an unusually high boiling point. Heavier molecules like H_2S, which do not exhibit hydrogen bonds, should have a higher boiling point than the much lighter water molecule. But despite being lighter, water has the higher boiling point because it exhibits hydrogen bonding and hydrogen sulfide does not. (See Figure 5.26.)

Think about this: Back in Chapter 2 a reason was offered for why "The watched pot never boils." With dipole interactions between the water molecules and hydrogen bonding present as well, no wonder it takes water so long to boil, high atmospheric pressure or not!

VAN DER WAALS (LONDON DISPERSION) FORCES

Even though nonpolar molecules are not thought of as having any attraction between them, there exists one very weak, temporary attraction called the *Van der Waals force.* You will often hear this force called the *dispersion force* as well. This force is a temporary force that comes about from the possibility of the electrons moving randomly and creating an uneven charge around the atom. The most important thing to know about the Van der Waals force is that, besides existing between nonpolar molecules, the force becomes stronger as the atomic masses of the nonpolar molecules becomes greater. This explains why the nonpolar diatomic molecular halogens exist in three different phases.

Diatomic Halogen	Molar Mass	Phase at STP
F_2	38 grams/mole	Gas
Cl_2	71 grams/mole	Gas
Br_2	160 grams/mole	Liquid
I_2	254 grams/mole	Solid

As the atomic mass increases, so do the Van der Waals forces between the molecules. This causes the molecules to be held together more tightly as the atomic masses increase. Iodine, the heaviest of the halogens listed, has the greatest mass and the greatest Van der Waals forces and exists as a solid.

PROBLEM: Of these three nonpolar substances, C_8H_{18}, $C_{20}H_{42}$, and CH_4, one of them is a solid, one is a liquid, and one is a gas at STP. Which is which?

Solution: $C_{20}H_{42}$ with the greatest mass of the three is expected to be the solid. CH_4 is the lightest in mass of the three and is expected to be a gas. C_8H_{18} by process of elimination is the liquid. Just to let the cat out of the bag, C_8H_{18} is called octane and it is the liquid that goes into the gas tanks of cars. $C_{20}H_{42}$ is called wax and exists as a solid. CH_4 is methane gas.

MOLECULE-ION ATTRACTION

When salts are placed in water, they dissolve (some more than others) and form ions. After a soluble salt has been dissolved in water there is going to be an attraction between the charged ions and the water molecules that are polar. This is called the *molecule-ion attraction*. Because of these forces of attraction, the hydrated ions in solution cause the water molecules to orient themselves in a particular fashion. The oxygen portion of the water molecule has a negative charge because of its high electronegativity. This negative part of the water molecule will be attracted to the cations in solution. Because the hydrogen atoms in water have a slight positive charge, they will orient themselves toward the anions in solution. Such is the case when NaCl is dissolved in water as shown in Figure 5.27.

PROBLEM: What type of bonding is found between the following molecules and atoms: $F_2(g)$, $Ne(g)$, $KBr(aq)$, $H_2O(l)$, $NH_3(l)$, and $CH_4(g)$?

Solution: Fluorine gas is a nonpolar gas and will experience Van der Waals forces. Ne is a noble gas and will experience Van der Waals forces as well. KBr(aq) is an ionic compound that has been dissolved in water. The forces of attraction between the ions and water molecules are molecule-ion forces. Water and ammonia molecules will experience both dipole forces and hydrogen bonding. Finally, because methane gas is a nonpolar gas it will experience Van der Waals forces between the molecules.

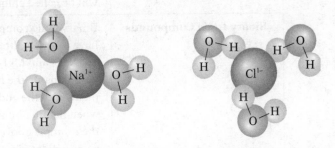

Figure 5.27 Molecule-Ion Attraction

This chart provides a handy summary of intermolecular bonding.

Dipole Forces	The molecule is polar if: • There exists polar bonds AND • The molecule is not symmetrical AND • There is no counterbalance of the dipole arrows. The molecule is nonpolar if: • There are no polar bonds at all OR • There exists polar bonds while the dipole arrows counterbalance because of symmetry in the molecule.
Hydrogen Bonding	Hydrogen bonding will exist if there are hydrogen atoms bonded to the atoms fluorine, oxygen, or nitrogen in a molecule. Remember: "FON."
Van der Waals (London Dispersion) Forces	Nonpolar molecules will experience temporary attractions as electrons move randomly. Van der Waals forces become stronger as the atoms and molecules become heavier and heavier.
Molecule-Ion Attraction	Ions dissolved in a polar substance like water cause the molecules of the polar substance to orient themselves so that the charges of the polar substance are attracted to the charges of the ions.

NAMING COMPOUNDS

The ability to name compounds and determine the chemical formula for a compound comes from the ability to distinguish between ionic and covalent compounds. The name of a compound depends heavily on the type of bond present between the atoms. Besides being able to identify certain types of bonds, when learning to name compounds it is best to remember the rules that apply to the type of bond in question. The rules for naming four common kinds of compounds are outlined below.

Single Atom Anions	Single atom anions, like Cl^{1-} end in -*ide*. This anion will be called chlor*ide*.
Binary Ionic Compounds	Binary ionic compounds are ionic compounds that have just two different elements present. Examples are $NaCl$ and MgI_2. When naming a binary ionic compound, name the metal first and then name the nonmetal with the ending -*ide*. There are no prefixes used when naming binary ionic compounds. The names of the example above are sodium chlor*ide* and magnesium iod*ide*.

Polyatomic Ions	Polyatomic ions are ions that have many atoms bonded together. It is best to become familiar with the names and charges of these compounds. A listing can be found in the reference tables in Appendix 4 in the back of this book. An example of a polyatomic ion that has been presented earlier in this chapter is NH_4^{1+}, the ammonium ion. The compound NH_4Cl would be called ammonium chloride.
Covalent Compounds	The names of covalent compounds differ from those of ionic compounds in that covalent compounds require the use of a prefix to indicate the number of atoms present. Examples are CO carbon *mon*oxide and CO_2 carbon *di*oxide. Other prefixes that can be used are *tri-* and *tetra-*.

PROBLEM: Name the following: S^{2-}, N^{3-}, CaF_2, K_2S, NaOH, Na_2SO_4, SO_3, and CCl_4.

Solution: The first two ions are single atom anions and are called sulfide and nitride. The next two are binary ionic compounds, calcium fluoride and potassium sulfide. The polyatomic ions hydroxide and sulfate are present in sodium hydroxide and sodium sulfate. Finally, the last two compounds are covalently bonded and are called sulfur trioxide and carbon tetrachloride.

■ DETERMINING CHEMICAL FORMULAS

This chapter has examined different ways for atoms to bond together. For certain problems it was noted that more than one atom of an element was needed to help complete the octet of another. Determining the chemical formula of a compound by putting together a Lewis structure is a tedious, laborious method. The crisscross method is a faster method to use provided that you understand how to find the charge of an ion. With the crisscross method, the number in the charge of one element becomes the number in the subscript in the other element. Here's how this works in determining the chemical formula for magnesium chloride. First, determine the charges of the ions present in Mg^{2+} and Cl^{1-}. Second, exchange the numbers in the ionic charge so that they become subscripts as in Figure 5.28.

Mg^{2+} ⟩⟨ Cl^{1-} Becomes Mg_1Cl_2

Figure 5.28 Using the Crisscross Method

Finally, use the lowest ratio of subscripts if the compound is ionic, $MgCl_2$. The crisscross method also works for compounds containing polyatomic ions. One note of caution, however: when there are multiple units of a polyatomic ion present, you must use parentheses to indicate this fact. Using this note of caution you can determine the chemical formula for calcium phosphate. Calcium, a group 2 metal, will take on a charge of 2+, and using the reference tables you see that phosphate has a 3+ charge. So you have Ca^{2+} and PO_4^{3-}. Next, crisscross the numbers in the charges as in Figure 5.29.

$$Ca^{2+} \diagdown\diagup PO_4^{3-} \quad \text{Becomes } Ca_3(PO_4)_2$$

Figure 5.29 Using the Crisscross Method

Notice how the charges are no longer written in the formula. Also take note of the parentheses used around the polyatomic phosphate ion.

PROBLEM: Give the chemical formulas for ammonium sulfate and potassium dichromate.

> *Solution:* Ammonium ion NH_4^{1+} and sulfate ion SO_4^{2-} can crisscross to become $(NH_4)_2SO_4$. Again, emphasis is placed on the use of parentheses. Potassium forms a 1+ ion and dichromate is $Cr_2O_7^{2-}$. After using the crisscross method the formula becomes $K_2Cr_2O_7$.

STOCK METHOD

The *stock method* for naming compounds helps clarify the names of compounds that contain the transition elements. For example, there are two types of iron chlorides, $FeCl_2$ and $FeCl_3$. How can you distinguish one from the other? In $FeCl_2$ the iron has an ionic charge of 2+, whereas in $FeCl_3$ the iron has an ionic charge of 3+. A Roman numeral in parentheses indicates the charge present on the cation. Therefore, $FeCl_2$ is called iron(II) chloride and $FeCl_3$ is called iron(III) chloride. After seeing the charge of the cation you can then apply the rules for the crisscross method.

PROBLEM: What are the chemical formulas for tin(IV) fluoride and lead(IV) oxide?

> *Solution:* Using the crisscross method, tin(IV) fluoride will be SnF_4. When using the crisscross method for lead(IV) oxide, initially the formula will look like Pb_2O_4 but remembering that all ionic compounds are written in the lowest ratios, the real formula is PbO_2.

▬▬ CHAPTER REVIEW QUESTIONS

1. Which substance has a polar covalent bond between its atoms?

   (A) K_3N
   (B) Ca_3N_2
   (C) NaCl
   (D) F_2
   (E) NH_3

2. Which kinds of bonding can be found in a sample of $H_2O(l)$?

   (A) Hydrogen bonds only
   (B) Nonpolar covalent bonds only
   (C) Ionic and nonpolar hydrogen bonds
   (D) Both polar covalent and hydrogen bonds
   (E) Metallic and ionic bonds

3. When an ionic compound is dissolved in water, the ions in solution can best be described as

   (A) hydrated molecules only
   (B) dehydrated ions and molecules
   (C) both hydrated molecules and hydrated ions
   (D) neither hydrated ions nor hydrated molecules
   (E) hydrated ions only

4. Which substance represents a molecule that can combine with a proton (H^{1+})?

   (A) NH_3
   (B) Na^{1+}
   (C) HCl
   (D) H_3O^{1+}
   (E) H

5. Which compound contains no ionic character?

   (A) NH_4Cl
   (B) CaO
   (C) K_2O
   (D) Li_2O
   (E) CO

6. The forces of attraction that exist between nonpolar molecules are called

   (A) Van der Waals / dispersion forces
   (B) ionic bonds
   (C) covalent bonds
   (D) electrovalent bonds
   (E) metallic bonds

7. Which substance is a network solid?

   (A) Li_2O
   (B) SiO_2
   (C) H_2O
   (D) CO_2
   (E) NaCl

8. Which molecule is a polar molecule?

   (A) N_2
   (B) H_2O
   (C) CH_4
   (D) CO_2
   (E) KCl

9. Which is the chemical formula for iron(III) sulfate?

   (A) Fe_2SO_4
   (B) Fe_3SO_4
   (C) $Fe(SO_4)_3$
   (D) $Fe_2(SO_4)_3$
   (E) Fe_2S_3

10. In which of the following compounds are hydrogen bonds between molecules the strongest?

    (A) HF
    (B) HCl
    (C) HBr
    (D) HI
    (E) HAt

11. When a salt dissolves in water, the water molecules are attracted by ions in solution. This attraction is called
    (A) atom-atom
    (B) molecule-molecule
    (C) molecule-ion
    (D) ion-ion
    (E) atom-ion

12. Which element is expected to have a "sea" of electrons?
    (A) Hydrogen
    (B) Nitrogen
    (C) Cobalt
    (D) Chlorine
    (E) Oceanium

13. In which of the following liquids are the Van der Waals forces of attraction between the molecules weakest?
    (A) Xe
    (B) Kr
    (C) Ar
    (D) Ne
    (E) He

14. Which molecule has both nonpolar intra-molecular and nonpolar intermolecular bonds?
    (A) CCl_4
    (B) CO
    (C) HF
    (D) HCl
    (E) F_2

15. The name of the compound $MgBr_2$ is
    (A) manganese bromite
    (B) manganese bromide
    (C) magnesium bromite
    (D) magnesium bromide
    (E) magnesium dibromide

16. The anion S^{2-} is called
    (A) sulfide
    (B) sulfite
    (C) sulphorus
    (D) sulfuron
    (E) sulfate

17. The compound PF_5 is called
    (A) monophorofluoride
    (B) phosphorus pentafluoride
    (C) pentaphosphoro fluoride
    (D) phosphorus tetrafluoride
    (E) potassium pentafluoride

18. Element X forms the compounds XCl_3 and X_2O_3. Element X would most likely belong to the group called
    (A) alkali metals
    (B) alkaline earth metals
    (C) group 13
    (D) halogens
    (E) noble gases

19. A nonmetal (X) reacts with a metal (M) to give the formula of M_2X. Which pairing below is most like elements represented by M and X?
    (A) M = Ca and X = N
    (B) M = Li and X = S
    (C) M = Si and X = O
    (D) M = Rb and X = F
    (E) M = Mg and X = Cl

20. How many sigma and pi bonds are found in the following molecule?
    $H-C\equiv C-CH_2-CH_2-CH=CH_2$
    (A) There are 3 pi bonds and 13 sigma bonds.
    (B) There are 12 sigma bonds and 5 pi bonds.
    (C) There are 12 sigma bonds and 2 pi bonds.
    (D) There are 2 pi bonds and 4 sigma bonds.
    (E) There are 8 sigma bonds and 2 pi bonds.

ANSWERS:

1. (E)	2. (D)	3. (E)
4. (A)	5. (E)	6. (A)
7. (B)	8. (B)	9. (D)
10. (A)	11. (C)	12. (C)
13. (E)	14. (E)	15. (D)
16. (A)	17. (B)	18. (C)
19. (B)	20. (A)	

CHAPTER 6

STOICHIOMETRY AND SOLUTION CHEMISTRY

IN THIS CHAPTER YOU WILL LEARN ABOUT . . .

Chemical Formulas
Balancing Chemical Equations
Mole Ratios
Moles, Mass, Volumes, and Molecules
Limiting and Excess Reagents
Percent Composition
Empirical Formulas from Percent Composition
Concentration and Dilutions
Colligative Properties
Solubility of Compounds
Net Ionic Equations

Stoichiometry is the branch of chemistry that deals with the amounts of products produced from certain amounts of reactants. Most of the chemistry discussed so far in this book has dealt with *what* is present (qualitative chemistry). The next step is to examine *how much* is present (quantitative chemistry). You may want to refer to Appendix 1, Mathematical Skills Review, in the back of this book.

CHEMICAL FORMULAS

Chemical formulas not only tell which elements are present in a compound but also how much of each element is present. There are three different types of chemical formulas that need to be examined. The chart below shows the major differences between them.

Symbols	You might think that chemical *symbols* tell only which element is present. For example, the chemical symbol C simply means the element carbon. In fact, the symbol C in an equation tells you not only that carbon is present, but also that one mole or one atom of carbon is present.

Molecular Formulas	*Molecular formulas* indicate the total number of atoms of each element that are present in a covalently bonded molecule. An example is CH_4, which indicates that there is one carbon atom and four hydrogen atoms in this covalently bonded molecule of methane.
Empirical Formulas	Ionic compounds form lattices that have an almost endless number of ions bonded together. Because it is impossible to count every ion in a sample, the lowest ratio of the elements present in the compound is used. A great example of this is NaCl. In every sample of NaCl there is one sodium ion to every chlorine ion. This does not mean that *empirical formulas* are limited to only ionic compounds. For example, if you examine the empirical formula of glucose $C_6H_{12}O_6$ and see that it is CH_2O, then you can better understand the term carbohydrate to mean "hydrated carbon."

Think about this: *If you had a 58.5-gram sample of NaCl and were reluctant to use an empirical formula, you would have to write it out as:* $Na_{602,000,000,000,000,000,000,000}Cl_{602,000,000,000,000,000,000,000}$. *Not exactly convenient.*

PROBLEM: A compound has an empirical formula of CH_2 and a molar mass of 70 grams/mole. What is the molecular formula of this compound?

> **Solution:** The empirical formula CH_2 has a mass of 14. Divide 70 by 14 to find that there are five units of CH_2 in this compound. This means that the molecular formula is C_5H_{10}.

BALANCING CHEMICAL EQUATIONS

Balancing a chemical equation requires an understanding of the *Law of Conservation of Mass*, which says that mass cannot be created or destroyed. The amount of mass in the reactants will be the amount of mass in the products. The credit for this discovery is given to Antoine Lavoisier, who took very careful measurements of the quantities of chemicals and equipment that he used. Conservation of mass also holds true when balancing equations. The number of atoms of each element in the reactants will be equal to the number of atoms of each element in the products. A useful mnemonic device for conservation of mass is "What goes in, must come out."

The two most important rules to remember when balancing equations are:

- You may change only the *coefficients*.
- You must use the lowest whole number coefficients.

Now add one more unofficial rule to those listed above:

- Leave the simplest substance until last.

Step by step you can now balance an equation. Try this example:

$$Al + O_2 \rightarrow Al_2O_3$$

Inspection shows that you should leave the reactant Al for last because it is the "simplest" and not bonded to any other elements. Inspection also shows that two oxygen atoms enter the reaction and three leave the reaction. The numbers two and three are factors of the number six. Therefore you can alter the coefficients that are before the substances containing oxygen and get:

$$Al + 3O_2 \rightarrow 2Al_2O_3$$

This now shows a total of six oxygen atoms as reactants and six oxygen atoms as products. The coefficients are multipliers and not only modify the number of oxygen atoms, but modify the number of aluminum atoms as well in aluminum oxide. Now you have four atoms of aluminum on the right side of the equation. To balance this, place a coefficient of 4 before the Al on the reactant side and get:

$$4Al + 3O_2 \rightarrow 2Al_2O_3$$

Does it all add up? Four aluminum atoms and six oxygen atoms are on the reactant side of the equation and four aluminum and six oxygen atoms are on the product side of the equation.

PROBLEM: Balance the following chemical equations:

1. $Zn + HCl \rightarrow H_2 + ZnCl_2$
2. $SiO_2 + HF \rightarrow SiF_4 + H_2O$
3. $SiCl_4 + Mg \rightarrow Si + MgCl_2$
4. $H_2 + N_2 \rightarrow NH_3$
5. $SO_3 \rightarrow S + O_2$

Solutions:

1. $Zn + 2HCl \rightarrow H_2 + ZnCl_2$
2. $SiO_2 + 4HF \rightarrow SiF_4 + 2H_2O$
3. $SiCl_4 + 2Mg \rightarrow Si + 2MgCl_2$
4. $3H_2 + N_2 \rightarrow 2NH_3$
5. $2SO_3 \rightarrow 2S + 3O_2$

In addition to balancing chemical equations, you can also classify the types of reactions that occur. There are four types of reactions: *synthesis, decomposition, single replacement,* and *double replacement*. Explanations and examples of each are as follows:

- In a synthesis reaction, many substances come together to form one compound:

$$A + B \rightarrow AB$$

- In a decomposition reaction, one compound breaks down into many substances:

$$YZ \rightarrow Y + Z$$

- In a single replacement reaction, one element replaces one other element.

$$AB + C \rightarrow CB + A$$

- In a double replacement reaction, two elements "switch partners":

$$AB + XY \rightarrow AY + XB$$

PROBLEMS: Classify the five balanced equations above as single replacement, double replacement, synthesis, or decomposition.

Solutions:

1. Single Replacement
2. Double Replacement
3. Single Replacement
4. Synthesis
5. Decomposition

MOLE RATIOS

A *mole ratio* is the ratio of the number of moles of one substance to the number of moles of another substance. Because coefficients can represent moles, molecules, or atoms, you can think of a mole ratio as a "coefficient ratio." For example, look at the equation for aluminum oxide, $4Al + 3O_2 \rightarrow 2Al_2O_3$. You can pick any two substances from the equation and determine their mole ratio. The mole ratio of Al to Al_2O_3 is 4:2 or 2:1, while the mole ratio of O_2 to Al_2O_3 is 3:2. This leads to another type of problem that you might encounter. Suppose you were asked to produce 1,000 moles of Al_2O_3 for a big chemical company. How much aluminum and oxygen would you need to purchase? Start with the balanced equation: $4Al + 3O_2 \rightarrow 2Al_2O_3$.

$$\text{Set up a proportion:} \quad \frac{4Al}{x\ Al} = \frac{2Al_2O_3}{1,000\ Al_2O_3}$$

Solve: $2x = 4,000.$ $x = 2,000$ moles of Al.

To solve for the number of moles of oxygen gas, use a dimensional analysis:

$$1,000 \text{ moles of } Al_2O_3 \frac{\left(3 \text{ moles of } O_2\right)}{\left(2 \text{ moles of } Al_2O_3\right)} = 1,500 \text{ moles of } O_2$$

PROBLEM: The Haber process is a method for making ammonia according to the equation $3H_2 + N_2 \rightarrow 2NH_3$. How many moles of nitrogen gas and hydrogen gas are needed to produce 700 moles of ammonia?

Solution: Start with the balanced equation: $3H_2 + N_2 \rightarrow 2NH_3$.

$$\text{Set up a proportion:} \quad \frac{3H_2}{x\ H_2} = \frac{2NH_3}{700\ NH_3}$$

Solve: $2x = 2,100.$ $x = 1050$ moles of H_2.

Doing the same for N_2 via a dimensional analysis reveals that 350 moles of N_2 are required too:

$$\frac{700 \text{ moles of } NH_3 \left(1 \text{ mole of } N_2\right)}{\left(2 \text{ moles of } NH_3\right)} = 350 \text{ moles of } N_2$$

MOLES, MASS, VOLUMES, AND MOLECULES

You have already encountered problems involving moles, molecules, and molar masses earlier in this book. There is still one other relationship that needs to be connected with the mole and that is *molar volume*. Once you make a connection between moles and volume, mass, and molecules you will be able to solve problems easily. One very helpful mnemonic device to use is the Mole-Go-Round. Some think of this method as a way of "cheating the system," but because the SAT II exam does not require you to show work, the Mole-Go-Round is a perfectly acceptable method for achieving better results.

So what is this Mole-Go-Round? It is a simple diagram that shows a pattern between the relationship of moles and other factors. The Mole-Go-Round is shown in Figure 6.1.

Moles are in the middle as they should be because once you know how many moles there are of a sample, then you can make conversions to find out other quantities. Note these two things:

1. When converting to moles the mathematical operation is division, while "eXiting" moles requires the operation to be multiplication. (The letter "X" has been emphasized so that you remember to multiply).
2. The factor between grams and molar mass will differ depending upon the molar mass of the compound in question.

PROBLEM: Given a 22.0-gram sample of $CO_2(g)$ at STP, how many liters will this sample occupy? How many molecules are present?

Solution: This problem starts out in the "mass" portion of the Mole-Go-Round. You have 22.0 grams of CO_2. (Notice how a number [22.0], units [gram], and substance [CO_2] are carefully recorded in each step of the problem.) The next step is to convert to moles by dividing by the molar mass. The molar mass for CO_2 is 44.0 grams/mole. This gives 0.50 moles of CO_2 as shown in Figure 6.2.

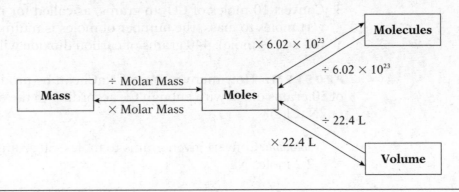

Figure 6.1 The Mole-Go-Round

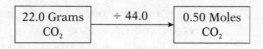

Figure 6.2 Converting Grams to Moles

The next step is to convert the 0.50 moles of CO_2 to molar volume and to the number of molecules. This is done by multiplying 0.50 moles by 22.4 liters and then multiplying 0.50 moles by 6.02×10^{23} as shown in Figure 6.3.

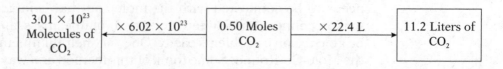

Figure 6.3 Converting Moles to Volume and Number of Molecules

Mass and Volume Problems

Knowing how to convert moles to mass and to volume opens up a range of other types of problems that can be solved. Earlier in this chapter, you looked at how many moles of reactants it would take to produce a certain number of products. But you might also be asked to produce a certain amount of product in grams instead of moles or to find the amounts of reactants needed in grams as well. This type of problem is called a mass-mass problem, and it can be solved with the help of the Mole-Go-Round and three simple steps.

Consider the following: $C(s) + O_2(g) \rightarrow CO_2(g)$. How many grams of carbon dioxide can be formed from the burning of 120 grams of carbon? Assume an abundant amount of oxygen.

1. 120 grams of C is converted to moles by dividing by the atomic mass of carbon. 120 grams divided by 12 is 10 moles of C(s).
2. Use the mole ratio from the balanced equation and substitute the new amount of reactants:

$$\frac{1 \text{ mole C}}{10 \text{ moles C}} = \frac{1 \text{ mole } CO_2}{x \text{ moles } CO_2}$$

10 moles of carbon dioxide are produced because for everyone one mole of carbon used, one mole of carbon dioxide is produced.
3. Convert 10 moles of CO_2 to grams, as called for in the problem. To convert moles to mass, the number of moles is multiplied by the molar mass (44 grams/mole). 440 grams of carbon dioxide will be produced.

PROBLEM: How many grams of NaCl can be produced from the reaction of 50 grams of Na with enough Cl_2 according to the equation $2Na(s) + Cl_2(g) \rightarrow 2NaCl(s)$?

Solution: Convert given grams to moles. 50 grams Na ÷ 23 grams/mole = 2.2 moles Na.

$$\text{Use the mole ratio } \frac{2 \text{ moles Na}}{2.2 \text{ moles Na}} = \frac{2 \text{ moles NaCl}}{x \text{ moles NaCl}}$$

and find that an equal number of moles of NaCl are produced; 2.2 moles of NaCl are produced. Convert 2.2 moles of NaCl to grams of NaCl by multiplying by the molar mass of 58.5 grams/mole = 128.7 grams of NaCl.

Coefficients can indicate still another quantity: *volume*. When reactions contain gases, provided that the conditions are the same (temperature and pressure), the coefficients can indicate the number of "volumes" of a gas that are present. Look at the burning of methane gas: $CH_4(g) + 2O_2(g) \rightarrow 2H_2O(g) + CO_2(g)$. The mole ratio for these gases is also their volume ratio. Let's consider the production of the famous greenhouse gas CO_2. How many liters of CO_2 can be produced from the burning of 50 liters of CH_4? Set up a proportion and solve:

$$\frac{1 \text{ "volume" } CH_4}{50 \text{ liters } CH_4} = \frac{1 \text{ "volume" } CO_2}{x \text{ liters } CO_2}$$

The proportion set up above dictates that 50 liters of carbon dioxide gas are produced.

PROBLEM: Given the reaction: $C_3H_8(g) + O_2(g) \rightarrow CO_2(g) + H_2O(g)$, balance the equation. If 100 liters of C_3H_8 are burned, how many liters of CO_2 are produced?

Solution: The balanced equation is $C_3H_8(g) + 5O_2(g) \rightarrow 3CO_2(g) + 4H_2O(g)$. (Did you place the coefficient for O_2 in the last step in balancing the equation?) Next, set up the volume proportion:

$$\frac{1 \text{ liter } C_3H_8}{100 \text{ liters } C_3H_8} = \frac{3 \text{ liters } CO_2}{x \text{ liters } CO_2}$$

and find that 300 liters of CO_2 are produced.

LIMITING AND EXCESS REAGENTS

Atoms and molecules react in specific ratios and amounts as shown throughout this chapter. At the heart of the proportions and ratios is the mole. What happens when reactants aren't measured out in specific amounts but are simply "thrown together"? Will all of the reactants react? This is not the case with chemical reactions. Reactants react in certain proportions and ratios and, at times, there will be excess reagents left over.

Consider the baloney sandwiches made for your lunch when you went to elementary school. Let's say that your favorite sandwich had two slices of bread and three slices of baloney. In the refrigerator there are four slices of bread and seven slices of baloney. You are in luck because now you can make two sandwiches from four slices of bread and six slices of baloney. But one of the original slices of baloney is now left over. One slice of baloney is in excess. The "leftovers" in a chemical reaction are called *excess reagents*. The substances that are used up completely are called *limiting reagents*. The limiting reagent for your lunch was the bread.

Consider the following reaction: $HCl + NaOH \rightarrow NaCl + H_2O$. The HCl and NaOH are consumed in a 1:1 ratio. If two moles of HCl reacted with one mole of NaOH, one mole of HCl is in excess because only one mole was needed to react with one mole of NaOH. Now consider the following: $2H_2 + O_2 \rightarrow 2H_2O$. How many grams of water can be made from 8.0 grams of H_2 and 96.0 grams of O_2? Start by converting to moles because the balanced equation shows a mole ratio and not a gram ratio.

1. 8.0 grams of H_2 ÷ 2 grams/mole = 4.0 moles H_2.
2. 96.0 grams of O_2 ÷ 32 grams/mole = 3.0 moles O_2.

In order to react 4.0 moles of H_2 you need 2.0 moles of O_2 because, as dictated by the balanced equation, two moles of hydrogen gas react with one mole of oxygen gas, a 2:1 ratio. Because 3.0 moles of oxygen gas are present and only 2.0 moles are needed, oxygen gas is in excess and the hydrogen gas is the limiting reagent. Here is another way to look at it: in order to react all 3.0 moles of oxygen gas, you would need 6.0 moles of hydrogen gas, an amount that is not available and again labels hydrogen gas as the limiting reagent.

PROBLEM: Potassium and bromine will react according to the equation: $2K + Br_2 \rightarrow 2KBr$. If 117 grams of potassium are reacted with 160 grams of bromine, how many moles of KBr can be produced?

Solution: Convert the known masses from grams to moles:

117 grams of K ÷ 39 grams/mole = 3.0 moles of K

160 grams of Br_2 ÷ 160 grams/mole = 1.0 mole Br_2

By inspection you see that three moles of K would require 1.5 moles of Br_2 because, according to the balanced equation $2K + Br_2 \rightarrow 2KBr$, the ratio of K to Br_2 is 2:1. The amount of bromine present is just 1.0 mole and this would make the bromine the limiting reagent. Because only 1.0 mole of Br_2 can react, only 2.0 moles of the potassium will be used. This would yield only 2.0 moles of KBr, as dictated by the balanced equation.

PERCENT COMPOSITION

Percent composition, also called percent by mass, is a useful piece of data to obtain when looking at the composition of certain substances. In percent composition problems you are asked to find the percent of the mass of an element in a compound as compared to the molar mass of the compound. A simple ratio will suffice and the result is multiplied by 100%. For $CaCl_2$, for example, what percent of this compound is made up of chlorine? The total mass is

(1 Ca atom × 40 = 40) + (2 Cl atoms × 35.5 = 71) = 111 grams/mol

The percent by mass of chlorine is $\frac{71}{111}$ × 100% = 64% chlorine.

PROBLEM: Find the percent by mass of each element in the compound $C_6H_{12}O_6$ (molar mass = 180).

Solution:

The percent carbon is $(72/180) \times 100\% = 40\%$

The percent hydrogen is $(12/180) \times 100\% = 6.7\%$

The percent oxygen is $(96/180) \times 100\% = 53.3\%$

EMPIRICAL FORMULAS FROM PERCENT COMPOSITION

In the previous section the problems presented a chemical formula and asked for the percent composition. There is a method for going from percent composition to chemical formula; however, you will obtain only the empirical formula from this. The three steps in determining the empirical formula of a compound from the percent composition are as follows:

1. Assume a 100-gram sample. This will allow the percent signs to be written as "grams."
2. Convert the number of grams of each element into moles of each element.
3. Divide the number of moles of each element by the number of moles that is the smallest of all the numbers.

Let's put theory into practice: A sample is found to contain 58.80% Ba, 13.75% S, and 27.45% O. What is the empirical formula of this substance?

1. Assume a 100-gram sample. This allows the percent signs to become "grams" and leaves us with 58.80 grams of Ba, 13.75 grams of S, and 27.45 grams of O.
2. Convert grams to moles:

    58.80 grams of Ba ÷ 137.34 grams/mole = 0.43 moles of Ba

    13.75 grams of S ÷ 32 grams/mole = 0.43 moles of S

    27.45 grams of O ÷ 16 grams/mole = 1.72 moles of O

3. As of now the compound looks like $Ba_{0.43} S_{0.43} O_{1.72}$. This is an absurd way of writing an empirical formula! So now the numbers of moles are divided by the lowest number of moles, 0.43. This leaves the empirical formula as $BaSO_4$.

PROBLEM: A compound is 14.6% C and 85.4% Cl by mass. This compound also has a molar mass of 166 grams/mole. What are the empirical and molecular formulas of this compound?

Solution: First find the empirical formula:

1. You were given 14.6% C and 85.4% Cl; assuming a 100-gram sample you have 14.6 grams of C and 85.4 grams of Cl.
2. 14.6 grams of C ÷ 12 grams/mole = 1.22 moles of C.
    85.4 grams of Cl ÷ 35.5 grams/mole = 2.41 moles of Cl.

3. Divide by the lowest number of moles $\frac{C_{1.22}}{1.22} = C_1$ and $\frac{Cl_{2.41}}{1.22} = Cl_2$

The empirical formula is CCl_2.

4. The mass of empirical formula CCl_2 is 83 grams/mole. This compound has a molar mass of 166 grams/mole. Dividing 166 by 83 tells that there are 2 units of the empirical formula CCl_2. The molecular formula of this compound is C_2Cl_4.

CONCENTRATION AND DILUTIONS

Two cups of tea are presented to two individuals who are enjoying their time together. Person A has ordered one cup (8 ounces) of tea with one lump of sugar. Person B has ordered a "double" and gets a cup of tea that is twice as large (16 ounces). Person B drops two lumps of sugar into the larger cup of tea. An argument ensues over whose tea is sweeter. Person B believes that the larger cup of tea is sweeter because it has twice as many lumps of sugar in it. What should Person A say to Person B?

Concentration can be expressed in many different ways. No matter which way concentration is expressed, it always has one thing in common. It is a ratio of solute to solvent. A solute is a substance that changes phase when dissolved; for example, $NaCl(s)$ becomes $NaCl(aq)$ when dissolved in water. The solvent does not change phase when something is dissolved in it.

The most common way of expressing concentration in chemistry is *molarity*. Molarity is the ratio of moles of solute to total liters of solution:

$$M = \frac{\text{number of moles of solute}}{\text{total liters of solution}} \text{ or } M = \frac{\text{moles}}{\text{liters}}$$

Notice the term "total liters of solution." This does not mean the volume of the solvent as the denominator considers the total volume of the solute and solvent together.

PROBLEM: What is the molarity of a solution that is made by dissolving 100 grams of NaCl in enough water to make 750 mL of solution?

Solution:

1. Convert the grams of NaCl to moles by dividing by the molar mass of NaCl, 58.5 g/mol.

$$100 \text{ grams} \div 58.5 \text{ grams/mole} = 1.71 \text{ moles of NaCl}$$

2. The 750 mL of solution need to be expressed in liters, so convert

$$750 \text{ mL } (1 \text{ liter}/1000 \text{ mL}) = 0.750 \text{ liters}$$

3. Solve:

$$M = \frac{1.71 \text{ moles NaCl}}{0.750 \text{ liters of sol'n}} = 2.28 \ M \text{ NaCl}$$

Now take the 2.28 M NaCl solution made above and add water to the solution. Has the concentration changed? Adding water alters the denominator and increases its value. A larger denominator means a lower overall value, and the molarity should decrease. If 0.250 liters of water were added to make the total volume 1.00 liters of solution, what would the new molarity be? Use the equation $M_1V_1 = M_2V_2$ to solve for the new molarity of a solution that has been diluted. The original molarity was 2.28 M (M_1), the original volume was 0.750 L (V_1) and the new volume is 1.00 L (V_2). The new molarity should decrease because of the addition of the water. So, $M_1V_1 = M_2V_2$ and substitution gives:

$$(2.28\ M)(0.750\ L) = (M_2)(1.00\ L)$$

Solving gives a new molarity = 1.71 M NaCl. The concentration has decreased as predicted.

PROBLEM: A 1.20-liter solution of 0.50 M HCl is diluted to make 2.0 liters of solution. What is the new molarity of this solution?

Solution: Using the equation $M_1V_1 = M_2V_2$, substitute and get:

$$(0.50\ M)(1.20\ L) = (M_2)(2.0\ L)$$

The final molarity is 0.30 M HCl.

COLLIGATIVE PROPERTIES

The directions for cooking pasta sometimes call for adding salt to the water in which the pasta is cooked. Adding a solute to a solvent changes the properties of the solvent. Some of the properties that change are boiling points, freezing points, and vapor pressure. The degree of change that can be brought about depends upon the concentration of the particles in solution. This way of expressing concentration is called *molality*.

Molality can be defined as: $m = \dfrac{\text{moles of dissolved particles}}{\text{kilograms of solvent}}$

Notice that the equation calls for "moles of dissolved particles." Ionic compounds can dissociate in solution and form a number of particles in solution. For example, if a 1.0-molal solution of NaCl were prepared, it would "act" as if it were 2.0 molal in nature. This is because every one mole of NaCl releases two moles of ions (Na^+ and Cl^-). This is a huge difference from a compound like glucose, $C_6H_{12}O_6$, where the atoms are covalently bonded and will not dissociate in solution. That means a 1.0-molal solution of glucose will be one molal in particles even though one mole of glucose has more atoms than one mole of NaCl has ions.

PROBLEM: 222 grams of $CaCl_2$ are dissolved in 2.50 kg of water. What is the resulting molality of this solution?

Solution: First convert the grams of $CaCl_2$ to moles by dividing by the molar mass (molar mass = 111). This yields 2.0 moles of $CaCl_2$. However, the solution does not have 2.0 moles of particles in it. Because calcium chloride is ionic and it contains 3 moles of ions per mole, the solution contains 6.0 moles of ionic particles. Substitution into the equation gives:

$$m = \frac{6.0 \text{ moles of dissolved particles}}{2.50 \text{ kilograms of solvent}} \text{ and } m = 2.4 \, m \, CaCl_2$$

Now that you can calculate the molality of a solution, you can further examine how much a *colligative property* can be changed once a certain amount of particles have been dissolved in solution. When looking at the effects on boiling point and freezing point, remember the following mnemonic device: "The rich get richer and the poor get poorer," meaning that the boiling point elevates while the freezing point depresses.

The boiling point of water will increase by a constant of 0.52°C for every 1 m of solute dissolved in solution. Consider the 2.4 m CaCl$_2$ solution made earlier. What will be the boiling point for this solution? A simple multiplication reveals that the increase will be 2.4 m (0.52°C / 1 m) = 1.25°C. But this is not the new boiling point of the solution. Remember that the original boiling point was 100°C. Now the boiling point is 1.25°C higher than 100°C and has become 101.25°C.

The freezing point of water will decrease by a constant of 1.86°C for every one 1 m of solute dissolved in solution. Again, consider the changes in freezing point for the water in the 2.4 m CaCl$_2$ solution. To calculate the change in freezing point, set up the following and multiply: 2.4 m CaCl$_2$ (1.86°C / 1 m) = 4.46°C lower freezing point. Because the original freezing point was 0°C, the new freezing point is –4.46°C.

Think about this: *People use a variety of substances to melt the ice on their sidewalks. Sometimes they use sodium chloride, sometimes calcium chloride. Which salt is a better choice?*

PROBLEM: 50 grams of AlCl$_3$ are dissolved in 1.1 kilograms of water. Find the molality of this solution. What is the boiling point of the solution?

Solution: Convert 50 grams of AlCl$_3$ to moles by dividing by the molar mass of 133.5 grams/mole. This gives 0.375 moles of AlCl$_3$. Because this is a soluble salt that yields four ions, the moles of the particles is 0.375 times 4 = 1.5 moles of particles. To find the molality, set up m = (1.5 moles/1.1 kg) = 1.36m AlCl$_3$. To find the increase in boiling point, multiply the molality by the constant: (1.36m)(0.52°C/1m) = 0.71°C. The new boiling point will be 100°C + 0.71°C = 100.71°C.

▨ SOLUBILITY OF COMPOUNDS

Even though it may seem like a good idea to use any salt to melt ice or change the boiling point of water, not every salt can be dissolved completely in water. One of the salts that will not dissociate 100% into its ions is AgCl. Just how much AgCl can dissolve in water will be examined later when we examine solubility products in Chapter 8. You should also know that the temperature and amount of solvent used to dissolve a salt also alter how much of the salt can be dissolved. Because different amounts of solvent can be used, a standard of 100 grams of water has been set as the norm on solubility curves.

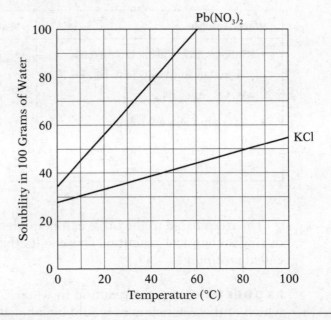

Figure 6.4 Solubility Curve

The maximum amount of a solute that can dissolve in 100 grams of water is called the solute's *solubility*. This amount is what makes a solution *saturated*. If 100 grams of a solvent have less than the maximum amount of solute dissolved, the solution is said to be *unsaturated*. If the solvent can be "tricked" into dissolving more solute than what it takes to make the solution saturated, then the solution is said to be *supersaturated*.

Gases and solids show different trends in solubility as the temperature of solution changes. In general, solids increase in solubility as temperature increases and gases decrease in solubility as temperature increases. The graph in Figure 6.4 shows the solubility of KCl and $Pb(NO_3)_2$ as the temperature of the 100-gram sample of water is heated from 0°C to 100°C.

Think about this: *Have you ever seen bubbles in a pot of water that is being heated? As the temperature of the water increases, the gases that are dissolved in the water become less soluble and can now be seen.*

While solubility curves can be quite detailed, there are some general rules that can be used to help determine if a solute is soluble in water. These rules are outlined below.

Water Soluble	Water Insoluble
• All nitrates, NO_3^{1-}, and acetates, $C_2H_3O_2^{1-}$, are soluble. • Salts of the halogens Cl, Br, and I are soluble except when they form salts with Hg, Pb, or Ag.	• Sulfides, S^{2-}, are insoluble except when they form salts with group 1 or group 2 metals or with ammonium ions. • Carbonates, CO_3^{2-}, are insoluble except when they form salts with group 1 metals or ammonium ions.

Water Soluble	Water Insoluble
• Sulfates, $SO_4{}^{2-}$, are soluble except when they form salts with Hg, Pb, Ag, Ca, Sr, or Ba. • All group I ions are soluble.	• Phosphates, $PO_4{}^{3-}$, are insoluble except when they form salts with group 1 metals or ammonium ions. • Hydroxides, OH^{1-}, are insoluble except when they form salts with group 1 metals or Ca, Sr, or Ba.

The rules listed in the table can help identify the insoluble salts that form during a chemical reaction. These salts that "settle out" of the solution are called *precipitates*.

PROBLEM: Given the reaction in water: $2KI + Pb(NO_3)_2 \rightarrow PbI_2 + 2KNO_3$, which of these substances would be labeled as aqueous (aq) or solid precipitates (s)?

Solution: KI is soluble in water; the halogen is not bonded to Hg, Pb, or Ag. $Pb(NO_3)_2$ is soluble because all nitrates are soluble. PbI_2 is not soluble in water because the iodide ion is bonded to Pb. Finally, KNO_3 is soluble because all nitrates are soluble. The final equation should look like this:

$$2KI(aq) + Pb(NO_3)_2(aq) \rightarrow PbI_2(s) + 2KNO_3(aq)$$

NET IONIC EQUATIONS

Now that you know the rules for solubility, you can take a look at exactly which substances take part in a reaction. While reactions are written to show the reactants and products in the overall reaction, not every substance plays a part in the reaction. These nonparticipating substances are called *spectators*. Consider the equation: $2KI(aq) + Pb(NO_3)_2(aq) \rightarrow PbI_2(s) + 2KNO_3(aq)$. There is a substance in this reaction that is a spectator and you can find it if you know the solubility rules and how to write a net ionic equation.

When writing a net ionic equation, first write out all soluble substances as ions in solution. Substances that are not soluble or do not dissociate into ions completely are written as shown in the overall equation. Return to the equation: $2KI(aq) + Pb(NO_3)_2(aq) \rightarrow PbI_2(s) + 2KNO_3(aq)$. First write out the soluble substances as ions in solution:

$$2K^{1+}(aq) \, 2I^{1-}(aq) + Pb^{2+}(aq) + 2NO_3{}^{1-}(aq) \rightarrow PbI_2(s) + 2K^{1+}(aq) + 2NO_3{}^{1-}(aq)$$

Next, find the substances that appear on both sides of the equation in equal amounts. This would be the potassium ions and nitrate ions. These ions appear exactly the same on both sides of the equation and they are the spectator ions in the reaction. These will cancel out and the net ionic reaction remains:

$$2I^{1-}(aq) + Pb^{2+}(aq) \rightarrow PbI_2(s)$$

PROBLEM: Give the net ionic reactions for the following reaction:

$$AgNO_3(aq) + NaCl(aq) \rightarrow AgCl(s) + NaNO_3(aq)$$

Solution: First we write out all of the aqueous ions in solution for the soluble substances:

$$Ag^{1+}(aq) + NO_3^{1-}(aq) + Na^{1+}(aq) + Cl^{1-}(aq) \rightarrow$$
$$AgCl(s) + Na^{1+}(aq) + NO_3^{1-}(aq)$$

Next we see that the sodium and nitrate ions appear on both sides of the equation so they will cancel out and we get a net ionic equation that looks like: $Ag^{1+}(aq) + Cl^{1-}(aq) \rightarrow AgCl(s)$.

PROBLEM: Give the net ionic equations for the following reaction:

$$2(NH_4)_3PO_4(aq) + 3CaCl_2(aq) \rightarrow 6NH_4Cl(aq) + Ca_3(PO_4)_2(s)$$

Solution: Although this looks complicated, the steps are still the same as in the previous two examples. Taking it one step at a time will ensure a correct answer. First write out the ions that are dissolved in solution:

$$6NH_4^{1+}(aq) + 2PO_4^{3-}(aq) + 3Ca^{2+}(aq) + 6Cl^{1-}(aq) \rightarrow$$
$$6NH_4^{1+}(aq) + 6Cl^{1-}(aq) + Ca_3(PO_4)_2(s)$$

The ions that are spectators are the ammonium and chloride ions. They will not appear in the net ionic equation: $2PO_4^{3-}(aq) + 3Ca^{2+}(aq) \rightarrow Ca_3(PO_4)_2(s)$.

▮▮ CHAPTER REVIEW QUESTIONS

1. What is the mass of 3.0×10^{23} atoms of neon gas?

   (A) 0.50 grams

   (B) 1.0 grams

   (C) 5.0 grams

   (D) 40.0 grams

   (E) 10.0 grams

2. A compound has a composition of 40% sulfur and 60% oxygen by mass. What is the empirical formula of this compound?

   (A) SO

   (B) S_2O_3

   (C) S_2O_7

   (D) SO_3

   (E) SO_2

3. What is the total number of atoms represented in one molecule of $(CH_3)_2NH$?

   (A) 5

   (B) 8

   (C) 9

   (D) 10

   (E) 12

4. A hydrocarbon has the empirical formula CH_3. A probable molecular formula for this compound could be

   (A) C_3H_3

   (B) C_2H_6

   (C) C_3H_8

   (D) C_4H_8

   (E) C_5H_{10}

5. The chemical symbol Ar could stand for

   (A) one mole of argon

   (B) one atom of argon

   (C) both a mole or an atom of argon

   (D) neither a mole or an atom of argon

   (E) one molecule of argon

6. Which salt has a solubility that is different from the other four?

   (A) AgCl

   (B) $PbBr_2$

   (C) $Ca_3(PO_4)_2$

   (D) Na_2CO_3

   (E) $Al(OH)_3$

7. A solution of a salt and 100 grams of water that can still dissolve more solute at a given temperature is classified as

   (A) unsaturated

   (B) supersaturated

   (C) saturated

   (D) anhydrous

   (E) hypertonic

8. The net ionic equation for the reaction between $CaCl_2$ and Na_2CO_3 to form calcium carbonate and sodium chloride would include all of the following except:

   (A) Ca^{2+}

   (B) CO_3^{2-}

   (C) $2Na^{1+}$

   (D) $CaCO_3$

   (E) All of the substances above would be in the net ionic equation.

9. Which solution listed below is going to have the highest boiling point?

   (A) $1.5\ m$ NaCl

   (B) $1.5\ m$ AgCl

   (C) $2.0\ m\ C_6H_{12}O_6$

   (D) $2.0\ m\ CaCl_2$

   (E) $1.0\ m\ Al_2(SO_4)_3$

10. Which equation is correctly balanced?

   (A) $Na + Cl_2 \rightarrow 2NaCl$

   (B) $CH_4 + 3O_2 \rightarrow CO_2 + H_2O$

   (C) $2KI + Pb(NO_3)_2 \rightarrow 2KNO_3 + PbI_2$

   (D) $H_2SO_4 + KOH \rightarrow K_2SO_4 + H_2O$

   (E) $C_6H_{12}O_6 + 6O_2 \rightarrow 6CO_2 + H_2O$

11. 110 grams of KF are dissolved in water to make 850 ml of solution. What is the molarity of the solution?

    (A) 0.129 *M*

    (B) 0.620 *M*

    (C) 0.002 *M*

    (D) 0.068 *M*

    (E) 2.23 *M*

12. Given one mole of $CH_4(g)$ as STP. Which statements are true?

    I. There are 6.02×10^{23} molecules present.

    II. The sample will occupy 22.4 l.

    III. The sample will weigh 16 g.

    (A) I only.

    (B) II only.

    (C) I and III only.

    (D) II and III only.

    (E) I, II, and III.

<u>Directions:</u> The following question consists of two statements. Determine whether statement I in the leftmost column is true (T) or false (F) and whether statement II in the rightmost column is true (T) or false (F).

	I		II
13.	The percent composition of oxygen in water is 33%	BECAUSE	one atom of oxygen makes up one-third of the mass of a water molecule.
14.	Gases become more soluble in water with an increase in temperature	BECAUSE	heating up water forces the water molecules closer together to hold gas molecules in solution.
15.	Calcium phosphate is a water-soluble compound	BECAUSE	all group I ions form water-soluble salts.

ANSWERS:

1. (E)	2. (D)	3. (D)	4. (B)
5. (C)	6. (D)	7. (A)	8. (C)
9. (D)	10. (C)	11. (E)	12. (E)
13. (F, F)	14. (F, F)	15. (F, T)	

CHAPTER 7

ENERGY AND CHEMICAL REACTIONS

IN THIS CHAPTER YOU WILL LEARN ABOUT . . .

Potential Energy Diagrams Revisited
Heat and Changes in Phase
Thermometry
Hess's Law
Bond Dissociation Energy
Entropy, Gibbs Free Energy, and Spontaneous Reactions

POTENTIAL ENERGY DIAGRAMS REVISITED

By definition, chemistry is the study of matter. However, matter alone does not govern the processes that take place in a chemical reaction. Energy is a huge factor that governs what will occur for all processes that occur in the universe. Changes in heat energy were touched upon in Chapter 1 with the examination of potential energy diagrams. This chapter will review these diagrams to help you put all the pieces together regarding heat and changes in heat energy.

The potential energy diagram is a diagram that can tell

- Whether a reaction is endothermic or exothermic
- The change in heat (ΔH) of a reaction
- The potential energies of the reactants and products (PER and PEP)
- The amount activation energy needed to start the reaction (E_a)
- The potential energy of the activated complex

These variables can be seen in the potential energy diagram in Figure 7.1:

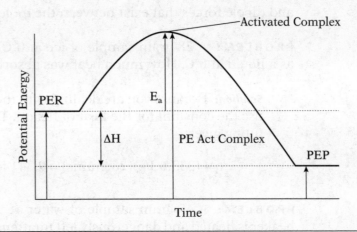

Figure 7.1 Potential Energy Diagram for an Exothermic Reaction

PROBLEM: In a chemical reaction the potential energy of the reactants is 40 kJ/mol, the potential energy of the products is 15 kJ/mol, and the activation energy is 25 kJ/mol. Find the potential energy of the activated complex and the heat of reaction.

Solution: The potential energy of the reactants is 40 kJ/mol. Add to this the activation energy, 25 kJ/mol, to find the potential energy of the activated complex, 65 kJ/mol. The heat of reaction is found using the equation $\Delta H = PEP - PER$. This comes out to be $\Delta H = 15$ kJ/mol $- 40$ kJ/mol $= -25$ kJ/mol. The negative signals that the reaction is exothermic.

HEAT AND CHANGES IN PHASE

Adding heat energy to a substance can change the temperature and phase of that substance. Removing heat can have the same effect. Chapter 2 presented the heating curve for water. In that curve there were parts of the graph that showed a change in temperature and no phase change and other parts of the curve where there was no temperature change while a phase change took place. The heat needed to make water undergo changes in phase and changes in temperature can be calculated.

When heating liquid water (a common process) you can calculate the amount of heat being absorbed by the water if you know the change in temperature. The equation $q = mc\Delta T$ is used to find this amount of heat energy. The variable q in the equation stands for heat absorbed or released, m is the mass of the sample, ΔT is the change in temperature (final temperature minus the initial temperature), and c is the specific heat of a substance (a constant of 4.18 J/g °C for water).

When there is a phase change, there is no temperature change. Therefore, finding out how much heat is needed to melt a sample of ice or boil a sample of water is not going to be feasible using the equation $q = mc\Delta T$ because ΔT will be zero. Instead, you use the equations $q = H_f m$ and $q = H_v m$ where H_f is the heat of fusion of water (333.6 J/g) and H_v is the heat of vaporization of water (2,259 J/g). Notice, from looking at the two constants, that it takes a larger amount of heat to vaporize one gram of water than to melt one gram of water from its solid state. This large value is due to the hydrogen bonds and dipole forces that exist between the molecules.

PROBLEM: A 28-gram sample of ice at 0°C completely melts and remains as a liquid at 0°C. How much heat was absorbed by the ice sample?

Solution: Because you are dealing with the melting of ice you will need to use the constant for the heat of fusion. The equation is $q = H_f m$. Substituting gives:

$$q = (333.6 \text{ J/g})(28 \text{ grams}) = 9,341 \text{ joules of heat energy}$$

PROBLEM: A 45-gram sample of water at 50°C is placed in a beaker. The beaker is heated and dangerously left unattended. In a few minutes the water has completely evaporated. How much energy was absorbed by this sample of water?

Solution: This problem requires two equations, one to calculate the amount of heat needed to heat the water to its boiling point and one to calculate the amount of heat absorbed during the change of phase from liquid to gas. First, calculate the heat needed to raise the temperature of the water from 50°C to 100°C. Set up and solve using $q = mc\Delta T$ to get:

$$q = (45 \text{ grams})(4.18 \text{ J/g °C})(100°C - 50°C) = 9,405 \text{ joules of heat}$$

Next, calculate the amount of heat needed to completely boil the water at 100°C. Use the equation $q = H_v m$. Substituting and solving gives:

$$q = (2,259 \text{ J/g})(45 \text{ grams}) = 101,655 \text{ joules of heat}$$

The total amount of heat needed to completely vaporize this sample of water is 101,655 J + 9,405 J = 111,060 joules of heat total.

THERMOMETRY

Figure 7.2 shows two individual blocks of the same material of equal mass but at different temperatures. What will happen to the temperature of the blocks if they are brought together to touch? Let's assume a closed system where heat cannot escape or enter.

Thermometry says that heat flows from a higher temperature to a lower temperature. When this happens, the amount of energy lost by one system is the same as the amount of energy that is gained by the other system.

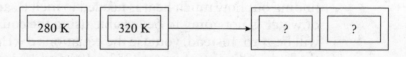

Figure 7.2 Transfer of Heat

In Figure 7.2, after the blocks make contact, they will eventually reach an equal temperature of 300 K.

Think about this: *What is the purpose of insulation in a home? Does it keep in the heat? Does it keep out the cold? Keeping your home warm in the winter is all about keeping in the heat. Hot air moves to where the air is colder. This means that when you experience a draft, it is because hot air is moving out of the house, not because cold air is moving into the house.*

HESS'S LAW

Many reactions require a number of other reactions to occur before the final products are formed. These reactions too have their own activation energies, heats of reaction, and so on. If you know the heats of reaction for the smaller, individual steps, you can calculate the final heat of the overall reaction using *Hess's law*. The final heat of reaction will be the sum of the heats of reaction for the individual steps in the overall reaction. Consider the following overall reaction:

A + B → 2E. This overall reaction has the following intermediate reactions that occur:

$$\mathbf{A + B \rightarrow} C \qquad \Delta H = -100 \text{ kJ}$$

$$C \rightarrow 2D \qquad \Delta H = -150 \text{ kJ}$$

$$\underline{2D \rightarrow \mathbf{2E}} \qquad \Delta H = +80 \text{ kJ}$$

$$\mathbf{A + B \rightarrow 2E} \qquad \Delta H = -170 \text{ kJ}$$

To find the overall heat of reaction simply add up the three heats of reaction. Also, when adding up the individual reactions to find the overall reaction, you will see that substances C and 2D would cancel out because they appear on both sides of the equation (as indicated by the italics).

Consider another reaction where 2A + 2B → D + E. This reaction has the following individual steps:

$$A + B \rightarrow C \qquad \Delta H = +90 \text{ kJ}$$

$$D + E \rightarrow 2C \qquad \Delta H = -300 \text{ kJ}$$

To find the heat of reaction it is necessary to make some adjustments to the way the individual steps are written and to adjust the heats of reaction accordingly. This can be done by first doubling the first reaction:

$$2A + 2B \rightarrow 2C \quad \Delta H = +180 \text{ kJ}$$

This was done because the overall equation calls for 2A and 2B to be reactants. Next, switch the reactants and products for the second equation:

$$2C \rightarrow D + E \qquad \Delta H = +300 \text{ kJ}$$

Here the sign for the heat of reaction has been changed. Now the two adjusted reaction steps can be added up to get the overall reaction:

$$2A + 2B \rightarrow 2C \qquad \Delta H = +180 \text{ kJ}$$

$$2C \rightarrow D + E \qquad \Delta H = +300 \text{ kJ}$$

$$2A + 2B \rightarrow D + E \qquad \Delta H = +480 \text{ kJ}$$

PROBLEM: Find the heat of reaction for $2C + 3H_2 + \frac{1}{2}O_2 \rightarrow C_2H_5OH$ if:

$$C_2H_5OH + 3O_2 \rightarrow 2CO_2 + 3H_2O \qquad \Delta H = -1,367 \text{ kJ}$$

$$C + O_2 \rightarrow CO_2 \qquad \Delta H = -393 \text{ kJ}$$

$$H_2 + \frac{1}{2}O_2 \rightarrow H_2O \qquad \Delta H = -285 \text{ kJ}$$

Solution: First reverse the first reaction to get C_2H_5OH on the right as a product.

$$2CO_2 + 3H_2O \rightarrow C_2H_5OH + 3O_2 \qquad \Delta H = +1,367 \text{ kJ}$$

The second reaction needs to be doubled so that 2C can enter the reaction:

$$2C + 2O_2 \rightarrow 2CO_2 \qquad \Delta H = -786 \text{ kJ}$$

The third reaction needs to be tripled so that $3H_2$ can enter the reaction:

$$3H_2 + \frac{3}{2}O_2 \rightarrow 3H_2O \qquad \Delta H = -855 \text{ kJ}$$

Add up the three individual reactions:

$$2CO_2 + 3H_2O \rightarrow C_2H_5OH + 3O_2 \qquad \Delta H = +1,367 \text{ kJ}$$

$$2C + 2O_2 \rightarrow 2CO_2 \qquad \Delta H = -786 \text{ kJ}$$

$$3H_2 + \frac{3}{2}O_2 \rightarrow 3H_2O \qquad \Delta H = -855 \text{ kJ}$$

$$2C + 3H_2 + \frac{1}{2}O_2 \rightarrow C_2H_5OH \qquad \Delta H = -274 \text{ kJ}$$

The amount of O_2 may be confusing. Review it carefully: $\frac{1}{2}O_2$ on the left and $3O_2$ on the right gives $\frac{1}{2}O_2$ on the left.

BOND DISSOCIATION ENERGY

The strength of a bond that has been formed depends upon how much energy was released when the bond was formed. Because a lower energy state is more stable and preferred, reactions that release greater amounts of energy will form more stable products that have stronger bonds. For example, consider the reaction of methane and oxygen gas. This reaction is an exothermic reaction that forms products that are more stable than the reactants. Proof of this lies in the fact that one of the products, carbon dioxide, has two double bonds in its molecule. The double bonds formed are stronger and more stable than the bonds found in the reactants. You can use the bond dissociation energies of the bonds found in the reactants and products to calculate the amount of heat absorbed and released in a reaction. The difference between the two will be the enthalpy of reaction.

In order to find the enthalpy of reaction, first look at a balanced equation of methane and oxygen gas reacting to form carbon dioxide and water: $CH_4 + 2O_2 \rightarrow CO_2 + 2H_2O$. Using the bond dissociation energies found in Appendix 4, Reference Tables, you can perform the calculation. There are 4 C—H bonds, 2 O—O bonds, 2 C=O bonds and 4 H—O bonds. The heat of reaction will be the sum of the energies of the bonds broken minus the sum of the bond energies of the bonds formed. Setting up to solve:

$$[4 \, (C-H) + 2 \, (O-O)] \text{ (Energies of the bonds broken)}$$

$$- [2 \, (C=O) + 4 \, (H-O)] \text{ (Energies of the bonds formed)}$$

Substituting gives:

$$[4 \, (412 \text{ kJ/mol}) + 2 \, (145 \text{ kJ/mol})] - [2 \, (798 \text{ kJ/mol}) + 4 \, (462 \text{ kJ/mol})]$$

$$[1{,}648 \text{ kJ/mol} + 290 \text{ kJ/mol}] - [1{,}596 \text{ kJ/mol} + 1{,}848 \text{ kJ/mol}] =$$

$$[1{,}938 \text{ kJ/mol}] - [3{,}448 \text{ kJ/mol}] = -1{,}510 \text{ kJ/mol}$$

PROBLEM: Find the heat of reaction for $2H_2 + O_2 \rightarrow 2H_2O$.

Solution: In this problem there are 2 H—H bonds, 1 O—O bond and 4 H—O bonds.

Set up to solve:

$$[2 \, (H-H) + 1(O-O)] - [4(H-O)]$$

$$[2 \, (435 \text{ kJ/mol}) + 1(145 \text{ kJ/mol})] - [4(462 \text{ kJ/mol})]$$

$$[(870 \text{ kJ/mol}) + (145 \text{ kJ/mol})] - [(1{,}848 \text{ kJ/mol})]$$

$$[1{,}015 \text{ kJ/mol}] - [1{,}848 \text{ kJ/mol}] = -833 \text{ kJ/mol}$$

Think about this: *Of the bond dissociation energies used above, the greatest value went to the carbon atom double bonded to the oxygen atom. This is because it takes more energy to break a double bond than a single bond. Also, because a double bond is more stable than a single bond, more energy will be released when a double bond is formed than when a single bond is formed.*

■■■ **ENTROPY, GIBBS FREE ENERGY, AND SPONTANEOUS REACTIONS**

Why is it that when an object is dropped it continues to fall without any further intervention? Do objects need to be coached in how to fall? Of course not; gravity "takes over" once you let go of an object and no further intervention is needed to make the object hit the floor. Processes that occur without added external energy or without additional intervention are called *spontaneous* processes. Now imagine that the very same object that was dropped to the floor suddenly leaves the floor and comes back to your hand (assume no springs, rockets, or legs to jump with). That would be quite a surprise! An object that has fallen does not come back to your hand because that would be a nonspontaneous process. There are many spontaneous processes in our universe; for example, the spontaneous decay of the nucleus of an atom. It should be noted that if a reaction is spontaneous, the reverse reaction will not be spontaneous. If we are going to make a nonspontaneous reaction occur, then energy and outside intervention will be needed.

You already know that lower energy states are preferred by nature. This is why excited electrons release light energy and return back to their ground states. The desire to achieve a lower energy state can be seen when an egg is dropped. The egg will fall to the floor instead of "falling to the ceiling." Something else happens to the egg once it hits the floor; the yolk and egg shells spread out all over making a big mess. Again, the egg shell and yolk do not come back together again, no matter what the nursery rhyme says. Nature prefers the egg's yolk to spread out and make a mess. Chaos, randomness, and disorder are all states that are preferred by nature. The term *entropy* is used to describe chaos, randomness, and disorder. Changes in entropy are designated by the symbol and letter, ΔS.

By determining the changes in energy (or enthalpy) in a reaction and the changes in entropy, you can then determine if a reaction is spontaneous. Spontaneity can be determined by using the *Gibbs Free Energy* equation: $\Delta G = \Delta H - T\Delta S$. Let's put the equation to work and see if we can determine the sign of ΔG for a spontaneous reaction.

Nature prefers a lower enthalpy state. To accomplish this, energy must be lost. This means that nature prefers a ΔH (–). Nature also prefers states of increased entropy and chaos. Therefore, nature prefers ΔS (+). Substituting into the Gibbs Free Energy equation (and remembering that the temperature is in Kelvin so that there can be no zeros or negative values):

$$\Delta G = \Delta H - T\Delta S$$

$$\Delta G = (-) - (+)(+)$$

$$\Delta G = (-) - (+)$$

Remember, when subtracting change the minus sign to a positive sign and change the sign of the second number so $\Delta G = (-) + (-) = (-)$. $\Delta G = (-)$ is what is needed for a reaction or process to be spontaneous.

If ΔH is positive and ΔS is negative, then you can safely conclude that the sign of ΔG would be positive. The process would be classified as nonspontaneous and would not take place. What if ΔG had a value of zero? In this scenario the reaction is said to be in a state of equilibrium, where the forward and reverse reactions proceed equally.

Let's look at one more possibility, ice melting at room temperature (about 293 K). This is a case where the entropy is increasing (ΔS is positive) because the molecules in the melting ice are spreading out, but at the same time the reaction is absorbing heat. This means that ΔH will be positive as well. If the signs for enthalpy and entropy are the same, then the temperature becomes the deciding factor in determining whether or not the reaction is spontaneous. At a very low temperature the ice will not melt, but at a higher temperature the ice will melt. These trends are summarized in the chart that follows:

ΔH	ΔS	ΔG
−	+	Will always be spontaneous (−).
+	−	Will always be nonspontaneous (+).
+	+	Will be spontaneous at a high temperature.
−	−	Will be spontaneous at a low temperature.

Think about this: *Should you be mandated to clean your room? Cleaning your room works against the laws of nature. Cleaning up means that there is less disorder (ΔS is negative), and it means that you have to put energy into doing work (ΔH is positive). According to this, cleaning your room is a nonspontaneous process that nature dictates should not happen.*

PROBLEMS: Give the signs for enthalpy and entropy for the following processes: The freezing of water, the big bang becoming the modern universe, and the submersion of potassium metal into water.

Solutions: The freezing of water requires that heat be released, so the enthalpy will be negative. The entropy will be negative as well because the molecules are more orderly in a solid. The universe has been expanding (entropy is positive) ever since the big bang (enthalpy is negative). Placing potassium metal in water will cause a violent reaction (enthalpy is negative) and the potassium atoms and water molecules will spread out as ions and a gas are formed (entropy is positive).

▰ CHAPTER REVIEW QUESTIONS

1. Two systems at different temperatures come in contact. The heat will flow from the system at

    (A) 30°C to a system at 317 K
    (B) 40°C to a system at 323 K
    (C) 50°C to a system at 303 K
    (D) 60°C to a system at 358 K
    (E) 70°C to a system at 370 K

2. How many joules of heat are released by a 150-gram sample of water that that cools from 25°C to 5°C? (c for H_2O is 4.18 J/gK)

    (A) 78,375 joules
    (B) 83.6 joules
    (C) 720 joules
    (D) 627 joules
    (E) 12,540 joules

3. Calculate the number of joules required to completely evaporate 18 grams of water at 98°C. (H_v = 2259 J/g and c = 4.18 J/gK)

    (A) 40,812 joules
    (B) 40,512 joules
    (C) 150 joules
    (D) 40,662 joules
    (E) 6.12×10^6 joules

4. Which process below has been described correctly for a temperature above 274K?

    (A) $H_2O(l) \rightarrow H_2O(s)$ is exothermic and spontaneous.
    (B) $H_2O(l) \rightarrow H_2O(s)$ is endothermic and spontaneous.
    (C) $H_2O(g) \rightarrow H_2O(l)$ is endothermic and spontaneous.
    (D) $H_2O(s) \rightarrow H_2O(l)$ is endothermic and spontaneous.
    (E) $H_2O(s) \rightarrow H_2O(l)$ is exothermic and spontaneous.

5. Based on Gibbs Free Energy equation $\Delta G = \Delta H - T\Delta S$, a process will occur spontaneously when

    (A) ΔG is positive and ΔS is positive
    (B) ΔH is positive and ΔT is negative
    (C) ΔH is negative and ΔS is positive
    (D) ΔH is negative and ΔS is negative
    (E) ΔG is positive and ΔS is negative

6. The overall reaction: $A + B + 1.5C \rightarrow D$ has three individual reactions that take place,

    Step 1: $A + 2B \rightarrow E$
    Step 2: $F \rightarrow B + C$
    Step 3: ?

    What is the reaction that takes place in Step 3?

    (A) $D + F \rightarrow C + E$
    (B) $E + \frac{5}{2}C \rightarrow D + F$
    (C) $D + \frac{3}{2}C \rightarrow A + E$
    (D) $B + C \rightarrow F + E$
    (E) $A + B + 1.5C \rightarrow D$

7. Calculate the heat for the overall reaction: $Mg(s) + \frac{1}{2}O_2 \rightarrow MgO(s)$ given the heats of reaction below:

    $Mg(s) + 2HCl(aq) \rightarrow MgCl_2(aq) + H_2(g)$
    $$\Delta H = -143 \text{ kJ}$$
    $MgO(s) + 2HCl(aq) \rightarrow MgCl_2(aq) + H_2O(g)$
    $$\Delta H = -216 \text{ kJ}$$
    $H_2(g) + \frac{1}{2}O_2 \rightarrow H_2O(l)$
    $$\Delta H = -285 \text{ kJ}$$

    (A) −644 kJ
    (B) −212 kJ
    (C) +644 kJ
    (D) −74 kJ
    (E) +74 kJ

ANSWERS:

1. (C)	2. (E)	3. (A)	4. (D)
5. (C)	6. (B)	7. (B)	

CHAPTER 8

REACTION RATES AND CHEMICAL EQUILIBRIUM

IN THIS CHAPTER YOU WILL LEARN ABOUT . . .

Reaction Rates

Reverse Reactions and Potential Energy Diagrams

Equilibrium

Le Châtelier's Principle

Equilibrium Constants and Mass Action Equations

Solubility Product Constants

Common Ion Effect

Reactions That Go to Completion

REACTION RATES

In sports there are certain times when contact is required between two players. Consider a situation in a game of football where a big linebacker wants to tackle a quarterback. Does the linebacker walk up to the quarterback and tap him lightly in hopes that the quarterback will fall down? Think about a hockey player who couldn't check the opposing players into the boards and keep them out of the play. That's not exactly a great way to defend one's own goal!

If substances are going to react with each other, there must be frequent contact between the substances. However, as in the two situations just mentioned, contact isn't always enough. Reactants need to collide effectively and frequently if a reaction is going to occur. That is, not only do they need to make contact, but they must also make contact with enough energy or at a certain angle or a certain orientation. As this occurs, the *rate* of the reaction will increase, meaning that there will be a greater change in the concentration of the reactants and products over time.

There are a number of ways conditions can be modified to make molecules collide more effectively. These cases can be examined in the outline below:

Effects of Temperature	Keeping in mind the definition of temperature as average kinetic energy, a higher temperature means that more molecules will have a greater speed. This will help the molecules move or collide more frequently and more effectively. An increase in temperature will increase the rate of reaction.

Effects of Surface Area	Consider a single large block of ice melting. Will it melt faster than a set of smaller ice cubes of equal mass? People who want the ice in their coolers to melt more slowly will use a larger block of ice. A larger block of ice has less surface area, and thus less of the ice is exposed to the warmer temperatures present in the cooler. Smaller ice cubes of the same mass will melt faster because they collectively have more surface area exposed. Chemists often avoid putting whole chunks of salts in their reactions. To increase the speed of the reaction, the solids are chopped up into a finer powder rather than left as larger pieces.
Effects of Pressure	Pressure has its effect on gases. As discussed in Chapter 2, increasing the pressure on a gas decreases the volume of the gas. The molecules become more tightly packed together. This increases the frequency of collisions between the gas molecules. Increasing the pressure on a gas will increase the rate of reaction between gas molecules.
Effects of Reactants	The nature of the reactants can play a part in the rate at which a reaction takes place. Organic chemists often reflux (a careful method of boiling flammable compounds) their reactions for a number of hours to get covalently bonded compounds to react. Reactions involving aqueous ionic compounds will react instantly. For example, if the colorless solutions $Pb(NO_3)_2(aq)$ and $KI(aq)$ are mixed, a yellow solid precipitates instantly. Covalently bonded substances have strong bonds that take time and energy to break in a controlled manner. Ionic compounds dissolved in solution have freed up their ions and they remain available to react with other substances in solution.
Effects of Concentration	An increase in concentration will increase the frequency of collisions between molecules and the rate of reaction. Look at Figure 8.1:

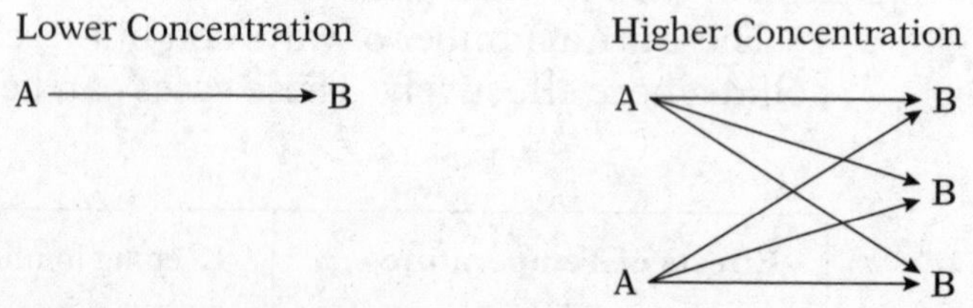

Lower Concentration

A ⟶ B

Higher Concentration

Figure 8.1 Concentration and Frequency of Collision

	In the example on the left, substance A can only collide with one molecule of substance B. On the right, substance A has more of a chance of colliding with substance B because there are more molecules of A and B available. The increase in collision will increase the rate.
Effects of a Catalyst	As seen in the previous chapter, a certain amount of energy is needed to start a reaction. This is called the activation energy of a reaction. Catalysts can increase the rate of a reaction by lowering the potential energy of the activated complex and the activation energy. The dashed line in Figure 8.2 shows the effect of a catalyst.
	The longer arrow on the left shows the activation energy barrier that exists before the catalyst was added. After the catalyst is added the new, lower activation energy is diagramed by the arrow on the right. Notice that the catalyst does not alter the potential energy of the reactants or products. This means that the change in heat for the reaction is not altered either. The only energies that are changed are the activation energies and the potential energy of the activated complex.

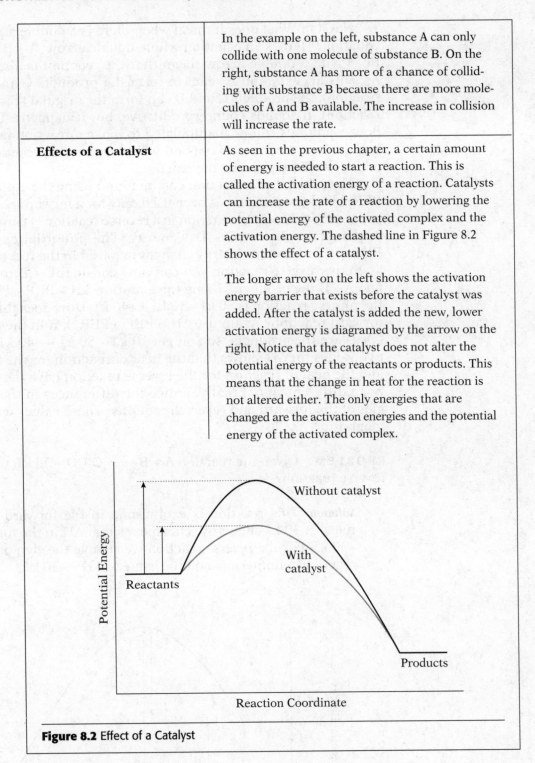

Figure 8.2 Effect of a Catalyst

Think about this: *It takes just seconds to make a hot cup of tea from boiling water and a tea bag. How long does it take to make a cold-brewed iced tea?*

REVERSE REACTIONS AND POTENTIAL ENERGY DIAGRAMS

The reactions considered up to this point have had only a single arrow in their chemical equation. An example is the following reaction: $2Na(s) + Cl_2(g)$

→ 2NaCl(s). What does it mean when there is a double arrow in a chemical equation? Here is an equation with a double arrow: A + B ←→ C + D. Reactions with a double arrow mean that the reaction is a *reversible reaction.* As substances A and B react to form the products C and D (the forward reaction), the products C and D can form the original reactants (the reverse reaction). It sounds counterproductive, but it happens. Later you will see how conditions can be manipulated to make a reaction favor the products, but for now the focus is only on what a reversible reaction means for the enthalpy changes during the reaction.

The enthalpy of a reaction can be found using the equation $\Delta H = PEP - PER$, but this equation has been used only for a forward reaction. What happens to the enthalpy of reaction in a reverse reaction? Consider the following reaction: A + B ←→ C + D + energy. The potential energy diagram in Figure 8.3 shows the energy changes involved in the reaction.

In the reverse reaction you can now consider C + D to be the reactants and A + B the products. Using the equation $\Delta H = PEP - PER$, the value for the reverse reaction is 50 kJ – 10 kJ = +40 kJ. How does this compare to the forward reaction? Again use $\Delta H = PEP - PER$, but in the forward reaction C + D are the products, so you get 10 kJ – 50 kJ = –40 kJ. Comparing the two values for ΔH shows that the heat of reaction for the forward reaction and the heat of reaction for the reverse reaction have the same magnitude but a different sign. Finally, notice the differences in the activation energies of the forward and reverse reactions. These values will be different in magnitude.

PROBLEM: Given the reaction A + B ←→ C + D + 30 kJ, what is ΔH for the reverse reaction?

> *Solution:* This reaction is exothermic in the forward direction and it releases 30 kJ of energy. This means that ΔH in the forward reaction is –30 kJ. For the reverse reaction we change the sign of ΔH, making the reaction endothermic and the value of ΔH = +30 kJ.

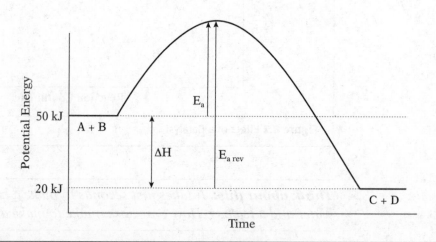

Figure 8.3 Potential Energy Diagram for a Reverse Reaction

EQUILIBRIUM

When two reactions oppose each other, they will eventually reach a point where the amount of product formed is equal to the amount of reactant formed. This situation of an equal "give and take" is called a state of *equilibrium*. Equilibrium is defined as a state of balance between two opposing reactions that are occurring at the same rate. Notice that the definition says nothing about the amounts or concentrations of any reactants or products. The only factors that are equal at equilibrium are the rates of the forward and reverse reactions.

Phase equilibrium is one type of equilibrium. Consider the following closed system at equilibrium (see Figure 8.4):

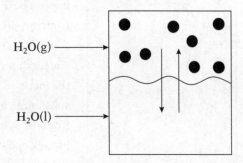

$H_2O(g)$

$H_2O(l)$

Figure 8.4 Phase Equilibrium

At first, the container is sealed with only liquid water and as time goes on, more water vapor begins to form. Eventually the space above the water becomes saturated with water vapor. The system now has the conditions for phase equilibrium to occur; the system is closed and saturation has been achieved. Once the space above the water becomes saturated with water vapor, for every water molecule that evaporates into the gas phase, one gas molecule will condense into liquid water. Because of this, the level of water in the container will remain constant.

There are a number of methods for identifying a chemical reaction that has reached equilibrium. One is to perform an experiment to measure the rates of reaction. Another method is to check if the concentrations are remaining constant. It is much easier, however, just to make sure that no further changes are occurring in temperature, pressure, volume, and/or color. What if a reaction does have a color change occurring? In this case equilibrium has not been established because there is a change in conditions that is causing a stress upon the point of equilibrium. A stress, or changes in conditions, can force the point of equilibrium to change. However, stresses can also be used to manipulate a reaction and force it in a particular direction.

LE CHÂTELIER'S PRINCIPLE

Le Châtelier's Principle regarding changes in the point of equilibrium was stated by Le Châtelier in the late 1800s. Le Châtelier's Principle states that if a stress or change in conditions is applied to a system at equilibrium, the point of equilibrium will shift in such a manner as to relieve the applied stress. There are several conditions that can change the point of equilibrium. These are detailed in the table below.

Effect of Changes in Concentration	Reactants are mixed together to achieve the goal of producing more product. Consider the reaction A + B ←→ C + D. If you want to make more C and D, you would add more A and B. If equilibrium was already established when more A and B were added, the equilibrium would shift to the right to produce more C and D and, at the same time, consume the additional A and B. What would happen here is that the equilibrium would shift away from the side of the reaction that had an increase in concentration. The same would hold true if more C or D were added: the equilibrium would shift to the left, consuming the increase and making more A and B. What if one of the substances in the reaction at equilibrium were removed from the system? In that case something would be missing as opposed to being in excess. As a result, the equilibrium would shift to produce more of what was missing.
Effect of Changes in Pressure	Before examining how pressure can change equilibrium, first recall Boyle's Law. Remember that as pressure increases, volume decreases. Remember too that the coefficients in a chemical equation can also stand for volumes of gases. Look at this simple reaction between two gases: $2A(g) \longleftrightarrow B(g)$. What happens if the pressure is increased? The left side has two volumes of gas A while the right side has just one volume of gas B. If the pressure is increased, the equilibrium would shift to the right because the right side of the equation has a smaller volume of gases.
Effect of Changes in Temperature	Before examining the effects of changes in temperature, remember that if heat energy is a product then the reaction is exothermic, whereas if heat energy is a reactant then the reaction is endothermic. Consider the following endothermic reaction: heat + A ←→ B. The heat energy is a reactant and should be treated as such. Now this situation is just like that of any other problem that involves a change in concentration. If the temperature of this reaction at equilibrium is increased, then you could say that there is an increase in the heat, or you can consider it an increase in one of the reactants. This would cause a shift to the right in order to consume the added heat.

(continued)

Effect of Adding a Catalyst	A catalyst will not shift the point of equilibrium of a reaction. However, a catalyst will help the reaction reach equilibrium faster. This is done again by lowering the potential energy of the activated complex and the activation energy.

PROBLEM: Given the following reaction: $3H_2(g) + N_2(g) \longleftrightarrow 2NH_3(g) + 22$ kcal heat energy. What is the heat of reaction for the reverse reaction? What would you do to the temperature, pressure, and concentrations of the reactants and products to shift the equilibrium so that more ammonia is made?

Solution: The ΔH for the reverse reaction will have a positive sign because the reverse reaction is endothermic. The ΔH is equal to $+ 22$ kcal. To shift the equilibrium to make more ammonia, start by adding more hydrogen gas and nitrogen gas because having more reactants present will make more products. Because the heat energy is a product, you do not want to add heat. You would have to lower the temperature to remove the heat so that the reaction will shift to the right in an effort to replace the heat. Because there is a total of four volumes of gas on the left and two volumes of gas on the right, an increase in pressure will favor the production of ammonia because it is the side with fewer volumes of gas.

EQUILIBRIUM CONSTANTS AND MASS ACTION EQUATIONS

Looking again at the equation: $A + B \longleftrightarrow C + D$, if C and D react to form the original reactants, and the original reactants form more products, who wins the battle? How can you tell if the reaction favors the formation of reactants or the formation of products? A mathematical expression can be set up to calculate just how much the reaction favors the formation of products. This mathematical expression is used to calculate the equilibrium constant for a particular reaction. For the reaction $aA + bB \longleftrightarrow cC + dD$ the mathematical expression can be written as

$$K_{eq} = \frac{[C]^c [D]^d}{[A]^a [B]^b}$$

This expression is called a *mass action equation*. The brackets in the equation represent the concentration of the products and reactants. Notice that the products are placed in the numerator while the reactants are placed in the denominator. Also notice that the coefficients are now the powers of the concentration of each substance. The mnemonic device used to remember these rules is "Products Over Reactants, Coefficients Become Powers." One other rule that you need to know is that you do not include substances that are in the liquid or solid phases. These concentrations are written as the number "1" in the equation.

Values for K_{eq} that are greater than 1.0 indicate a reaction that favors the products, as the value of the numerator (products) outweighs that of the denominator (reactants). Values less than 1.0 indicate that the denominator is greater in value than the numerator and the reactants are favored.

PROBLEM: Write the mass action equation for W(aq) + X(s) $\longleftrightarrow$ 3Y(aq) + 2Z(l). At equilibrium the concentration of W is 0.1 M and the concentration of Y is 0.1 M. What is the value of the equilibrium constant? Does this reaction favor the formation of reactants or products?

Solution: Remember the rule, "Products Over Reactants, Coefficients Become Powers." Leave out the solids and liquids from the reaction and get:

$$K_{eq} = \frac{[Y]^3}{[W]}$$

Substitution and solving gives:

$$K_{eq} = \frac{[0.1\ M]^3}{[0.1\ M]} = \frac{0.001}{0.1} = 0.01 \text{ meaning that the formation of reactants is favored}$$

PROBLEM: At equilibrium, a 1.00 liter flask is found to contain 0.01 moles $H_2(g)$ and 0.02 moles of $I_2(g)$ that have reacted according to the equation: $H_2(g) + I_2(g) \longleftrightarrow 2HI(g)$. The K_{eq} for this reaction at a particular temperature is 50.5. What is the concentration of HI inside the flask?

Solution: Because the gases are in a 1.00-liter flask, the concentrations of the reactants are $[H_2] = 0.01\ M$ and $[I_2] = 0.02\ M$. Write the mass action equation as:

$$K_{eq} = \frac{[HI]^2}{[H_2][I_2]} \text{ Substituting gives: } 50.5 = \frac{[HI]^2}{[0.01\ M][0.02\ M]}$$

Solving gives $50.5[0.01][0.02] = x^2$ and $x = 0.10\ M$.

PROBLEM: Given the following reaction: A + B $\longleftrightarrow$ C. Calculate the equilibrium constant in a reaction where the initial concentration of A is 1.5 M, the initial concentration of B is 1.3 M, and, after A and B react and come to equilibrium, the concentration of C is 1.2 M.

Solution: Unlike in the previous problem you are not given the concentrations of reactants at equilibrium. Instead you are given the initial concentrations of reactants.

Set up a chart to organize the changes in concentration as the reaction proceeds. First put in the initial concentrations of A and B. Put in C also because you know its final concentration.

	A	B	C
Initial Concentrations	1.5 M	1.3 M	0
Change in Concentrations			
Final Concentrations			+1.2 M

Next, add the changes in concentration. These changes are in the same 1:1 ratio as the coefficients in the balanced equation.

	A	B	C
Initial Concentrations	1.5 M	1.3 M	0
Change in Concentrations	–1.2 M	–1.2 M	+1.2 M
Final Concentrations			+1.2 M

Calculate the final concentrations.

	A	B	C
Initial Concentrations	1.5 M	1.3 M	0
Change in Concentrations	–1.2 M	–1.2 M	+1.2 M
Final Concentrations	0.3 M	0.1 M	+1.2 M

Place the final concentrations into the mass action equation and calculate the equilibrium constant:

$$K = \frac{[C]}{[A][B]} = \frac{[1.2]}{[0.3][0.1]} = 40$$

showing that this reaction greatly favors the formation of the product, C.

■ SOLUBILITY PRODUCT CONSTANTS

Recall from previous chapters that solubility is the amount of a solid that it takes to saturate the solution that it is in. Also remember that not every salt is completely soluble in water. The *solubility product constant* helps determine how soluble a salt is in solution at a particular temperature. The values for these constants can be found in Appendix 4, Reference Tables, in the back of this book.

One salt to examine is $PbSO_4$. Lead sulfate dissolves according to the following equation: $PbSO_4(s) \longleftrightarrow Pb^{2+}(aq) + SO_4^{2-}(aq)$. The solubility product constant for lead sulfate can be written as $K_{sp} = [Pb^{2+}][SO_4^{2-}] / 1$. Notice that the solid was not included in the equilibrium constant expression. What concentration of lead and sulfate ions will you find in a saturated solution of lead sulfate at 298 K? The solubility product constant for lead sulfate is 1.6×10^{-8}. This means that the concentration of lead ions and sulfate ions is each 1.26×10^{-4} M.

PROBLEM: Magnesium hydroxide has a solubility product constant of 1.8×10^{-11} at 298 K. Write the equilibrium constant expression for this salt. What is the concentration of magnesium and hydroxide ions in a saturated solution of magnesium hydroxide? Which salt is more soluble in water, $Mg(OH)_2$ or $PbSO_4$?

> **Solution:** The equation looks like $Mg(OH)_2(s) \leftrightarrow Mg^{2+}(aq) + 2OH^{1-}(aq)$. The solubility product constant expression is written as: $K_{sp} = [Mg^{2+}][OH^{1-}]^2/1$. Because for each magnesium ion formed, two hydroxide ions form, substitution gives: $1.8 \times 10^{-11} = [x][x]^2 = x^3$.
>
> Solving for x gives $x = 2.6 \times 10^{-4}\ M$
>
> The $[Mg^{2+}] = 2.6 \times 10^{-4}\ M$ and the $[OH^{1-}] = 6.8 \times 10^{-8}\ M$

$PbSO_4$ is more soluble because it has a greater solubility product constant than that of magnesium hydroxide.

COMMON ION EFFECT

Consider the chemical equation for AgCl dissolved in water to make a saturated solution: $AgCl(s) \leftrightarrow Ag^{1+}(aq) + Cl^{1-}(aq)$. At 298 K the solubility product constant is 1.8×10^{-10}, which indicates that is a slightly soluble salt. There is a way of making AgCl even less soluble, via the *common ion effect*. Consider the following, when an ion that is already present is added to the solution, the equilibrium will shift to consume the increase in concentration of the ion.

If NaCl(aq) were added to a solution of silver chloride, the NaCl would introduce more Cl^{1-} ions and shift the equilibrium to the left. This would favor the production of more AgCl(s). Adding the common ion, Cl^{1-}, actually decreases the solubility of AgCl by forming more of its solid.

REACTIONS THAT GO TO COMPLETION

Why isn't every reaction reversible? Why don't reactions such as $Mg + 2HCl \rightarrow MgCl_2 + H_2$ have a reversible reaction? Inserting the phases into the chemical equation will help to answer this question: $Mg(s) + 2HCl(aq) \rightarrow MgCl_2(aq) + H_2(g)$. This reaction produces a gas—a gas that, according to the laws of entropy, loves to escape from the reaction so that it can spread out. Because the reaction is constantly losing one of its products to the atmosphere, the reactants work feverishly to replace the missing hydrogen gas. This attempt to replace the missing hydrogen gas drives the reaction completely to the right.

In addition to reactions that form a gas, reactions that form a precipitate also go to completion. Consider the following reaction: $Pb(NO_3)_2(aq) + 2KI(aq) \rightarrow PbI_2(s) + 2KNO_3(aq)$. This reaction forms a solid precipitate that removes itself from the reaction and settles to the bottom of the reaction vessel. Again, because the reaction is constantly losing one of its products, the reactants work feverishly to replace the missing lead iodide. This attempt to replace the missing lead iodide drives the reaction completely to the right. Finally, reactions that form water tend to go to completion.

CHAPTER REVIEW QUESTIONS

1. Given the reaction: $Zn(s) + 2HCl(aq) \rightarrow ZnCl_2(aq) + H_2(g)$

   Why is the reaction slower when a single piece of zinc is used than when powdered zinc of the same mass is used?

   (A) The powdered zinc is more concentrated.
   (B) The single piece of zinc is more reactive.
   (C) The powdered zinc requires less activation energy
   (D) The powdered zinc generates more heat energy.
   (E) The powdered zinc has a greater surface area.

2. Which takes place when a catalyst is added to a reaction at equilibrium?

   (A) The point of equilibrium is shifted to the right.
   (B) The point of equilibrium is shifted to the left.
   (C) The forward and reverse reactions rates are increased unequally.
   (D) The forward and reverse reactions rates are increased equally.
   (E) The value of ΔH has the same magnitude but a different sign.

3. As the frequency and the number of effective collisions between reacting particles increases, the rate of the reaction

   (A) increases
   (B) decreases
   (C) remains the same
   (D) approaches zero
   (E) none of the above

4. Which factors are equal in a reversible chemical reaction that has reached equilibrium?

   (A) The number of moles of the reactants and products.
   (B) The potential energies of the reactants and products.
   (C) The activation energies of the forward and reverse reactions.
   (D) The rates of reaction for the forward and reverse reactions.
   (E) The concentrations of the reactants and products.

5. A catalyst is added to a system at equilibrium. The concentration of the reactants will then

   (A) decrease
   (B) increase
   (C) remain the same
   (D) approach zero
   (E) none of the above

6. Given the following reaction that has reached equilibrium: $NaCl(s) \longleftrightarrow NaCl(aq)$. For the phase equilibrium to exist, the $NaCl(aq)$ must be a solution that is

   (A) concentrated
   (B) saturated
   (C) dilute
   (D) heated
   (E) unsaturated

7. In an effort to speed up a reaction between a solid and a gas one would not:

   (A) make an effort to concentrate the reactants as best as possible
   (B) add a catalyst
   (C) cool the reaction down
   (D) increase the pressure on the system
   (E) use a powdered solid instead of one big lump of the same solid

8. Which reaction below is expected to go to completion?

I. $Zn + HCl$

II. $HCl + NaOH$

III. $Ag^{1+}(aq) + Cl^{1-}(aq)$

(A) II only.

(B) III only.

(C) I and II only.

(D) II and III only.

(E) I, II, and III.

9. Which salt listed in Appendix 4 of this book has the greatest solubility in water under equal conditions?

(A) Lead iodide

(B) Lead sulfate

(C) Magnesium hydroxide

(D) Silver chloride

(E) The salts are all equally soluble.

Directions: The following question consists of two statements. Determine whether statement I in the leftmost column is true (T) or false (F) and whether statement II in the rightmost column is true (T) or false (F).

I		II

10. Lead iodide is more soluble than   BECAUSE   the K_{sp} value for lead iodide is lower than that of silver chloride   silver chloride

11. Refer to the following chart and choices for the reaction $A + B \rightarrow C$:

	A	B	C
Initial Concentrations	0.5 M	0.4 M	0
Change in Concentrations			
Final Concentrations			+0.3 M

I. The final concentration of B is 0.1 M.

II. The change in concentration of C is 0.3 M.

III. The final concentration of A is 0.3 M.

Which of the above statements are correct?

(A) I only

(B) II only

(C) II and III only

(D) I and II only

(E) I, II, and III

ANSWERS:

1. (E)	2. (D)	3. (A)
4. (D)	5. (C)	6. (B)
7. (C)	8. (E)	9. (B)
10. (T, F)	11. (D)	

CHAPTER 9

ACIDS AND BASES

IN THIS CHAPTER YOU WILL LEARN ABOUT . . .

Naming Acids and Bases
Operational Definitions of Acids and Bases
Conceptual Definition of Acids and Bases
Arrhenius Acids and Bases
Brønsted-Lowry Acids and Bases and Conjugate Pairs
Lewis Acids and Bases
Titration and Neutralization
Hydrolysis
K_a and the Strength of Acids
pH and the pH Scale
Buffers

Why does lemon juice sting when it gets on an open cut? How does an antacid work? Why can HF be used to etch glass?

NAMING ACIDS AND BASES

By this point in this book you should have come across a few familiar acids and bases. Either you recognized the molecular formula of an acid or a base or you saw the name of a familiar acid or base. Acids and bases can be defined in a number of ways, as you will see shortly. It's a good idea to see how acids and bases are named before you focus on the differences between them. You will see that the names do not follow the same rules used for other covalently bonded substances.

Here is a simple outline of how to name acids that have certain anions bonded within them.

Example	Name	Name as an Acid
HCl	Hydrogen chlor*ide*	This acid ends in -*ide*. For its name as an acid, add the prefix *hydro-*. Then drop the ending -*ide* and add -*ic acid*. The name of this acid is *hydro*chlor*ic acid*.

(continued)

Example	Name	Name as an Acid
H_2SO_4	Hydrogen sulfate	This acid ends in -ate. For its name as an acid, drop the -ate and add the suffix -ic acid. This acid is called sulfuric acid. Notice that, unlike the previous acid, there is no mention of the hydrogen atom in the name.
H_2SO_3	Hydrogen sulfite	This acid ends in -ite. For its name as an acid, drop the -ite and add the suffix -ous acid. This acid is called sulfurous acid. Again, there is no mention of the hydrogen atom in the name.

PROBLEMS: Name the following: HBr, H_3PO_4, and $HClO_2$.

Solution: HBr is called hydrobromic acid because the ending to the compound is -ide, hydrogen bromide. H_3PO_4 is called phosphoric acid because of the phosphate ion present. $HClO_2$ is called chlorous acid because of the chlorite ion present.

OPERATIONAL DEFINITIONS OF ACIDS AND BASES

As mentioned previously, acids and bases can be defined in a number of ways. One way to define an acid or base is by what you "see" when an acid or base reacts with other substances. For example, your senses can help you identify an acid or base because acids taste sour and bases taste bitter. The sour taste of lemons can be attributed to the citric acid found in the lemon juice. In addition, bases have a slippery feel to them (please, do not touch or taste acids or bases).

There are also a few reactions that can help define acids and bases. Here are some examples:

- Acids and bases react to form a salt and water. This process is called *neutralization*. Proof lies in the fact that the water formed can be evaporated and the salt will remain behind. An example of this reaction is: $HCl(aq) + NaOH(aq) \rightarrow NaCl(aq) + H_2O(l)$
- Acids react with active metals, such as Zn or Mg, to form a salt and hydrogen gas. As the reaction proceeds, you can see the hydrogen gas being given off in the reaction. An example is: $2HCl(aq) + Mg(s) \rightarrow MgCl_2(aq) + H_2(g)$
- Acids can be formed from a reaction of nonmetal oxides and water; for example: $CO_2(g) + H_2O(l) \rightarrow H_2CO_3(aq)$
- Bases can be formed from a reaction between metal oxides and water; for example: $Na_2O(s) + H_2O(l) \rightarrow 2NaOH(aq)$

Finally, acids and bases can change the colors of certain *indicators*. Some examples are *phenolphthalein* and *litmus*. The chart below summarizes the colors of these two important indicators. A good mnemonic device to help you remember litmus indicators is BRA: "**B**lue turns **R**ed in **A**cid." Other indicators that can be used are methyl orange and bromothymol blue.

Indicator	Color in Acid	Color in Base
Phenolphthalein	Colorless	Pink/purple/magenta
Litmus	Red	Blue

CONCEPTUAL DEFINITION OF ACIDS AND BASES

Acids and bases can further be defined by three conceptual definitions as outlined below.

Theory	Acids	Bases
Arrhenius Definition	Acids are substances that yield H^{1+} ions (or H_3O^{1+} ions) as the only positive ions in solution.	Bases are substances that yield OH^{1-} ions as the only negative ion in solution.
Brønsted-Lowry Definition	Acids are proton (H^{1+}) donors.	Bases are proton (H^{1+}) acceptors.
Lewis Definition	Acids are electron pair acceptors.	Bases are electron pair donors.

ARRHENIUS ACIDS AND BASES

Arrhenius's definition of acids and bases is the definition that most people know. Acids can be recognized because their chemical formulas have an "H" at the beginning, like HBr and HNO_3. Bases are easy to recognize because their chemical formulas end with "OH," like NaOH or KOH. There are a few strong acids and bases that you should be familiar with. These strong acids and bases ionize completely in solution and dissociate into as many hydronium ions and hydroxide ions as are available.

Strong Acids	Strong Bases
HCl	Group 1 hydroxides and the
HBr	heavier group 2 hydroxides
HI	such as NaOH, KOH, and
H_2SO_4	$Ca(OH)_2$
HNO_3	
$HClO_4$	

Strong acids and bases are also known to be strong *electrolytes*. That is, when they dissolve in solution they form ions that can carry an electrical current. This phenomenon is not limited to acids and bases; aqueous salts and molten salts have mobile ions that are also capable of carrying an electrical current.

Think about this: Why is it dangerous to drop an electric appliance such as a hairdryer or radio into a bathtub of water?

BRØNSTED-LOWRY ACIDS AND BASES AND CONJUGATE PAIRS

The Brønsted-Lowry definition of acids and bases does not replace the Arrhenius definition, but extends it. The Brønsted-Lowry definition of acids and bases requires you to take a closer look at the reactants and products of an acid-base reaction. In this case, acids and bases are not easily defined as having hydronium and hydroxide ions. Instead, you are asked to look and see which substance has lost a proton and which has gained the very same proton that was lost.

In the example: $HCl + H_2O \rightarrow H_3O^{1+} + Cl^{1-}$, which substance is the acid and which is the base? Look at the compound that contains chlorine. As a reactant the chlorine has one proton, but as a product the chlorine is an ion by itself. The chlorine compound has lost a proton and the HCl can be labeled as an acid. The water, or oxygen-containing compound, has two hydrogen atoms as a reactant but now has three hydrogen atoms/ions as a product. Therefore, the water is labeled as a base. (See Figure 9.1.)

Let's look at another situation involving the reaction $H_2O + NH_3 \rightarrow OH^{1-} + NH_4^{1+}$. The water lost a proton and became a hydroxide ion. The ammonia gained a proton and became an ammonium ion. In this case, the water is an acid and the ammonia is a base. (See Figure 9.2.)

This brings up an interesting situation. Isn't water a neutral substance? How can it react as an acid in one reaction and as a base in another? Water is a substance that can gain or lose a proton depending upon the environment it is in. Water is what is called an *amphoteric* substance because it can act as either an acid or a base.

Think about this: An amphibian can live on BOTH water and land. An amphitheater can be BOTH indoors or outdoors. What can be said about someone who can write with BOTH the left hand or right hand?

PROBLEM: Label the acid and base in the reaction: $HCO_3^{1-} + HSO_4^{1-} \rightarrow SO_4^{2-} + H_2CO_3$

Solution: The compound containing the carbonate ion starts with one H^{1+} ion bonded to it and ends up with two H^{1+} ions. The HCO_3^{1-} ion has accepted a proton and is the base. The compound with the sulfate ion started with one H^{1+} ion and ended up without any. The HSO_4^{1-} ion has

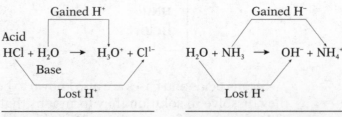

Figure 9.1 Proton Transfer $HCl + H_2O \rightarrow H_3O^{1+} + Cl^{1-}$

Figure 9.2 Proton Transfer $H_2O + NH_3 \rightarrow OH^{1-} + NH_4^{1+}$

lost a proton and is labeled as an acid. Notice that in this example both reactants have "H" in their chemical formulas. Despite this, one of them behaved like a base.

What if an acid-base reaction was reversible? Could the products act as acids and bases in a reversible reaction? Look at the following: $HF + NH_3 \longleftrightarrow F^{1-} + NH_4^{1+}$. In the forward reaction, the HF is the acid and the NH_3 is the base. The fluoride ion is called the *conjugate base* because it was formed from the loss of a proton and in the reverse reaction the fluoride ion is what accepts a proton. HF and F^{1-} are what we call a *conjugate pair*, a pair of substances that differ by an H^{1+} ion.

$$\begin{array}{cc} & \textit{Conj.} \\ \textit{Acid} & \textit{Base} \\ HF + NH_3 & \longleftrightarrow F^{1-} + NH_4^{1+} \end{array}$$

The opposite holds true for ammonia and the ammonium ion. Ammonia is classified as a base because it accepted a proton. The ammonium ion in the reverse reaction will be the proton donor. Because the ammonium ion was also formed from a base, the ammonium ion is called the *conjugate acid*. NH_3 and NH_4^{1+} are another conjugate pair, a pair of substances that differ by an H^{1+} ion.

$$\begin{array}{cc} & \textit{Conj.} \\ \textit{Base} & \textit{Acid} \\ HF + NH_3 & \longleftrightarrow F^{1-} + NH_4^{1+} \end{array}$$

PROBLEM: What are the conjugate acids and bases for HS^{1-}?

Solution: The conjugate base is formed from a substance that acts like an acid. If HS^{1-} acts like an acid and loses a proton, then S^{2-} remains as the conjugate base. The conjugate acid is formed from a substance that acts like a base. If HS^{1-} acts like a base and gains a proton, then H_2S remains as the conjugate acid.

LEWIS ACIDS AND BASES

Lewis acids and bases, because of their complexity, shall be examined briefly. Consider the structure of ammonia with its free pair of electrons. If the free pair of electrons were to make a bond with boron trifluoride, which substance is labeled as an acid and which one is the base? Because the boron accepted a pair of electrons it is considered to be the Lewis acid. Ammonia is the substance that donated the electron pair and is classified as the Lewis base. (See Figure 9.3.)

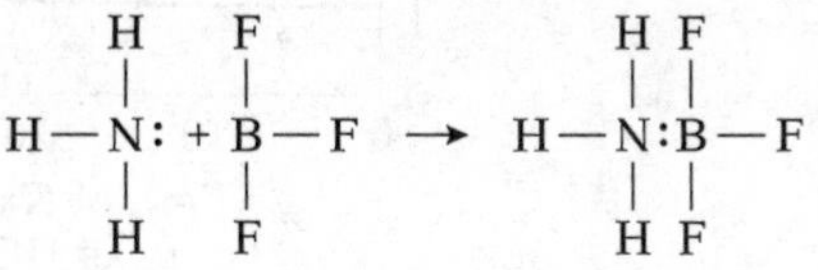

$$\begin{array}{ccc} H \quad F & & H \ F \\ | \quad | & & | \ | \\ H-N: + B-F & \rightarrow & H-N:B-F \\ | \quad | & & | \ | \\ H \quad F & & H \ F \end{array}$$

Figure 9.3 Lewis Acids and Bases

TITRATION AND NEUTRALIZATION

As discussed previously, neutralization is the process by which an acid is reacted with a base to form a salt and water. The general reaction for this process can be simplified by showing how the H^{1+} and OH^{1-} ions combine to form water. Perhaps you have seen a commercial on TV where an antacid promises to neutralize more acid than another antacid. Perhaps you have seen a situation in the laboratory where an acid was spilled. The laboratory specialist should have neutralized the spill with baking soda. These are situations where the amounts of acid and base being neutralized are not completely measured out. There is a process where acids and bases can be measured out in exact quantities so that they neutralize each other exactly and do so without any excess acid or base. This process is called a *titration*.

In a titration, you are presented with an acid or base of an unknown molarity. You then use an acid or base of a known molarity to neutralize the unknown. The process is done slowly with the use of burets to deliver exact amounts of acid and base. Finally, you use an indicator to help determine the *end point* of the reaction. When the indicator changes color, the end point has been reached and the acid and base have neutralized each other. Titrations need to be so carefully controlled that often just one drop of acid or base added to the reaction flask can mean the difference between the solution's being acidic or basic. This can be seen by the steep slope of the line in the graph in Figure 9.4.

The graph in Figure 9.4 shows that it took 50.00 mL of 0.100 M NaOH to neutralize 50.00 mL 0.100 M HCl. The equation used to make the necessary calculations in a titration is as follows: $M_aV_a = M_bV_b$, where M_a and M_b are the molarities of the acid and base and V_a and V_b are the volumes of acid and base used in the titration process.

PROBLEM: A student is performing a titration in the laboratory. She delivers 15.00 mL of HCl of an unknown molarity to a flask via a buret. She then adds

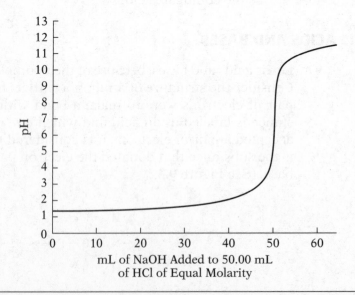

Figure 9.4 Titration Curve

one drop of phenolphthalein indicator to the acid in the flask and the solution remains colorless. The student then begins the titration and delivers 0.100 M NaOH to the flask drop by drop. She finds that it takes 22.10 mL of the NaOH to make the indicator change from colorless to pink. What is the molarity of the HCl that was delivered into the flask?

Solution: The volume of the acid delivered was 15.00 mL (V_a). The volume of the 0.100 M NaOH (M_b) used was 22.10 mL (V_b). You are asked to find the molarity of the acid (M_a). First write the equation used to solve titrations:

$$M_a V_a = M_b V_b \text{ then substitute:}$$

$$M_a(15.00 \text{ mL}) = (0.100 \text{ } M)(22.10 \text{ mL}) \text{ and solve:}$$

$$M_a = \frac{(0.100 \text{ } M)(22.10 \text{ mL})}{15.00 \text{ mL}} = 0.147 \text{ } M\text{HCl}$$

Think about this: In the problem above, the concentration of the acid was greater than that of the base. Because the base was less concentrated, it took a greater volume of base to neutralize the acid. What would happen if the resulting solution was carefully heated to evaporate all of the water?

HYDROLYSIS

Literally meaning "water splitting," *hydrolysis* can be thought of as the opposite of neutralization. This is represented by the diagram shown in Figure 9.5.

When a salt undergoes hydrolysis, it reacts with water to form an acid and a base. The acid and base formed can be considered to be the original reactants from which the salt was formed in a neutralization reaction. By determining the strengths of the original acid and base from which the salt was formed you can then determine if the salt is an acid, a base, or a neutral salt.

Consider NaCl, for example. To determine if this salt is acidic or basic, first locate the anion and cation in the salt. Na^{1+} is a cation and Cl^{1-} is the anion. Then draw a line between the Na and Cl in the compound to indicate the split by the water: Na|Cl. The cation will bond with OH^{1-} and the anion will bond with H^{1+}. This leads to the fact that the base used to form NaCl is NaOH and the acid used to form NaCl is HCl. Because NaOH is a strong base and HCl is a strong acid, the salt they form, NaCl, is a neutral salt.

$$\text{Acid + Base} \underset{\text{Hydrolysis}}{\overset{\text{Neutralization}}{\rightleftarrows}} \text{Salt + Water}$$

Figure 9.5 Neutralization versus Hydrolysis

PROBLEM: Determine if the salt $NaC_2H_3O_2$(s) is acidic, basic, or neutral.

Solution: Determine the cation and the anion present in the salt: Na|$C_2H_3O_2$.

The cation was the cation from a base, in this case, NaOH. The anion was the anion from an acid, in this case, $HC_2H_3O_2$. NaOH is a strong base and $HC_2H_3O_2$ is a weak acid. Therefore, this salt is a basic salt that was derived from a strong base and a weak acid.

K_a AND THE STRENGTH OF ACIDS

Weak acids do not dissociate completely in solution and an equilibrium is established between the acid and the ions that it forms. This means that an equilibrium constant expression can be written and the concentration of the hydronium ions in solution can be found. For example, for acetic acid, which is a weak acid, you can write the equation: $HC_2H_3O_2(aq) \longleftrightarrow H^{1+}(aq) + C_2H_3O_2^{1-}(aq)$ and write the equilibrium constant expression as:

$$K_a = \frac{\left[H^{1+}\right]\left[C_2H_3O_2^{1-}\right]}{\left[HC_2H_3O_2\right]}$$

Notice that the $[H^{1+}]$ is located in the numerator of the equation. If the value for K_a is greater than 1, then there is a greater concentration of H^{1+} present in the solution and the solution is more acidic. Therefore, a greater K_a value means that the acid is stronger. The reference tables in Appendix 4 show the K_a values of some common acids. Examination of these constants will show, for example, that hydrofluoric acid, used in the etching of glass, is a stronger acid than acetic acid, the acid that is responsible for the sour taste of vinegar.

PROBLEM: The K_a for HClO at 298 K is 3.0×10^{-8}. What is the concentration of hydronium ions in a 0.2 M solution of HClO?

Solution: The reaction proceeds as: $HClO(aq) \longleftrightarrow H^{1+}(aq) + ClO^{1-}(aq)$. The equilibrium constant expression can be set up as:

$$K_a = \frac{\left[H^{1+}\right]\left[ClO^{1-}\right]}{\left[HClO\right]}$$

and substitution gives [remember that there is an equal concentration of $H^{1+}(aq)$ and $ClO^{1-}(aq)$]:

$$3.0 \times 10^{-8} = \frac{[x][x]}{[0.2]} \quad \text{and} \quad 3.0 \times 10^{-8} = \frac{[x]^2}{[0.2]} \quad \text{and} \quad 6.0 \times 10^{-9} = [x]^2$$

Solving gives $[H^{1+}] = 7.7 \times 10^{-5}$ M.

pH AND THE pH SCALE

The *pH* of a solution is a measure of the solution's acidity. Before pH can be calculated, you must first know the concentration of H^{1+} in solution, thus the reason for calling it pH. The pH of a solution can be found by using the equation $-\log[H^{1+}]$. Consider the 0.2 M HClO from the previous section. The $[H^{1+}]$ was 7.7×10^{-5} M in that solution. Now you can calculate the pH of the solution. By substituting into the equation for pH you get: $-\log[7.7 \times 10^{-5}$ $M]$. The log of 7.7×10^{-5} M is -4.1. Negating this as called for by the pH equation, you find that the pH is 4.1.

Besides finding the pH, the concentration of H^{1+} can help find the concentration of OH^{1-} in solution as well. Pure water ionizes to form equal amounts of H^{1+} and OH^{1-} as shown in the following equation: $H_2O(l) \longleftrightarrow$ $H^{1+}(aq) + OH^{1-}(aq)$. The equilibrium constant expression for this equation can be written as $K_w = [H^{1+}][OH^{1-}]/1$ (don't include liquids in the equation). The equilibrium constant for the autoionization of water at 298 K is 1×10^{-14}. This constant will help find the concentrations of H^{1+} and OH^{1-} in acid and basic solutions. First, look at the pH of pure water.

In a solution of pure water the $[H^{1+}]$ is equal to $[OH^{1-}]$. So if $K_w = [H^{1+}][OH^{1-}]$ and $[H^{1+}]$ is equal to $[OH^{1-}]$, you can rewrite the equation as $1 \times 10^{-14} = [x][x] = x^2$. Taking the square root of both sides of the equals sign, you get $x = 1 \times 10^{-7}$. If the $[H^{1+}]$ is 1×10^{-7} M, then the pH of pure water is $-\log [1 \times 10^{-7} M]$ which equals 7. This is why a pH of 7 is neutral on the pH scale.

PROBLEM: A solution of a strong acid has an $[H^{1+}]$ equal to 1.0×10^{-2} M. What is the $[OH^{1-}]$ in this solution? What is the pH of this solution? Will the solution be acidic or basic?

Solution: Use the equation: $K_w = [H^{1+}][OH^{1-}]$ and substitute to get $1 \times 10^{-14} = [1.0 \times 10^{-2} M][x]$ and $x = 1.0 \times 10^{-12}$ M. The $[OH^{1-}]$ is 1.0×10^{-12} M. To find the pH use the equation $-\log[H^{1+}]$ and substitute to get $-\log[1.0 \times 10^{-2} M]$, which is equal to a pH of 2. To determine if the solution is acidic or basic, compare the concentrations of H^{1+} and OH^{1-}. Because the $[OH^{1-}]$ is 1.0×10^{-12} M and the $[H^{1+}]$ is 1.0×10^{-2} M you can see that there is a greater concentration of H^{1+} ions present. This tells you that a pH of 2 means that the solution is acidic.

With the help of the previous three problems, you can now set up the pH scale. The pH scale extends from numbers 1 through 14 with each increment meaning a difference in the $[H^{1+}]$ by a power of 10 (remember that you are using logarithms). You have already determined that a pH of 7 is neutral and that a pH of 2 is acidic. This means that the pH values above 7 will indicate a solution that is more basic. You can summarize these trends with the pH scale shown in Figure 9.6.

BUFFERS

How is it that human blood can maintain a pH that is just above 7 considering all the factors that could change the pH of human blood? What is it that causes the resistance to the change in pH? Chemists, and the human body, prepare solutions called *buffers* (or buffer solutions if you wish) that help counter changes in pH. Buffers contain substances that are available to counter any H^{1+} and OH^{1-} ions that may be added to a solution. This is achieved by preparing a solution that is equimolar of a weak acid or weak base and a salt of that weak acid or weak base. A common example is a buffer prepared from acetic acid ($HC_2H_3O_2$) and the salt sodium acetate ($NaC_2H_3O_2$).

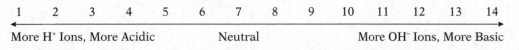

| 1 | 2 | 3 | 4 | 5 | 6 | 7 | 8 | 9 | 10 | 11 | 12 | 13 | 14 |

More H⁺ Ions, More Acidic   Neutral   More OH⁻ Ions, More Basic

Figure 9.6 The pH Scale

When the following equilibrium is established: $HC_2H_3O_2(aq) \longleftrightarrow H^{1+}(aq) + C_2H_3O_2^{1-}(aq)$ and a base is introduced, the H^{1+} is consumed but the acetic acid will form more H^{1+} ions as the equilibrium shifts to the left. If an acid were introduced, then the acetate ion is available to consume the added H^{1+} and shift the equilibrium to the left. These shifts in equilibrium are what cause the pH to remain stable. There is one catch, however. For the buffer to be effective, the concentrations of the acid and its conjugate base must have a concentration far greater than that of any added acid or base.

CHAPTER REVIEW QUESTIONS

1. A stronger base

   (A) is also a stronger acid
   (B) is also a stronger electrolyte
   (C) tastes sour
   (D) yields fewer OH^{1-} ions in solution
   (E) is easier to neutralize

2. When HCl(aq) reacts with Zn(s) the products formed are

   (A) water and a salt
   (B) an acid and a base
   (C) a salt and hydrogen gas
   (D) a nonmetal oxide
   (E) a metal oxide

3. A substance is added to a solution containing two drops of phenolphthalein. The solution then turns pink. Which substance would produce this color change?

   (A) HCl
   (B) H_2CO_3
   (C) KOH
   (D) CH_3CH_2OH
   (E) CH_3OH

4. Litmus is red when the H^{1+} concentration in the solution is

   (A) $1 \times 10^{-11} M$
   (B) $1 \times 10^{-9} M$
   (C) $1 \times 10^{-7} M$
   (D) $1 \times 10^{-5} M$
   (E) $1 \times 10^{-14} M$

5. A substance is dissolved in water and the only positive ions in the solution are H^{1+} ions. This substance is

   (A) KOH
   (B) NaH
   (C) H_2SO_4
   (D) NH_3
   (E) CH_4

6. Which is true about a solution that is acidic?

   (A) $[H^{1+}]$ equals zero.
   (B) $[OH^{1-}]$ equals $[H^{1+}]$.
   (C) $[H^{1+}]$ is less than $[OH^{1-}]$.
   (D) $[H^{1+}]$ is greater than $[OH^{1-}]$.
   (E) $K_w = 1 \times 10^{-7}$.

7. According to the Brønsted-Lowry theory, a base can

   (A) donate a proton
   (B) yield H^{1+} ions
   (C) donate an electron pair
   (D) accept an electron pair
   (E) accept a proton

8. What volume of 0.200 M NaOH(aq) is needed to neutralize 40.0 mL of a 0.100 M HCl(aq)?

   (A) 100.0 mL
   (B) 80.0 mL
   (C) 40.0 mL
   (D) 20.0 mL
   (E) 10.0 mL

9. As an acidic solution is titrated with drops of base, the pH value of the solution will

   (A) increase
   (B) decrease
   (C) remain the same
   (D) approach zero
   (E) none of the above

10. Which pH value demonstrates a solution with the greatest concentration of OH^{1-} ions?

   (A) 1
   (B) 7
   (C) 10
   (D) 12
   (E) 14

11. The reaction: $HI(aq) + LiOH(aq) \rightarrow H_2O(l) + LiI(aq)$ is classified as

    (A) a single replacement

    (B) a neutralization reaction

    (C) the process of hydrolysis

    (D) a synthesis reaction

    (E) an oxidation-reduction reaction

12. How many times stronger is an acid with a pH of 2 than an acid with a pH of 5?

    (A) A pH of 2 is three times as strong.

    (B) A pH of 2 is one thousand times as strong.

    (C) A pH of 2 is three times as weak.

    (D) A pH of 2 is one thousand times as weak.

    (E) A pH of 5 is three thousand times as strong.

13. Which substance below is expected to be the strongest electrolyte?

    (A) Chlorous acid

    (B) Water

    (C) Acetic acid

    (D) Hydrofluoric acid

    (E) Hypochlorous acid

14. Which of the following statements is true?

    (A) NaCl is a neutral salt.

    (B) $KC_2H_3O_2$ is an acidic salt.

    (C) KOH is an acid.

    (D) HCl and KOH react to form hydrogen gas and water.

    (E) NaBr is basic salt.

15. Which pairing is not a set of conjugates?

    (A) OH^{1-} and H_2O

    (B) $HC_2H_3O_2$ and $C_2H_3O_2^{1-}$

    (C) HCl and Cl^{1-}

    (D) NH_3 and NH_4^{1+}

    (E) H_2SO_4 and SO_4^{2-}

16. Which reaction below is incorrect based upon the reactants given?

    (A) $HF + LiOH \rightarrow H_2O + LiF$

    (B) $2HCl + Zn \rightarrow H_2O + ZnCl_2$

    (C) $SO_2 + H_2O \rightarrow H_2SO_3$

    (D) $K_2O + H_2O \rightarrow 2KOH$

    (E) All of the above reactions are correct.

17. Which compound below is not correctly paired with its name?

    (A) KOH is potassium hydroxide.

    (B) H_2SO_3 is sulfurous acid.

    (C) HI is hydroiodic acid.

    (D) $HClO_2$ is chloric acid

    (E) H_3PO_4 is phosphoric acid.

18. Which of the following pairs would be a good choice for preparing a buffer solution?

    (A) Acetic acid and sodium acetate

    (B) Hydrochloric acid and sodium chloride

    (C) Perchloric acid and sodium chlorate

    (D) Sulfuric acid and sodium sulfate

    (E) Water and sodium metal

ANSWERS:

1. (B)	2. (C)	3. (C)
4. (D)	5. (C)	6. (D)
7. (E)	8. (D)	9. (A)
10. (E)	11. (B)	12. (B)
13. (A)	14. (A)	15. (E)
16. (B)	17. (D)	18. (A)

CHAPTER 10

REDOX AND ELECTROCHEMISTRY

IN THIS CHAPTER YOU WILL LEARN ABOUT . . .

Redox Defined
Oxidation Numbers
Writing Half Reactions
Strength of Oxidizing and Reducing Agents
Balancing Half Reactions
The Voltaic Cell
E° Values and Spontaneous Reactions
The Electrolytic Cell

Have you ever bitten into an apple and set it down on a countertop? After a while the apple turns brown. The same thing happens to a potato that has been peeled; it too will turn brown. Why does this occur? Have you ever used a cleaning agent that has the term "ox" in it? Read on for the answers to these questions.

REDOX DEFINED

Redox, a term not often found in the spell-check option in word processors, is short for *reduction* and *oxidation.* The previous chapter looked at the loss and gain of the protons. This chapter focuses on the movement of electrons. Just like protons, electrons can be gained and lost in a chemical reaction. The loss and gain of electrons, of course, will involve ions as well. Before you go any further, you will need to learn a set of rules about *oxidation numbers.*

OXIDATION NUMBERS

Oxidation numbers can be thought of as the charge of an ion or the charge that an atom "feels." You can expect ionic compounds to contain charged particles, but the atoms in covalently bonded compounds do not normally carry charges as ions do. However, the nonmetals in a neutral covalently bonded compound can still have an oxidation state or charge that they "feel." You can find the oxidation state of an element by understanding and applying the following rules:

1. Free elements have an oxidation number of zero. That is, any element bonded to itself or by itself has no oxidation number. This is because there

will be an equal sharing of the electrons in the compound. Examples are Cl_2, Na, O_3, and S_8.

2. Group 1 metals will have an oxidation number of 1+ when bonded in compounds. For example, Na in NaCl, K in K_2O, and Li in LiOH all have oxidation states of 1+.

3. Group 2 metals will have an oxidation number of 2+ when bonded in compounds. For example, Ca in $CaCl_2$ and MgO both have oxidation states of 2+.

4. Hydrogen will have an oxidation state of 1+ when bonded in compounds. For example, H in HCl will have an oxidation state of 1+. An exception to this rule is when hydrogen is bonded in a metal hydride. Examples are NaH and KH. In a metal hydride the hydrogen will have an oxidation state of 1–.

5. Fluorine will have an oxidation state of 1– when bonded in compounds. This is the only oxidation state (other than zero) for fluorine.

6. Oxygen will have an oxidation state of 2–. An example is CaO. In this compound the oxygen has an oxidation state of 2– to balance the 2+ charge of Ca. When bonded to fluorine, oxygen has a positive oxidation number. When bonded in a peroxide, general formulas of X_2O_2, oxygen will have an oxidation state of 1–. Examples of this are H_2O_2 and Na_2O_2.

7. The oxidation number of an ion is the charge that the ion carries. For example, Fe^{3+} has an oxidation state of 3+.

PROBLEMS: Find the oxidation numbers for O in H_2O, K in KI, Ca in CaS, H in MgH_2, F in HF, and Fe in Fe(s).

Solutions: Oxygen will be 2– as it is quite often. Potassium is located in group 1 and will have an oxidation number of 1+. Calcium is in group 2 and will have a charge of 2+. Hydrogen is bonded with a metal in a metal hydride and will have an oxidation number of 1–. Fluorine will have an oxidation number of 1–. Iron is a lone element and will have an oxidation state of zero.

What if you were asked to find the oxidation state of an element that has not been listed in the rules above? For example, how do you know the oxidation state of Cr in $K_2Cr_2O_7$? This is a case where *conservation of charge* comes into play. Although you do not have oxidation state rules for Cr, you do know rules for K and O and you can apply conservation of charge to help find the oxidation state for chromium. Looking at the compound potassium chromate, you see that there is no overall charge on the compound. This means that the sum of all the oxidation states is zero. You can set up a solution that allows you to mathematically work out the oxidation states for the elements. A helpful method for solving for an unknown oxidation state is to write out the compound, find the oxidation numbers for the elements in which it is known, and work out the math to agree with the Law of Conservation of Charge. The oxidation states are always written above the elements in the compound and the math is worked out below the elements. Start by putting in the oxidation numbers for oxygen and potassium:

Ox   1+   2–

$K_2Cr_2O_7$

Math   = 0

Then multiply the oxidation state by the number of atoms of the element present:

$$\text{Ox} \quad 1+ \quad 2-$$
$$K_2Cr_2O_7$$
$$\text{Math} \quad +2 +? -14 = 0$$

Next solve for the unknown number in the equation:

$$\text{Ox} \quad 1+ \quad 2-$$
$$K_2Cr_2O_7$$
$$\text{Math} \quad +2 +12 -14 = 0$$

Finally, solve for the oxidation number for Cr. Because two Cr ions contribute a total of +12 to the math, each individual Cr ion has a charge of 6+:

$$\text{Ox} \quad 1+ \; 6+ \; 2-$$
$$K_2Cr_2O_7$$

When finding the oxidation state of an element in a polyatomic ion, remember that, mathematically, the oxidation states must equal the total charge on the polyatomic ion. For example, SO_4^{2-} must be set up as:

$$\text{Ox}$$
$$SO_4^{2-}$$
$$\text{Math} \qquad\qquad = -2$$

Now include the oxidation number and math for oxygen:

$$\text{Ox} \qquad 2-$$
$$SO_4^{2-}$$
$$\text{Math} \quad ? -8 \quad = -2$$

Finally, solve for the oxidation number for the sulfur:

$$\text{Ox} \quad 6+ \; 2-$$
$$SO_4^{2-}$$
$$\text{Math} \quad +6 -8 = -2$$

The oxidation state for sulfur is 6+.

PROBLEMS: Find the oxidation states for Mn in $KMnO_4$, Cl in $HClO_4$, P in H_3PO_3, and C in HCO_3^{1-}.

Solutions: In short, the oxidation states are as follows:

Ox	1+ 7+ 2–	Ox	1+ 7+ 2–	Ox	1+ 3+ 2–	Ox	1+ 4+ 2–
	$KMnO_4$		$HClO_4$		H_3PO_3		HCO_3^{1-}
Math	+1 +7 –8 = 0	Math	+1 +7 –8 = 0	Math	+3 +3 –6 = 0	Math	+1 +4 –6 = –1

■ WRITING HALF REACTIONS

Once you have a firm grasp on the rules for oxidation numbers and solving problems involving oxidation numbers, you can use oxidation numbers to determine the substance that undergoes a reduction and an oxidation in a redox reaction. You can tell that a substance has been reduced if it has gained electrons (electrons are a reactant). A substance that has been oxidized has lost electrons (electrons are a product). After the substances that have changed oxidation states in a redox reaction are identified, you then write separate *half reactions*. Half reactions are two separate reactions that show the oxidation and reduction reactions separately. An example follows.

When determining which substances were oxidized and reduced in a redox reaction, first assign the oxidation numbers to all the elements in the reaction. The formation for magnesium chloride serves as an example:

$$0 \quad\quad 0 \quad\quad +2 \quad -1$$
$$Mg + Cl_2 \rightarrow MgCl_2$$

In this equation, the free elements have oxidation states of zero and the oxidation states have been assigned to Mg and Cl in magnesium chloride. Next, separate the two elements that have undergone a change in oxidation state:

$$Mg^0 \rightarrow Mg^{2+}$$
$$Cl_2^0 \rightarrow 2Cl^-$$

Next, add electrons to balance the charges and obey the Law of Conservation of Charge. A student once pointed out to me that the electrons always go on the more positive side of the equation. To this day, my students are taught this rule religiously. This means that two electrons will be written as a product for the magnesium half reaction and two electrons will be written as a reactant for the chlorine half reaction:

$$Mg \rightarrow Mg^{2+} + 2e^-$$
$$2e^- + Cl_2 \rightarrow 2Cl^-$$

Because the Mg lost electrons, it was the substance oxidized and it is called the *reducing agent*. Because the Cl gained electrons, it was the substance reduced and it is called the *oxidizing agent*. A good mnemonic device to use is "LEO the lion says GER" (**L**oss of **E**lectrons is **O**xidation, **G**ain of **E**lectrons is **R**eduction). The so-called agents are always one of the reactants in a redox reaction. Finally, notice that the number of electrons lost is the same number of electrons gained. You might think that four electrons were transferred in this reaction. Actually, only two electrons were transferred because oxidation and reduction occur simultaneously. The two electrons lost are the same two that were gained.

Write the two half reactions for: $2H_2O \rightarrow 2H_2 + O_2$. First, assign oxidation numbers:

$$1+ \; 2- \quad\quad 0 \quad\quad 0$$
$$2H_2O \rightarrow 2H_2 + O_2$$

Next, separate the reactions for the elements that changed oxidation numbers:

$$4H^{1+} \rightarrow 2H_2$$

$$2O^{2-} \rightarrow O_2$$

Now add electrons to balance the charges. Electrons go on the more positive side of the half reaction.

$$4e^- + 4H^{1+} \rightarrow 2H_2$$

$$2O^{2-} \rightarrow O_2 + 4e^-$$

The substance oxidized (reducing agent) is O^{2-} because it is the reactant that lost electrons. The substance reduced (oxidizing agent) is H^{1+} because it is the reactant that gained electrons.

PROBLEM: Write two half reactions for $2Na + Cl_2 \rightarrow 2NaCl$. Which substances were oxidized and reduced?

Solution: Assign oxidation numbers and separate the substances that had a change in oxidation state:

$$2Na \rightarrow 2Na^{1+}$$

$$Cl_2 \rightarrow 2Cl^{1-}$$

Now write in the electrons to the more positive side of the equation to balance the charge:

$$2Na \rightarrow 2Na^{1+} + 2e^-$$

$$2e^- + Cl_2 \rightarrow 2Cl^{1-}$$

Na was oxidized and is the reducing agent. Cl_2 was reduced (notice the oxidation number was reduced from 0 to 1–) and is the oxidizing agent.

STRENGTH OF OXIDIZING AND REDUCING AGENTS

It is often difficult to find the strongest or weakest oxidizing or reducing agent in a redox reaction. To answer such questions, go back to the basics: electronegativity. A highly electronegative element like fluorine will gain electrons (will be reduced). What does this say about fluorine? Fluorine is the best oxidizing agent. At the other end of the spectrum is lithium. Lithium has a very low electronegativity and will give away its electrons (will be oxidized). Therefore, lithium is a very good reducing agent.

Think about this: Lithium is a popular metal used in batteries. What does this say about lithium's ability to be a reducing agent?

BALANCING HALF REACTIONS

Look at the following redox reaction and write two half reactions for it:

$$Cu + Ag^{1+} \rightarrow Ag + Cu^{2+}$$

The Cu and Ag have oxidation states of 0 and the ions have oxidation numbers equal to their charge:

$$Cu^0 \rightarrow Cu^{2+}$$

$$Ag^{1+} \rightarrow Ag^0$$

Finally, add in the electrons:

$$Cu^0 \rightarrow Cu^{2+} + 2e^-$$

$$1e^- + Ag^{1+} \rightarrow Ag^0$$

Here is a situation where the number of electrons lost is not equal to the number of electrons gained—processes that must occur simultaneously with an equal number of electrons. To remedy the situation, you must call upon the distributive property of mathematics. Multiply the silver half reaction by 2 to get an equal number of electrons lost and gained, and at the same time, correct the coefficients for the silver atom and ion in the equation.

$$2(1e^- + Ag^{1+} \rightarrow Ag^0) = 2e^- + 2Ag^{1+} \rightarrow 2Ag^0$$

Next, add up the two reactions:

$$Cu^0 \rightarrow Cu^{2+} + 2e^-$$

$$\frac{2e^- + 2Ag^{1+} \rightarrow 2Ag^0}{2e^- + 2Ag^{1+} + Cu^0 \rightarrow 2Ag^0 + Cu^{2+} + 2e^-}$$

Because the electrons appear on both sides of the equation in equal amounts, you can cancel them out and have a balanced equation: $Cu + 2Ag^{1+} \rightarrow 2Ag + Cu^{2+}$.

Just in case you were wondering, Cu^0 was oxidized and is the reducing agent, while Ag^{1+} was reduced and is the oxidizing agent.

PROBLEM: Balance the following: $Zn + Fe^{3+} \rightarrow Zn^{2+} + Fe$

Solution: After placing the electrons on the more positive side of the half reactions, the half reactions will look like this:

$$Zn \rightarrow Zn^{2+} + 2e^-$$

$$3e^- + Fe^{3+} \rightarrow Fe$$

Multiply to get:

$$3(Zn \rightarrow Zn^{2+} + 2e^-) \quad = \quad 3Zn \rightarrow 3Zn^{2+} + 6e^-$$

$$2(3e^- + Fe^{3+} \rightarrow Fe) \quad = \quad 6e^- + 2Fe^{3+} \rightarrow 2Fe$$

Add up the two equations and drop out the electrons that are in equal amounts on both sides of the equation to get: $3Zn + 2Fe^{3+} \rightarrow 3Zn^{2+} + 2Fe$.

Balancing other types of redox reactions can be more complicated and require more work. Fortunately, there is a set method for solving the complex balancing of redox reactions. These steps must be learned if you are to be successful in balancing a complex redox reaction. We will first consider the reaction: $As + ClO_3^{1-} \rightarrow H_3AsO_3 + HClO$.

1. First, write separate half reactions for the As and Cl. It is important to carry along the H and O that appear in the compounds:

$$As \rightarrow H_3AsO_3 \quad \text{and} \quad ClO_3^{1-} \rightarrow HClO$$

2. Then balance the elements other than H and O. The As and Cl atoms are already balanced in the two half reactions, so you still have:

$$As \rightarrow H_3AsO_3 \quad \text{and} \quad ClO_3^{1-} \rightarrow HClO$$

3. Add water molecules to the half reactions to balance the oxygen atoms:

$$3H_2O + As \rightarrow H_3AsO_3 \quad \text{and} \quad ClO_3^{1-} \rightarrow HClO + 2H_2O$$

4. Add H^{1+} ions to the half reactions to balance the hydrogen atoms:

$$3H_2O + As \rightarrow H_3AsO_3 + 3H^{1+} \quad \text{and} \quad 5H^{1+} + ClO_3^{1-} \rightarrow HClO + 2H_2O$$

5. Obey conservation of charge by adding electrons to the more positive side of the half reactions:

$$3H_2O + As \rightarrow H_3AsO_3 + 3H^{1+} + 3e^- \quad \text{and}$$

$$4e^- + 5H^{1+} + ClO_3^{1-} \rightarrow HClO + 2H_2O$$

6. Use the distributive property of mathematics to multiply the half reactions so that the number of electrons lost is equal to the number gained:

$$(3H_2O + As \rightarrow H_3AsO_3 + 3H^{1+} + 3e^-)4 \quad = \quad 12H_2O + 4As \rightarrow 4H_3AsO_3$$
$$+ 12H^{1+} + 12e^-$$

$$3(4e^- + 5H^{1+} + ClO_3^{1-} \rightarrow HClO + 2H_2O) \quad = \quad 12e^- + 15H^{1+} + 3ClO_3^{1-} \rightarrow$$
$$3HClO + 6H_2O$$

7. Finally, add the two half reactions and cancel out substances that appear on both sides of the equation:

$$12H_2O + 4As \rightarrow 4H_3AsO_3 + 12H^{1+} + 12e^-$$
$$12e^- + 15H^{1+} + 3ClO_3^{1-} \rightarrow 3HClO + 6H_2O$$

becomes:

$$12H_2O + 4As \rightarrow 4H_3AsO_3 + 12H^{1+}$$
$$15H^{1+} + 3ClO_3^{1-} \rightarrow 3HClO + 6H_2O$$

becomes:

$$6H_2O + 4As \rightarrow 4H_3AsO_3$$
$$\underline{3H^{1+} + 3ClO_3^{1-} \rightarrow 3HClO}$$
$$3H^{1+} + 6H_2O + 4As + 3ClO_3^{1-} \rightarrow 4H_3AsO_3 + 3HClO$$

Redox reactions can be balanced in basic solutions as well. This task will require an additional step but it can still be done following the steps shown above. To balance the redox reaction: $H_2O_2 + ClO_3^{1-} \rightarrow ClO_2^{1-} + O_2$ in a basic solution, do as follows:

1. First write separate half reactions for the O and Cl. It is important to carry along the H and O that appear in the compounds:

$$H_2O_2 \rightarrow O_2 \quad \text{and} \quad ClO_3^{1-} \rightarrow ClO_2^{1-}$$

2. Then balance the elements other than H and O. The Cl atoms are already balanced, so you still have:

$$H_2O_2 \rightarrow O_2 \quad \text{and} \quad ClO_3^{1-} \rightarrow ClO_2^{1-}$$

3. Add water molecules to the half reactions to balance the oxygen atoms:

$$H_2O_2 \rightarrow O_2 \quad \text{and} \quad ClO_3^{1-} \rightarrow ClO_2^{1-} + H_2O$$

4. Add H^{1+} ions to the half reactions to balance the hydrogen atoms:

$$H_2O_2 \rightarrow O_2 + 2H^{1+} \quad \text{and} \quad 2H^{1+} + ClO_3^{1-} \rightarrow ClO_2^{1-} + H_2O$$

5. Obey conservation of charge by adding electrons to the more positive side of the half reactions:

$$H_2O_2 \rightarrow O_2 + 2H^{1+} + 2e^- \quad \text{and} \quad 2e^- + 2H^{1+} + ClO_3^{1-} \rightarrow ClO_2^{1-} + H_2O$$

6. Because hydroxide ions neutralize hydronium ions in a basic solution, replace all of the H^{1+} ions with water and add an equal number of OH^{1-} ions to the other side of the half reaction:

$$2OH^{1-} + H_2O_2 \rightarrow O_2 + 2H_2O + 2e^- \quad \text{and} \quad 2e^- + 2H_2O + ClO_3^{1-} \rightarrow$$
$$ClO_2^{1-} + H_2O + 2OH^{1-}$$

7. Balance the number of electrons by using the distributive property of mathematics:

$$2OH^{1-} + H_2O_2 \rightarrow O_2 + 2H_2O + 2e^- \quad \text{and} \quad 2e^- + 2H_2O + ClO_3^{1-} \rightarrow$$
$$ClO_2^{1-} + H_2O + 2OH^{1-}$$

8. Finally, cancel out substances that appear on both sides of the equation:

$$2OH^{1-} + H_2O_2 \rightarrow O_2 + 2H_2O + 2e^- \quad \text{and} \quad 2e^- + H_2O + ClO_3^{1-} \rightarrow$$
$$ClO_2^{1-} + 2OH^{1-}$$

becomes: $2OH^{1-} + H_2O_2 \rightarrow O_2 + 2H_2O \quad$ and $\quad H_2O + ClO_3^{1-} \rightarrow$
$$ClO_2^{1-} + 2OH^{1-}$$

becomes: $H_2O_2 \rightarrow O_2 + 2H_2O \quad$ and $\quad H_2O + ClO_3^{1-} \rightarrow ClO_2^{1-}$

and the two half reactions are added:

$$\begin{array}{c} H_2O_2 \rightarrow O_2 + H_2O \\ ClO_3^{1-} \rightarrow ClO_2^{1-} \\ \hline H_2O_2 + ClO_3^{1-} \rightarrow O_2 + H_2O + ClO_2^{1-} \end{array}$$

PROBLEM: Balance the following reaction in an acidic solution: $Cu + NO_3^{1-} \rightarrow NO_2 + Cu^{2+}$.

Solution:

1. First write separate half reactions for the Cu and N. It is important to carry along the H and O that appear in the compounds:

$$NO_3^{1-} \rightarrow NO_2 \quad \text{and} \quad Cu \rightarrow Cu^{2+}$$

2. Then balance the elements other than H and O. The Cl and N atoms are already balanced, so you still have:

$$NO_3^{1-} \rightarrow NO_2 \quad \text{and} \quad Cu \rightarrow Cu^{2+}$$

3. Add water molecules to the half reactions to balance the oxygen atoms:

$$NO_3^{1-} \rightarrow NO_2 + H_2O \quad \text{and} \quad Cu \rightarrow Cu^{2+}$$

4. Add H^{1+} ions to the half reactions to balance the hydrogen atoms:

$$2H^{1+} + NO_3^{1-} \rightarrow NO_2 + H_2O \quad \text{and} \quad Cu \rightarrow Cu^{2+}$$

5. Obey conservation of charge by adding electrons to the more positive side of the half reactions:

$$1e^- + 2H^{1+} + NO_3^{1-} \rightarrow NO_2 + H_2O \quad \text{and} \quad Cu \rightarrow Cu^{2+} + 2e^-$$

6. Balance the number of electrons by using the distributive property of mathematics:

$$2e^- + 4H^{1+} + 2NO_3^{1-} \rightarrow 2NO_2 + 2H_2O \quad \text{and} \quad Cu \rightarrow Cu^{2+} + 2e^-$$

7. Finally, cancel out substances that appear on both sides of the equation:

$$4H^{1+} + 2NO_3^{1-} \rightarrow 2NO_2 + 2H_2O \quad \text{and} \quad Cu \rightarrow Cu^{2+}$$

Adding the two equations together we find that:

$$4H^{1+} + 2NO_3^{1-} + Cu \rightarrow 2NO_2 + 2H_2O + Cu^{2+}$$

THE VOLTAIC CELL

Because redox reactions can make electrons move from one substance to another, it is possible to create a setup so that the electrical energy produced in a redox reaction can be channeled to do work. There is a way to harvest the electrons produced by a redox reaction. Today these devices are called batteries. The first device that could do this was called a *voltaic cell*. In a voltaic cell a redox reaction occurs spontaneously so that the electrons can be used to do work. A typical voltaic cell is shown in Figure 10.1.

The wire in the voltaic cell carries the electrons from one *half cell* to another. The *salt bridge* allows ions to migrate from one half cell to the other so that there is no buildup of charge as the electrons are transferred from one half cell to the other. The *electrodes* are the sites of oxidation and reduction in the voltaic cell. These processes will occur on the surfaces of the *cathode* (electrode where reduction occurs) and the *anode* (electrode where oxidation occurs).

In order to understand in which direction electrons will flow in the voltaic cell, you must first get an idea of the activity of certain elements. This will tell you which elements want to lose their electrons more than others and force their electrons on these other elements. Figure 10.2 shows select metals and nonmetals sorted from most active to least active.

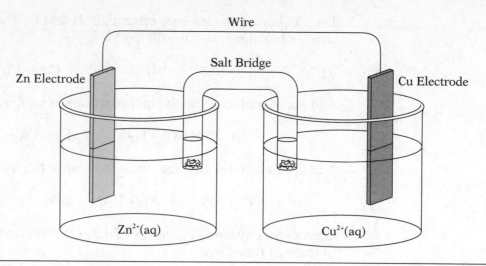

Figure 10.1 The Voltaic Cell

A more careful examination of the elements' electronegativity values shows a correlation (for the most part) between electronegativity and ability to lose or gain electrons.

Look back at Figure 10.1 showing the Zn and Cu electrodes set up in the voltaic cell. For simplicity, the setup can be abbreviated by writing: Zn / Zn^{2+} // Cu / Cu^{2+}. The electrons will flow spontaneously from the Zn electrode to the Cu electrode because according to the activity series, Zn is a more active metal than Cu. This means that Zn behaves more like a metal and loses electrons easily. The half reaction for Zn in this half cell will be: $Zn \rightarrow Zn^{2+} + 2e^-$.

The electrons lost from the Zn half cell travel through the wire (and possibly an electrical device) and remain on the surface of the Cu electrode. The negatively charged electrons will attract the positively charged Cu^{2+} ions and react to give: $Cu^{2+} + 2e^- \rightarrow Cu$.

Because the Zn electrode lost electrons (oxidation), it is called the anode (remember the mnemonic device "AN OX"). The Cu electrode gained electrons (reduction) and is called the cathode (remember the mnemonic device "RED CAT"). The anode is considered to be the negative electrode in the voltaic cell and the cathode is considered to be the positive electrode.

Finally, there is the issue of ion movement. The cations in the salt bridge will move toward the cathode half cell while the anions in the salt bridge

Activity Series for Metals and Nonmetals

Most Active Metals (Lose Electrons Best)   Least Active Metals

| Li | K | Na | Mg | Al | Zn | Cr | Fe | Co | Ni | Sn | Pb | H_2* | Cu | Ag | Au |

Most Active Nonmetals (Gain Electrons Best)   Least Active Nonmetals

| F_2 | | Cl_2 | | Br_2 | | I_2 |

H_2*—The hydrogen half cell is an arbitrary standard, which all other activities are measured against.

Figure 10.2 Activity Series

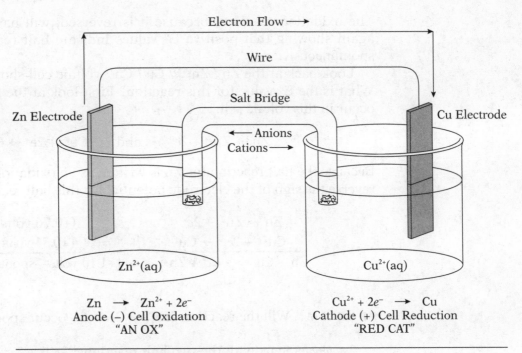

Figure 10.3 The Zinc/Copper Voltaic Cell

move toward the anode half cell. Figure 10.3 shows the voltaic cell labeled to reflect the discussion above.

PROBLEM: For the voltaic cell $Al/Al^{3+}//Co/Co^{2+}$, write the two half reactions and determine which electrode is the anode and which is the cathode.

Solution: Looking at the activity series shows that Al is a more active metal than Co. This means that Al will lose electrons and Co will have to gain them. The half reaction for the oxidation is $Al \rightarrow Al^{3+} + 3e^-$. The half reaction for the reduction is $2e^- + Co^{2+} \rightarrow Co$. The Al electrode is the anode and the Co electrode is the cathode.

E° VALUES AND SPONTANEOUS REACTIONS

The activity series tells the likelihood of a metal or a nonmetal to lose or gain electrons, but it does not quantify exactly how much more likely a metal or a nonmetal is to lose or gain an electron. You can examine the *electrode potentials* of certain elements to determine exactly how much voltage can be produced from the loss and gain of electrons between two substances. The electrode potentials can be found in Appendix 4, Reference Tables. For convenience, all reactions have been written as reduction reactions. Upon inspection of these values, you will notice a highly positive value for fluorine gas to gain electrons. Because of its high electronegativity, fluorine will gain electrons. This means that highly positive potential values ($E°$) for a half reaction are spontaneous half reactions. Looking under Metals in the reference table in Appendix 4 you see that the gaining of an electron by lithium has an E° value of –3.05. This indicates that the reduction of a lithium ion will not occur. Instead the opposite is most likely to happen, $Li \rightarrow Li^{1+} + 1e^-$.

The oxidation reaction, because it is reversed, will have an E° of +3.05, again showing that positive E° values indicate half reactions that occur spontaneously.

Look back at the Zn / Zn²⁺ // Cu / Cu²⁺ voltaic cell shown in Figure 10.3. What is the E° value for this reaction? First look at the half reactions that occur in this voltaic cell:

$$Zn \rightarrow Zn^{2+} + 2e^- \quad \text{and} \quad Cu^{2+} + 2e^- \rightarrow Cu$$

Because the half reaction for Zn is written as an oxidation, you will need to reverse the sign of the electrode potential for this half reaction:

$$
\begin{array}{ll}
Zn \rightarrow Zn^{2+} + 2e^- & \text{(+0.76 volts)} \\
\underline{Cu^{2+} + 2e^- \rightarrow Cu} & \underline{\text{(+0.34 volts)}} \\
Zn + Cu^{2+} \rightarrow Cu + Zn^{2+} & \text{(+1.10 volts—spontaneous)}
\end{array}
$$

PROBLEM: Will the reaction $2KCl \rightarrow 2K + Cl_2$ occur spontaneously?

Solution: Start with the two half reactions: $2K^{1+} + 2e^- \rightarrow 2K$ and $2Cl^{1-} \rightarrow 2e^- + Cl_2$. Now find the electrode potentials for each (remember to reverse the sign for the oxidation reaction):

$$
\begin{array}{ll}
2K^{1+} + 2e^- \rightarrow 2K & \text{(–2.93 volts)} \\
\underline{2Cl^{1-} \rightarrow 2e^- + Cl_2} & \underline{\text{(–1.36 volts)}} \\
2KCl \rightarrow 2K + Cl_2 & \text{(–4.29 volts—nonspontaneous)}
\end{array}
$$

One last word about the electrode potentials before moving on. Notice that when you multiply the half reaction by a coefficient to balance the equation, you do not multiply the electrode potential for the reaction by the coefficient. The electrode potential stays the same as printed in the reference tables and is never multiplied by any coefficient.

Think about this: *If you stumbled upon gold while mining, would you find it as a gold salt or as a solid piece of pure gold? Gold is found as solid nuggets in nature, as suggested by the half reaction $Au^{3+} + 3e^- \rightarrow Au(s)$ and accompanying E° value of +1.50 V. This clearly shows that you do not find gold as an ion occurring naturally.*

As redox reactions occur, the reactants are used and eventually reach completion. This is better known as "batteries dying." When this happens, the E° value is zero and the reaction has reached equilibrium.

Because conditions under which a battery is operating (such as temperature and concentration of reactants) can change, different potentials can be reached. The new electrode potential, E, can be calculated using the Nernst equation:

$$E = E° - \frac{2.30\,RT}{n\boldsymbol{f}}(\log Q)$$

This equation includes the standard electrode potential E° and the familiar constant R. The Q in the equation is the concentration of the products divided by the concentration of the reactants. The variable n stands for the number of moles of electrons transferred. The equation also introduces a new symbol, the *faraday*. A faraday, f, is the charge on one mole of electrons, or about 96,352 coulombs of charge. For convenience it is rounded off as 96,500 C of charge. Because R and f are constants and most reactions occur at 298 K, the equation can be simplified to look like this:

$$E = E° - \frac{0.0591 \text{ V}}{n}\left(\log Q\right)$$

Consider the familiar voltaic cell: $Zn + Cu^{2+} \rightarrow Zn^{2+} + Cu$. If you create a battery that has $[Cu^{2+}] = 3.0\ M$ and $[Zn^{2+}] = 0.1\ M$, what will be the potential of this battery? Start with the Nernst equation and substitute:

$$E = E° - \frac{0.0591 \text{ V}}{n}\left(\log Q\right) \quad \text{becomes} \quad E = 1.10 \text{ V} - \frac{0.0591 \text{ V}}{2}\left(\log[0.1]/[3.0]\right)$$

$$\text{becomes} \quad E = 1.10 \text{ V} - \frac{0.0591 \text{ V}}{2}(-1.48)$$

The new voltage becomes +1.144 volts, a voltage greater than that at standard conditions.

THE ELECTROLYTIC CELL

Earlier you saw that the decomposition of $2KCl \rightarrow 2K + Cl_2$ is not a spontaneous reaction. Does that mean that it is impossible to achieve this reaction? Remember that any reaction that is nonspontaneous can be made to be spontaneous if enough energy is added. The device that can help achieve this is called the *electrolytic cell*. While the voltaic cell spontaneously generates its own electricity, the electrolytic cell requires an outside source of current to make a nonspontaneous reaction occur. If the KCl undergoes *electrolysis* (literally meaning "electricity split") the K and Cl can be split up into their original elements.

The electrolytic cell is set up with an external source of current. The electrons will flow from the negative terminal in the battery into the electrode. If the KCl is going to carry a current, it must be molten so that the ions can move freely and carry the current needed to complete the electrolysis. This can be seen in Figure 10.4. Once the current flows into the molten KCl electrons on the negative electrode, reduction will occur at this electrode, the cathode. The equation for the reaction between the negative electrons and the positive potassium ions is $K^{1+} + 1e^- \rightarrow K(s)$. To complete the circuit, the electrons must return to the power source. This is done by the loss of electrons from the chlorine: $2Cl^{1-} \rightarrow 2e^- + Cl_2$.

As shown in Figure 10.4, the anode is still the site of oxidation while the cathode is the site of reduction. One major change is that in the electrolytic cell the anode is now positive and the cathode is negative.

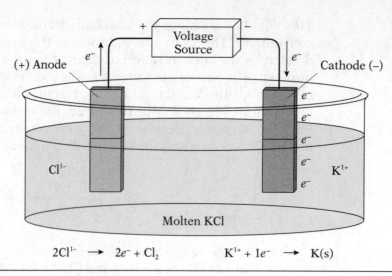

$$2Cl^{1-} \rightarrow 2e^- + Cl_2 \qquad\qquad K^{1+} + 1e^- \rightarrow K(s)$$

Figure 10.4 The Electrolysis of KCl

The electrolytic cell can also be used for *electroplating* an object with a metal. In Figure 10.5 the ring is being plated with silver. As the silver ions in solution are being attracted to the electrons on the ring to coat the ring with silver, the silver electrode is replenishing the silver that is being taken from the solution.

The reaction at the ring (the cathode) is $Ag^{1+} + 1e^- \rightarrow Ag(s)$. (Why does this make sense?) The reaction at the silver electrode (the anode) is $Ag(s) \rightarrow Ag^{1+} + 1e^-$.

The final example of electrolysis is that of water. When water undergoes hydrolysis, hydrogen gas and oxygen gas are the products: $2H_2O \rightarrow 2H_2 + O_2$. What is so special about this reaction is how the hydrogen and oxygen can be separated and captured. When the electricity from the external source

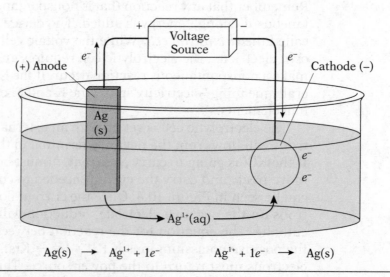

$$Ag(s) \rightarrow Ag^{1+} + 1e^- \qquad\qquad Ag^{1+} + 1e^- \rightarrow Ag(s)$$

Figure 10.5 Electroplating with Silver

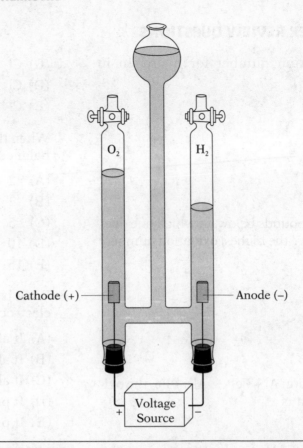

Figure 10.6 Electrolysis of Water

reaches the cathode, reduction occurs and hydrogen gas is formed: $4e^- + 4H^{1+}$ $\rightarrow 2H_2$. At the anode the oxygen from the water loses electrons to form oxygen gas: $2O^{2-} \rightarrow 4e^- + O_2$. This is shown in Figure 10.6.

The apparatus also allows the gases to be captured for everyday use and experimentation.

CHAPTER REVIEW QUESTIONS

1. The oxidation number for hydrogen in NaH is

   (A) 1+
   (B) 2+
   (C) 0
   (D) 1–
   (E) 2–

2. Of the compounds below, in which one does chlorine have the highest oxidation number?

   (A) HCl
   (B) $KClO_3$
   (C) $HClO_2$
   (D) $KClO_4$
   (E) $CaCl_2$

3. In the reaction $Al + Fe^{3+} \rightarrow Al^{3+} + Fe$, the oxidizing agent is

   (A) Al
   (B) Fe
   (C) Al^{3+}
   (D) Fe^{3+}
   (E) none of the above

4. In the chemical cell reaction $2Cr + 3Ni^{2+} \rightarrow 2Cr^{3+} + 3Ni$, which species is reduced?

   (A) Cr
   (B) Ni^{2+}
   (C) Cr^{3+}
   (D) Ni
   (E) none of the above

5. When Fe^{2+} is oxidized to Fe^{3+}, the Fe^{2+} ion

   (A) loses 1 electron
   (B) loses 1 proton
   (C) gains 1 electron
   (D) gains 1 proton
   (E) gains 1 neutron

6. Which half reaction demonstrates conservation of mass and conservation of charge?

   (A) $Cl_2 + e^- \rightarrow Cl^{1-}$
   (B) $Cl_2 + 2e^- \rightarrow Cl^{1-}$
   (C) $Cl_2 \rightarrow 2Cl^{1-} + e^-$
   (D) $Cl_2 + e^- \rightarrow 2Cl^{1-}$
   (E) $Cl_2 + 2e^- \rightarrow 2Cl^{1-}$

7. When the equation $Co + Ni^{2+} \rightarrow Co^{3+} + Ni$ is balanced, the sum of the coefficients is

   (A) 2
   (B) 3
   (C) 5
   (D) 10
   (E) 15

8. What is the purpose of the salt bridge in an electrochemical cell?

   (A) It allows ion migration.
   (B) It allows neutron migration.
   (C) It allows electron migration.
   (D) It prevents ion migration.
   (E) It prevents neutron migration.

9. Making reference to electronegativity values, which substance is most easily reduced?

   (A) Br_2
   (B) Cl_2
   (C) F_2
   (D) I_2
   (E) At_2

10. When nonspontaneous redox reactions occur by use of an external current, the process is called

    (A) neutralization
    (B) esterification
    (C) electrolysis
    (D) hydrolysis
    (E) voltaic ion

ANSWERS:

1. (D)	2. (D)	3. (D)
4. (B)	5. (A)	6. (E)
7. (D)	8. (A)	9. (C)
10. (C)		

CHAPTER 11
ORGANIC CHEMISTRY

IN THIS CHAPTER YOU WILL LEARN ABOUT . . .

Carbon
Hydrocarbons
Naming Organic Compounds
Naming Hydrocarbons and Their Isomers
Cyclic Hydrocarbons
Aromatics
Alcohols and Ethers
Carbonyl-Containing Compounds
Compounds Containing Nitrogen
Organic Reactions with Halogens

Organic chemistry: The words alone will make any premed student cringe. Why would anyone dedicate an entire year in college toward this one topic? Simply because carbon is special!

CARBON

Why is carbon so special? Why does it merit an entire chapter of its own? Carbon is special because of the millions of compounds that it can form. Some of these compounds are naturally occurring; others are prepared synthetically. The chart below contains the names of some natural and synthetic compounds. Notice how many of these compounds support life as we know it.

Natural	Synthetic
• Proteins	• Plastics
• Carbohydrates	• Polystyrene
• Lipids	• Rubber
• Steroids	• Nylon
• Nucleic acids	• Nonstick coatings

HYDROCARBONS

A big part of organic chemistry is the recognition and naming of the *functional groups* that are found in organic compounds. The following chart shows the major classes of *hydrocarbons: alkanes, alkenes,* and *alkynes.* Also, note the general formulas and saturation of these functional groups.

	Alkanes	Alkenes	Alkynes
Functional Group	All C-to-C single bonds	Have one double bond between two carbons	Have one triple bond between two carbons
General Formula	C_nH_{2n+2}	C_nH_{2n}	C_nH_{2n-2}
Saturation	Saturated— cannot have more H added	Unsaturated— can have more H added	Unsaturated—can have more H added

PROBLEM: Given these molecular formulas of two organic compounds, C_6H_{12} and C_8H_{14}, which functional groups are present? Are these compounds saturated or unsaturated?

Solution: C_6H_{12} has the general formula of C_nH_{2n}. It is an alkene and, because it is considered to be unsaturated, this compound has the potential to bond more hydrogen atoms to the molecule. C_8H_{14} has the general formula of C_nH_{2n-2}. Because of this, it is classified as an alkyne and it too is unsaturated.

NAMING ORGANIC COMPOUNDS

The naming of organic compounds depends upon two things:

1. The number of carbon atoms present in the longest chain of carbon atoms.
2. The presence of double or triple bonds.

A prefix is given to organic compounds to indicate the number of carbons present in the longest chain of the compound. These prefixes are shown in the chart below. A suffix is also given to the name of the compound to indicate the type of bonds present in the compound. The suffixes for hydrocarbons can be either *-ane, -ene* or *-yne.* Putting a prefix and suffix together gives the name of the compound. Notice that the compounds below follow the general formulas of their respective functional groups.

Prefix and Number of Carbon Atoms	Alkane	Alkene	Alkyne
One carbon atom *meth-*	Methane CH_4	None	None
Two carbon atoms *eth-*	Ethane C_2H_6	Ethene C_2H_4	Ethyne (Acetylene) C_2H_2
Three carbon atoms *prop-*	Propane C_3H_8	Propene C_3H_6	Propyne C_3H_4
Four carbon atoms *but-*	Butane C_4H_{10}	Butene C_4H_8	Butyne C_4H_6
Five carbon atoms *pent-*	Pentane C_5H_{12}	Pentene C_5H_{10}	Pentyne C_5H_8

NAMING HYDROCARBONS AND THEIR ISOMERS

Butane, C_4H_{10}, can be drawn as CH_3—CH_2—CH_2—CH_3. There is another compound with the same molecular formula of C_4H_{10} but with a different structure. Because this compound has the same molecular formula but a different structure it is said to be an *isomer* of butane, as seen in Figure 11.1. If this compound is not butane, what is it called? Follow the two rules presented above about naming organic compounds. First, find the longest chain and then look at the types of bonds between the carbon atoms. The compound in Figure 11.1 has a longest chain of three carbon atoms. That means the compound's name will be propane (three carbons in the longest chain and the bonds are all single).

The compound also has a CH_3 group or branch off of the longest chain. Because the branch has just one carbon it is called a meth*yl* group. The suffix *-yl* indicates that the branch looks like meth*ane* except that one hydrogen atom is missing. Because the methyl group is on the second carbon of the longest chain, it is given a number that tells its location in the molecule. Therefore the name of this compound is 2-methylpropane (see Figure 11.1).

PROBLEM: Name the compound in Figure 11.2.

Solution: The longest chain in this compound is five carbon atoms long and they are all single bonded. This part of the molecule is called pentane. There are two methyl groups, and they will be named as "dimethyl." Finally, the locations of the branches must be given the lowest numbers possible, in this case 2 and 3. The name of this compound is 2,3-dimethylpentane.

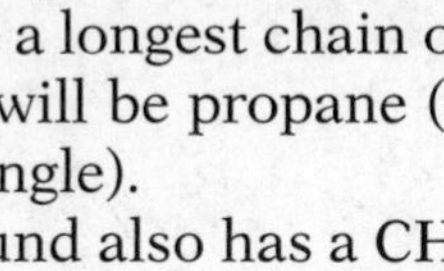

$$CH_3-CH-CH_3 \qquad\qquad CH_3-CH_2-CH-CH-CH_3$$
$$\;\;\;\;\;\;\;\;\;| \qquad\qquad\qquad\qquad\qquad\qquad | \;\;\; |$$
$$\;\;\;\;\;\;\;CH_3 \qquad\qquad\qquad\qquad\qquad\;\; CH_3 \;\; CH_3$$

Figure 11.1 Isobutane 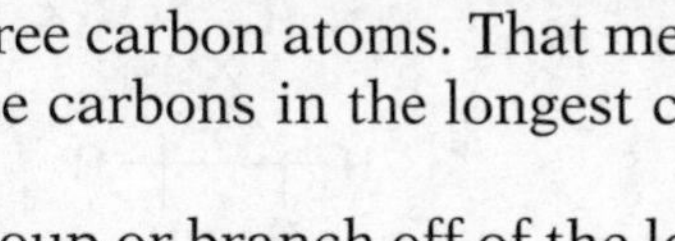   **Figure 11.2**

PROBLEM: Name: $CH_3-CH=CH-CH_3$ and $CH_3-C\equiv C-CH_3$.

Solution: Both compounds have four carbon atoms and will have the prefix *but-*. The first compound has a double bond and will end in *-ene* and the second one will end in *-yne* because of its triple bond. Both the double bond and triple bond are located between the second and third carbon atoms in their molecules. Remembering to use the lowest number possible, you can name these compounds 2-butene and 2-butyne.

CYCLIC HYDROCARBONS

These previous examples of hydrocarbons show chains of carbon atoms and branches on the chains. There are also classes of hydrocarbons that are arranged in rings. For example, a molecule of cyclopropane can be represented as shown in Figure 11.3.

This cumbersome structure can be simplified by drawing a simple triangle. Each "corner" or "bend" in the drawing represents a carbon atom and the single lines tell that the carbon atoms are single bonded. (See Figure 11.4.)

PROBLEM: Name the ringed compounds in Figure 11.5:

Solution: The first ringed compound has six carbon atoms and they are joined by single lines indicating all single bonds. This compound is called cyclohexane. The same holds true for the second compound except it has eight carbon atoms. Its name will be cycloctane.

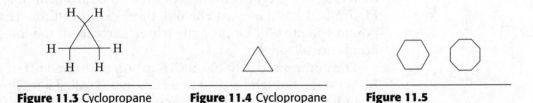

Figure 11.3 Cyclopropane   **Figure 11.4** Cyclopropane   **Figure 11.5**

AROMATICS

There is a special class of hydrocarbons called the aromatic compounds. The aromatics are characterized by their six carbon rings and their alternating double and single bonds. The simplest of these compounds is a compound called benzene. Because the double and single bonds alternate, there are two resonance structures (thanks to delocalized pi bonds) that are illustrated in Figure 11.6.

To simplify the concept of the two resonance structures, benzene can also be drawn as shown in Figure 11.7.

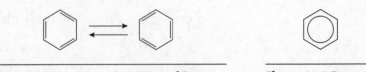

Figure 11.6 Resonance Structures of Benzene   **Figure 11.7** Benzene

ALCOHOLS AND ETHERS

Organic compounds are not limited to having only the elements carbon and hydrogen. Many times oxygen atoms may also be present in organic compounds. This opens the door to a new set of functional groups. Two of these functional groups are called alcohols and ethers. *Alcohols* and *ethers* are characterized by the presence of a singly bonded oxygen atom in an organic compound. However, there is a major difference between the two.

Alcohols are identified by the presence of an —OH or hydroxyl group. Don't be fooled into thinking that alcohols are bases. The —OH group in an alcohol is covalently bonded and will not form a hydroxide ion to make a solution basic. The general formula of an alcohol is R—OH, where R is a carbon chain. Three common alcohols are shown in Figure 11.8.

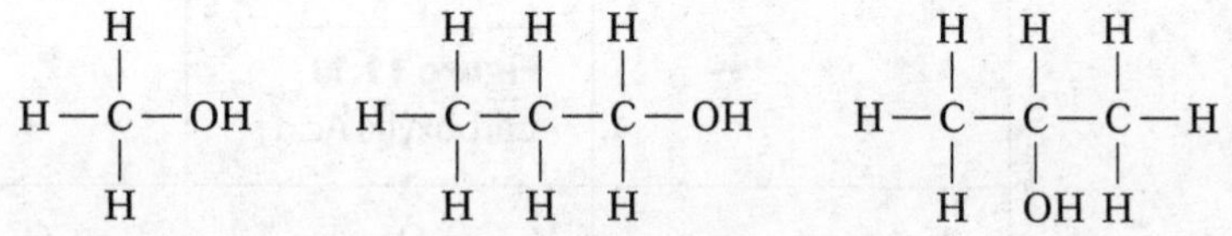

Figure 11.8 Three Alcohols

Notice that the —OH group is located in various parts of these molecules. Each of these alcohols is given a different name based upon the number of carbon atoms present and the location of the hydroxyl group.

When naming alcohols, take the name of the parent alkane and drop the final "e" from the parent name. Then add the suffix *-ol* to the end of the parent name. For example, the first alcohol in Figure 11.8 has one carbon atom and four single bonds. It looks like methane, but it is an alcohol. The name of this compound will be methanol. The second compound has three carbon atoms and is an alcohol. This compound is called 1-propanol, the "1" indicating that the functional group is on the "first" carbon atom. The alcohol on the right is called 2-propanol. Because it differs from 1-propanol only by the location of the —OH group, it can be considered to be an isomer of 1-propanol.

Ethers are similar to alcohols except that the oxygen atom will be bonded to two carbon chains, instead of one as in alcohols. An example of an ether is dimethyl ether, shown in Figure 11.9.

PROBLEM: Dimethyl ether has a molecular formula of C_2H_6O. Give the name of an alcohol that is an isomer of dimethyl ether.

Solution: The only alcohol that can be made from two carbon atoms is ethanol, as shown in Figure 11.10.

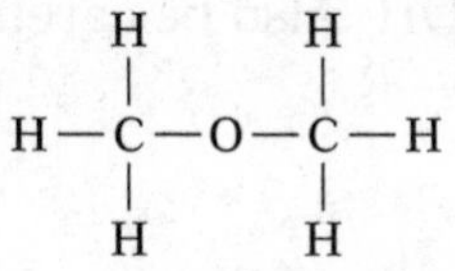

Figure 11.9 Dimethyl Ether 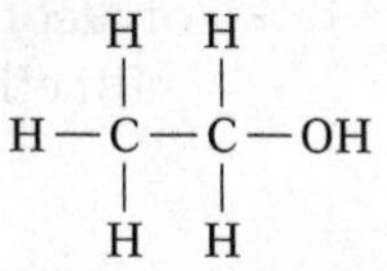   **Figure 11.10** Ethanol

CARBONYL-CONTAINING COMPOUNDS

The *carbonyl group* found in organic compounds is characterized by the double bond between a carbon atom and an oxygen atom, $C=O$. Depending upon the location of the carbonyl group and other arrangements found in an organic compound, there are a number of functional groups that can be formed. These basic functional groups can be found in Figures 11.11 through 11.14.

Functional Group	General Structure	Shorthand Structure	Example
Carboxylic Acids	$$\begin{array}{c} O \\ \parallel \\ R - C - OH \end{array}$$ **Figure 11.11** Carboxylic Acid	R—COOH	CH_3—CH_2—COOH *Propanoic acid*
Ketones	$$\begin{array}{c} O \\ \parallel \\ R - C - R' \end{array}$$ **Figure 11.12** Ketones	R—CO—R	CH_3—CO—CH_3 2-Propan*one*
Aldehydes	$$\begin{array}{c} O \\ \parallel \\ R - C - H \end{array}$$ **Figure 11.13** Aldehydes	R—CHO	CH_3—CH_2—CHO Propan*al*
Esters	$$\begin{array}{c} O \\ \parallel \\ R - C - O - R' \end{array}$$ **Figure 11.14** Esters	R—COO—R	CH_3—CO—O—CH_2—CH_3 Ethylethan*oate*

Two points should be made to help clarify the structures of these four functional groups.

1. The carboxylic acid and aldehyde functional groups are located at an end of the molecule, whereas ketones and esters have their functional group located in the interior of the molecules.
2. Do not confuse the shorthand structure of aldehydes, R—CHO, with that of alcohols, R—COH. Also be careful not to confuse ketones, R—CO—R, with ethers, R—O—R.

PROBLEMS: Name the following: CH_3—CH_2—CH_2—COOH, CH_3CH_2— CO—CH_3, CH_3—CH_2—CH_2—CHO, and CH_3—CO—O—CH_2—CH_2—CH_3.

Solutions: The first compound has four carbon atoms and the functional group is called a carboxylic acid. The name is of this compound is butanoic acid. The second compound has four carbon atoms and a ketone group on the second carbon (using the lowest number). This is 2-butanone. The third compound is butanal, the ending -*al* tells that it is an aldehyde. The final compound has three carbon atoms bonded to the single bonded oxygen. This part is called propyl. Because there are two carbon atoms in a chain bonded to the carbonyl group this part is called ethanoate. This molecule is called propylethanoate.

COMPOUNDS CONTAINING NITROGEN

The next set of functional groups contain the element nitrogen. These important compounds are called *amines* and *amides* as shown in Figures 11.15 and 11.16. This section will examine their structures and their importance in the role of protein production.

Functional Group	General Structure	Shorthand Structure	Example
Amines	C — N — H | H **Figure 11.15** Amines	R—NH₂	CH₃—CH₂—NH₂ ethan*amine*
Amides	O ‖ R — C — N **Figure 11.16** Amides	R—CO—NH₂	CH₃—CO—NH₂ ethan*amide*

The simple building blocks of proteins are called amino acids. The terms "amino" and "acids" refer to the functional groups called amines and carboxylic acids that have been examined in this chapter. When these functional groups interact, they form the amide functional group and build a longer chain of amino acids, which will eventually be long enough to be called a protein. This reaction can be seen in Figure 11.17.

In this reaction the H and OH highlighted in bold have been removed to form a molecule of water. The carbonyl of the amino acid on the left formed a bond with the nitrogen atom in the amino acid on the right. The amide functional group was formed in this dehydration synthesis reaction. Notice the ends of the newly formed dipeptide. There exists another amine group and carboxylic acid group at the ends of this chain to join with other amino acids and lengthen the chain.

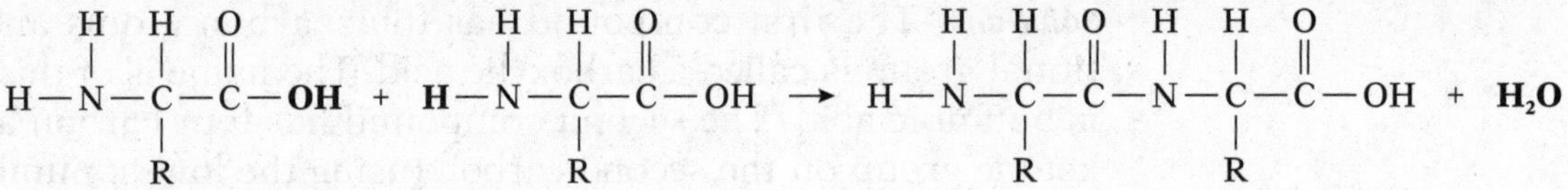

$$H-\overset{\overset{\displaystyle H}{|}}{N}-\overset{\overset{\displaystyle H}{|}}{\underset{\underset{\displaystyle R}{|}}{C}}-\overset{\overset{\displaystyle O}{\|}}{C}-OH \;+\; H-\overset{\overset{\displaystyle H}{|}}{N}-\overset{\overset{\displaystyle H}{|}}{\underset{\underset{\displaystyle R}{|}}{C}}-\overset{\overset{\displaystyle O}{\|}}{C}-OH \;\rightarrow\; H-\overset{\overset{\displaystyle H}{|}}{N}-\overset{\overset{\displaystyle H}{|}}{\underset{\underset{\displaystyle R}{|}}{C}}-\overset{\overset{\displaystyle O}{\|}}{C}-\overset{\overset{\displaystyle H}{|}}{N}-\overset{\overset{\displaystyle H}{|}}{\underset{\underset{\displaystyle R}{|}}{C}}-\overset{\overset{\displaystyle O}{\|}}{C}-OH \;+\; H_2O$$

Figure 11.17 Dipeptide

ORGANIC REACTIONS WITH HALOGENS

Think about this: One of the "wonders" of studying organic chemistry is the ability to create new compounds from existing ones. In this day and age, all compounds are compounds that can be manipulated in the laboratory. This includes the naturally occurring compounds that were mentioned at the beginning of this chapter. This is the reason why biotechnology has come a long way, especially in the field of genetics.

This section introduces the *halogens* into the mix and names compounds classified as *alkyl halides*. The prefixes used for the halogen are as follows: F = fluoro-, Cl = chloro-, Br = bromo-, and I = iodo-.

The first type of reaction is called the substitution reaction. In this reaction a hydrogen atom from an alkane is replaced by a halogen. For example, when methane reacts with diatomic chlorine gas, the major organic product is chloromethane: $CH_4 + Cl_2 \rightarrow CH_3Cl + HCl$.

Another reaction is the addition reaction. In this reaction a double or triple bond breaks to accommodate more atoms and the resulting compound contains all single bonds. For example, when ethene is reacted with hydrogen gas, you get: $CH_2{=}CH_2 + H_2 \rightarrow CH_3{-}CH_3$, ethane. Addition reactions work for diatomic halogen molecules as well, as shown in Figure 11.18. The product in this case is called 2,3-dibromobutane.

Notice that in this reaction the bromine atoms were added to the carbon atoms that had the double bond between them. The bromine atoms must be placed on the carbon atoms that contained the double bond because these were the carbon atoms that contained the functional group.

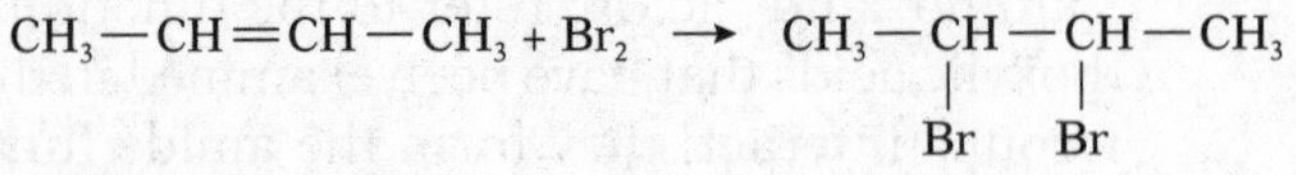

$$CH_3{-}CH{=}CH{-}CH_3 + Br_2 \;\rightarrow\; CH_3{-}\overset{\overset{\displaystyle }{}}{\underset{\underset{\displaystyle Br}{|}}{CH}}{-}\underset{\underset{\displaystyle Br}{|}}{CH}{-}CH_3$$

Figure 11.18 2,3-Dibromobutane

▅▅▅ CHAPTER REVIEW QUESTIONS

1. Which hydrocarbon will undergo a substitution reaction with a halogen?

   (A) Pentyne

   (B) Ethene

   (C) Propyne

   (D) Butane

   (E) Propene

2. Which type of organic reaction is represented by the equation $C_3H_6 + H_2 \rightarrow C_3H_8$?

   (A) Addition

   (B) Substitution

   (C) Condensation

   (D) Polymerization

   (E) Dehydration synthesis

3. When the amine group of one amino acid reacts with the carboxylic acid group of another amino acid, the resulting functional group formed is called

   (A) an amine

   (B) an amide

   (C) an ester

   (D) a plastic

   (E) a polymer

4. Which one of the following polymers is synthetic?

   (A) Nucleic acids

   (B) Plastic

   (C) Proteins

   (D) Cellulose

   (E) Starch

5. Which two compounds are not isomers of each other?

   (A) n-pentane and 2-methylbutane

   (B) CH_3CH_2OH and CH_3OCH_3

   (C) CH_3COOH and CH_3CH_2COOH

   (D) CH_3COCH_3 and CH_3CH_2CHO

   (E) $CH_3CH_2CH_2Cl$ and $CH_3CHClCH_3$

6. A carbonyl group is present in all of these functional groups except:

   (A) ketones

   (B) aldehydes

   (C) esters

   (D) amides

   (E) ethers

7. An organic compound has a molecular formula of C_3H_4. Which compound below belongs to the same class of hydrocarbons?

   (A) C_2H_6

   (B) C_3H_6

   (C) C_4H_8

   (D) C_2H_2

   (E) CH_4

8. Which statement is false?

   (A) $CH_3CH_2NH_2$ is ethanamine.

   (B) $CH_3CHBrCHBrCH_3$ is 2,3-dibromo-butane.

   (C) CH_3CH_2OH is an ether.

   (D) Cyclopentane and 2-pentene have a molecular formula of C_5H_{10}.

   (E) Alkenes and alkynes are unsaturated.

ANSWERS:

1. (D)	2. (A)	3. (B)
4. (B)	5. (C)	6. (E)
7. (D)	8. (C)	

CHAPTER 12
NUCLEAR CHEMISTRY

IN THIS CHAPTER YOU WILL LEARN ABOUT...

Radioactive Isotopes
Radioactive Emanations
Separating Nuclear Emanations
Nuclear Fusion and Fission
Half-Life ($t_{1/2}$)
Dangers and Benefits of Using Radioisotopes

RADIOACTIVE ISOTOPES

Most of this book has been concerned with the elements up through atomic number 38 and with isotopes that have a stable nucleus. This chapter focuses on the bottom of the periodic table where there are elements with no stable isotope and with nuclei that can undergo a *transmutation* to form a new element from the disintegration of the nucleus. Most of the credit for the discovery of radioactivity and the study of the breakdown of the nucleus is given to Marie Curie, Pierre Curie, and Henri Becquerel. Hans Geiger is responsible for the *Geiger Counter,* a device used to detect and measure the activity of radioactive particles. It is now known that the instability of the nucleus stems from the ratio of the number of neutrons to the number of protons in the nucleus. When the ratio of neutrons to protons begins to exceed a 1.5:1 ratio, there tends to be instability in the nucleus. All elements with the atomic number of 84 or greater have no stable isotope.

RADIOACTIVE EMANATIONS

When the nucleus of an atom breaks down, three types of emanations are ejected from the unstable nucleus. These three particles are called *alpha particles*, *beta particles*, and *gamma rays*. These particles can be examined side by side in the chart below:

Type of Radiation	Alpha Particles	Beta Particles	Gamma Rays
Symbol	^4_2He or α	$^0_{-1}e$ or β^-	γ
Description	Helium nucleus (not helium atom).	Same properties as an electron but was ejected from the nucleus.	Not a particle at all. Gamma rays are high-energy radiation.

Type of Radiation	Alpha Particles	Beta Particles	Gamma Rays
Mass	4 AMU (2 protons and 2 neutrons).	1/1836 AMU.	No mass.
Charge	2+ charge from the two protons present.	1– charge.	No charge.
Speed	Travels at one-tenth the speed of light.	Travels at nearly the speed of light.	Travels at the speed of light (3.0×10^8 m/s).
Penetrating Power	Weak and can easily be stopped by thinner materials.	Can be stopped by a thicker material.	Requires several inches of lead to be stopped.
Damage Inflicted	Ionizes gas molecules.	Ionizes gas molecules.	Breaks down larger molecules.
Example	$^{226}_{88}Ra \rightarrow {}^{222}_{86}Rn + {}^4_2He$ Ra-226 undergoes alpha decay to form Rn-222.	$^{214}_{82}Pb \rightarrow {}^{214}_{83}Bi + {}^0_{-1}e$ Pb-214 undergoes beta decay to form Bi-214.	$^{99m}_{43}Tc \rightarrow {}^{99}_{43}Tc + \gamma$ "Metastable" Tc-99 gives off gamma radiation.

From the examples in the chart above, you can see that the decay of the nucleus obeys conservation of mass and charge. The mass numbers to the left of the arrow add up to the mass numbers to the right of the arrow. Also, recall that the atomic number can also tell the nuclear charge of an isotope. Notice how these charges add up to equal each other on the left and right sides of the arrows. Making sure that the mass numbers and nuclear charges add up will ensure that you correctly solve problems involving decay. For example, suppose you are given the following: $^{234}_{91}Pa \rightarrow {}^{234}_{90}Th + X$. This nuclear reaction has a mass number of 234 already on both sides of the equation. Therefore the mass of X is 0. The nuclear charge of X must be 1 because a nuclear charge of 91 is on the left side of the equation and a nuclear charge of 91 must be present on the right side. X can be written as ^0_1X. This particle has no mass but a charge of 1+. This is called a *positron* and is represented by $^0_{+1}e$. The full equation looks like this: $^{234}_{91}Pa \rightarrow {}^{234}_{90}Th + {}^0_{+1}e$.

PROBLEMS: What are the particles represented by the letter X in the following nuclear reactions?

$$^{14}_6C + {}^1_1H \rightarrow {}^{14}_7N + X \qquad \text{and} \qquad {}^{35}_{17}Cl + {}^1_0n \rightarrow {}^{35}_{16}S + X$$

Solutions: The first problem has a total mass of 15 on the left. X has a mass of 1 so that the right side of the equation has a total mass of 15. The charge to the left of the arrow is +7. X must have no charge so that the right side can have a total of +7. X can be written as ^1_0X. A particle with a mass of 1 AMU but no charge is a neutron, ^1_0n.

The second problem indicates that the total mass on the left and right must be 36. X has a mass of 1. To make the total charge on the right side of the equation equal 17, X must have a charge of 1+. X can be written as ^1_1X. A particle that has a mass of 1 AMU and a charge of 1+ is a proton, ^1_1H.

The two problems presented above show a type of transmutation not yet encountered in this book. In these problems an isotope is being bombarded with a particle to trigger the transmutation. This is called an *artificial transmutation*. This is different from the first four reactions examined, in which the isotopes underwent a *natural transmutation* and did not need to be bombarded with other particles to undergo a transmutation. The natural transmutations that Th-232 undergoes can be seen in the *decay series* shown in Figure 12.1.

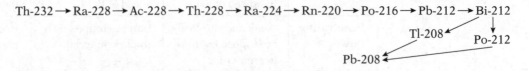

Figure 12.1 The Decay Series of Thorium-232

In the decay series, a series of alpha and beta decays occur. The type of decay that has occurred can be determined by looking at the decrease in the mass numbers from one isotope to the next. A drop in 4 AMU indicates alpha decay; no drop in the mass number indicates beta decay. These decays will continue until a stable isotope is produced (usually Pb-206, Pb-207, or Pb-208).

In addition to alpha particles, beta particles, gamma rays, and positrons, there is one more particle that you should become familiar with. Protons and neutrons make up the nucleus of an atom. Protons and neutrons are made up of even smaller particles called *quarks*. Quarks can have charges of +2/3 or –1/3. Combinations of three quarks and their charges can produce a proton and its 1+ charge or produce a neutron and its neutrality.

SEPARATING NUCLEAR EMANATIONS

Experiments involving radioactive particles can call for the particles to be attracted or deflected in an electric field. One such example takes place in a particle accelerator. Alternating voltages in the particle accelerator take advantage of the charges of particles to move them through the accelerator and eject them into the experiment being performed. An example of particles being attracted and deflected through a field can be seen in Figure 12.2.

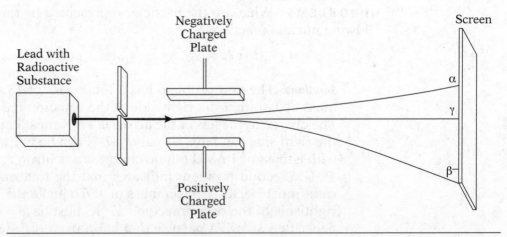

Figure 12.2 Separation of Nuclear Decay

Notice that the alpha particles were attracted toward the negatively charged plate while the beta particles were attracted toward the positively charged plate. Because the alpha particles are greater in mass, they were not as affected by the electric field as were the much lighter beta particles. The gamma rays were not deflected or attracted because they have no charge. The same would be true in the case of neutrons.

NUCLEAR FUSION AND FISSION

The concept of artificial transmutation can be applied to reactions that involve nuclear *fission* or the splitting of nuclei. An example of this is the fission reaction that occurs in an atomic bomb. When U-235 is bombarded with a neutron, the uranium is split according to the reaction:

$$^{235}U + {}^1_0n \rightarrow {}^{141}Ba + {}^{92}Kr + 3\ {}^1_0n + \text{energy}$$

This fission reaction starts with the splitting of just one U-235 nucleus. This reaction releases three neutrons that can then split three uranium nuclei, which give off nine neutrons, which split nine uranium nuclei, and so on. This process is called a *chain reaction*, and once started, it continues until all of the uranium is split. With each step in the chain reaction, three times more energy is released than in the previous step in the reaction. This explains why a nuclear weapon releases so much energy.

When controlled carefully, fission can be used to produce electricity. In nuclear power plants, moderators are used to slow down the neutrons produced during nuclear fission. It is common to have moderators made from graphite or with heavy water, ^2H_2O. Control rods are also placed in the nuclear reactor to absorb neutrons and slow down the rate at which the fission takes place.

Fusion is a different type of nuclear reaction. In fusion, there is a joining of nuclei to form a larger nucleus. Fusion is the reaction that allows the sun to burn so brightly. In the sun the following takes place:

$$4\ {}^1_1H \rightarrow {}^4_2He + 2\ {}^0_{+1}e + \text{energy}$$

Albert Einstein found that with a nuclear reaction there is a *mass defect;* that is, some of the mass of the particles involved in the nuclear reaction is converted to energy! The mass of the reacting nuclei will be greater than the mass of the nuclei of the products. The energy created from this conversion of mass to energy can be calculated by using Einstein's famous equation for binding energy, $E = mc^2$.

HALF-LIFE ($t_{1/2}$)

The rate at which an unstable nucleus disintegrates varies from isotope to isotope. Some isotopes can decay over a number of years; others may completely disintegrate in just a couple of seconds. Chemists often look at the *half-life* of an isotope to determine how long a sample of the particular isotope will remain unchanged over a certain period of time. The half-life of an isotope is the amount of time it takes for half of a radioactive substance to decay. Each nuclide has its own half-life that does not change.

Imagine that you have 100 grams of a radioactive substance that has a half-life of eight days. If you return to check up on your 100-gram sample after eight days you would see that the sample gradually decayed and 50 grams remained. If you came back in another eight days you would find that just 25 grams would be left. This process is outlined in Figure 12.3.

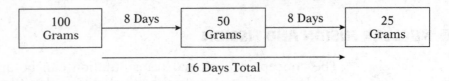

Figure 12.3 Half-Life

The diagram in Figure 12.4 is another way of showing the decay of this sample.

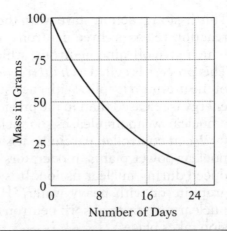

Figure 12.4 Graph for the Decay of a Substance

PROBLEM: The half-life of I-131 is eight days. After 32 days, how many grams of a 40-gram sample will remain?

Solution: The half-life of this isotope is eight days. After 32 days, four half-life periods have passed. This means that the 40-gram sample has been halved four times. This can be diagrammed as:

40 grams → 20 grams → 10 grams → 5 grams → 2.5 grams

2.5 grams of this isotope will remain.

PROBLEM: A 16-gram radioactive sample decays to two grams over a period of 30 years. What is the half-life of this substance?

Solution: This sample underwent three half-life periods as shown here:

16 grams → 8 grams → 4 grams → 2 grams

If the total time for three half-life periods is 30 years, then each half-life period, as represented by the arrows in the diagram, is 10 years.

DANGERS AND BENEFITS OF USING RADIOISOTOPES

Quite often you hear only negative stories about nuclear reactions and radioactivity. Radioactivity can mutate DNA molecules and cause cancer. The use of nuclear reactors to produce energy can create nuclear waste, which can harm the environment. Nuclear power plants have been known to have accidents and expose many people to radioactive particles. Radioactive radon gas can be found in the homes that people live in. Nuclear warheads and nuclear weapons can cause mass destruction. On the other hand, there are many uses for radioisotopes that can be beneficial to our lives. In order for a radioisotope to be effective, it must be used properly and in the proper dosages. Some benefits of radioisotopes are described in the following chart.

Some Benefits of Radioisotopes

C-14 Dating	C-14 is used to determine the age of formerly living organisms or of artifacts containing formerly living organisms. Because the half-life of C-14 is 5,730 years, it can be used to date objects that are up to 50,000 years old. C-14 is not a good way to date older objects because the amount of radioactivity in objects more than 50,000 years old is too light to be measured. Wines can be dated back about 100 years by detecting tritium, H-3. Finally, U-238, which has a half-life of 4,500,000,000 years, can be used to date the oldest rocks ever formed on Earth!
Food Preservation	Because radiation can kill living organisms, it is used to kill organisms that feed on food meant for human consumption. Radiated food has a longer shelf life because the radiation eliminates the organisms that cause the food to decay.
Radiotracers	Radiotracers are radioactive isotopes of elements that human bodies use to maintain normal function. Radiotracers are useful in medicine. Doctors can introduce them into a patient's body, and the body will use them in the same way it uses normal elements. The radiotracers can then be used to produce images of the internal organs. For example, the human thyroid gland normally makes use of the element iodine. When the gland malfunctions, the radioisotope I-131 can be introduced, and just like regular iodine, it will collect in the thyroid. A scan can then be done to get an image of the gland and its function. The inert gas Xe-133 can be used to image the lungs. Radiotracers have a short half-life, which minimizes the damage done to cells by the radioactive emanations.

(continued)

Radiation Therapy   If radiation can cause cancer, how can radiation therapy treat cancer and destroy cancer cells? The answer is that radiation in the right doses and used properly can be beneficial. When radiation is used to fight cancer, it is carefully focused on the cancerous cells, reducing the amount of exposure to normal, healthy cells.

CHAPTER REVIEW PROBLEMS

1. As a nucleus of a particular isotope disintegrates, another nuclide is formed. This change in the nucleus to form a new nuclide is called

    (A) binding energy
    (B) transmutation
    (C) stability
    (D) generation
    (E) synthesis

2. Which element has no known stable isotope?

    (A) Carbon
    (B) Silver
    (C) Radon
    (D) Phosphorus
    (E) Lead

3. In the artificial transmutation $^9_4Be + X \rightarrow$ $^6_3Li + {}^4_2He$, the particle represented by the letter X is a(n)

    (A) beta particle
    (B) positron
    (C) deuteron
    (D) proton
    (E) alpha particle

4. Which pair below would not be deflected or attracted by the charged plates in an electric field?

    (A) An alpha particle and a neutron
    (B) A beta particle and a positron
    (C) A quark and a deuteron
    (D) A proton and gamma radiation
    (E) Gamma radiation and a neutron

5. After 62.0 hours, 1.0 gram remained unchanged from a sample of ^{42}K (half-life is 12.4 hours). What was the mass of ^{42}K in the original sample?

    (A) 64 grams
    (B) 32 grams

    (C) 16 grams
    (D) 8 grams
    (E) 4 grams

6. The energy released by the detonation of an atomic bomb is NOT related to

    (A) fission of the atom's nucleus
    (B) fusion of the atom's nucleus
    (C) a chain reaction
    (D) the release of many neutrons
    (E) the uncontrolled speed of many neutrons

7. The joining of many hydrogen nuclei in the nuclear reaction that occurs in stars is called a

    (A) mass defect
    (B) sunburn
    (C) fusion reaction
    (D) fission reaction
    (E) helium reaction

8. Iodine-131 is an excellent radioisotope for diagnosing health problems of the

    (A) kidneys
    (B) heart
    (C) lungs
    (D) thyroid
    (E) bone marrow

9. In determining the age of an artifact, an archaeologist is most likely to examine the percentage of

    (A) carbon-14
    (B) phosphorus-31
    (C) hydrogen-3
    (D) chlorine-37
    (E) bromine-81

10. Which equation is an example of an artificial transmutation?

(A) $^{238}U \rightarrow {}^{4}He + {}^{234}Th$

(B) $^{27}Al + {}^{4}He \rightarrow {}^{30}P + {}^{1}_{0}n$

(C) $^{14}C \rightarrow {}^{14}N + {}^{0}_{-1}e$

(D) $^{226}Ra \rightarrow {}^{4}He + {}^{222}Rn$

(E) $^{99m}_{43}Tc \rightarrow {}^{99}_{43}Tc + \gamma$

11. When Li-7 is bombarded with a proton, two alpha particles are released along with energy. It turns out that the mass of the two alpha particles actually weighs less than the original products in the reaction. The mass that was converted to energy is called the

(A) Einstein conversion

(B) mass defect

(C) Theory of Relativity

(D) natural transmutation

(E) chain reaction

12. Radioactive emanations can be detected by using

(A) a person's DNA

(B) a block of lead

(C) a Geiger Counter

(D) an x-ray machine

(E) graphite and heavy water

13. One reason why certain isotopes have an unstable nucleus is because the number of

(A) protons outweigh the number of neutrons

(B) electrons outweigh the number of protons

(C) neutrons outweigh the number of electrons

(D) protons outweigh the number of electrons

(E) neutrons outweigh the number of protons

Questions 14–17 Refer to the following:

(A) Alpha particle

(B) Beta particle

(C) Gamma radiation

(D) Neutron

(E) Position

14. Has a negative charge

15. Has no mass and no charge

16. Has the greatest positive charge

17. Is very similar to an electron

ANSWERS:

1. (B)	2. (C)	3. (D)	4. (E)
5. (B)	6. (B)	7. (C)	8. (D)
9. (A)	10. (B)	11. (B)	12. (C)
13. (E)	14. (B)	15. (C)	16. (A)
17. (B)			

CHAPTER 13
LABORATORY SKILLS

Even though it is fun to learn chemistry in the classroom with the use of a textbook, nothing beats the hands-on laboratory experience!

LABORATORY SAFETY

The laboratory experience can be all the fun of taking a course in chemistry. This chapter will review a number of experiments and procedures. But before going into any detailed laboratory setups or experiments, it is important to first review laboratory safety. Although it is impossible to cover every safety rule for every experiment, here are some common safety rules everyone should know.

- Wear goggles, footwear that covers the entire foot, gloves, and lab coats or aprons at all times.
- Never eat or drink in the lab.
- Pour all chemicals over a sink.
- Perform experiments under a fume hood at all times.
- Never point a heated test tube at yourself or others.
- Never heat a corked test tube.
- Know where the fire extinguisher, eyewash station, and fire blankets are located.
- Read all labels carefully.
- Never rest bottle tops or stoppers on the table top.
- Always add acid to water and never water to an acid.
- Never push glass tubing or a thermometer through a stopper or cork.
- Handle hot materials with tongs or a test tube holder.

- Chemicals that come in touch with your skin must be flushed with water immediately.
- Note the odors of substances cautiously.

THE MENISCUS

Because adhesive forces can exist between liquids and glassware, a pronounced curvature of a liquid can form when the liquid is in a narrow cylinder or burette. This curvature is called the *meniscus*. A few important things to remember when reading the meniscus are that the meniscus should always be read at eye level while the glassware and setup is on a level surface. Also, make sure that you read the bottom of the meniscus (if it is curved downward) as shown in Figure 13.1.

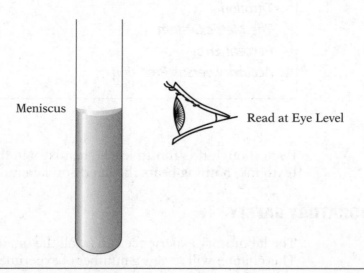

Figure 13.1 Reading a Meniscus

In some cases the meniscus can be curved upward. Should this happen, read the top of the meniscus. Finally, when using dark liquids, it may be impossible to read the meniscus. In this case, read the top of the level of the liquid.

COLLECTING GASES

Gases for use in experiments may be collected in a variety of ways. Two things that should be kept in mind when collecting a gas are the *density* and the polarity of the gas. For example, you would not collect a polar gas (such as ammonia) in water because water and ammonia have like polarities that allow the two substances to mix.

Hydrogen and oxygen gases can be collected by water displacement as shown in Figure 13.2. The reaction that produces the gases takes place in the bottle on the left. An outlet tube leads to the collection bottle on the right that has first been filled with water, turned upside down, and placed so that the mouth of the bottle is under water in a trough. The oxygen and hydrogen gas molecules are nonpolar molecules and will not dissolve very well in the water.

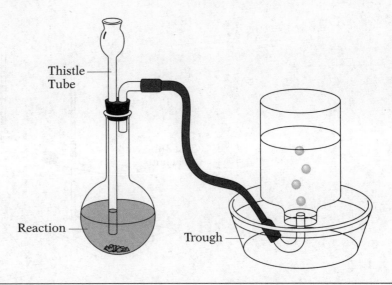

Figure 13.2 Collection of Hydrogen and Oxygen by Water Displacement

After collection, the collection bottle is stoppered or a glass plate is slid under the mouth of the bottle. This must be done while the mouth is still under water, facing downward.

Carbon dioxide is a heavy gas, so you do not need to collect it by water displacement. Instead it can be collected by an upward displacement of air, as shown in Figure 13.3. In this case the tubing from the reaction bottle leads to the bottom of the collection bottle. As the carbon dioxide gas is collected, the air already in the collection bottle is displaced upward.

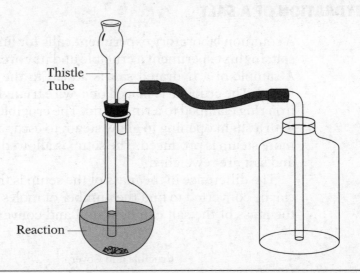

Figure 13.3 Collection of Carbon Dioxide by Upward Displacement of Air

If a small flame, such as a lit match, is held just inside the mouth of the collection bottle and the flame is extinguished, this means that the carbon dioxide has filled the bottle.

Ammonia is collected by downward displacement of air as shown in Figure 13.4. Because the preparation of ammonia gas requires a gentle heating of the reactants, the reaction chamber cannot be an ordinary bottle. Instead a test tube must be used. As the ammonia is collected, the air already in the

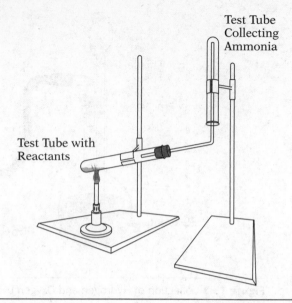

Figure 13.4 Collection of Ammonia by the Downward Displacement of Air

collection tube is displaced downward. Water displacement is not a suitable method for collecting a polar substance like ammonia because the ammonia will dissolve in the water.

To test for a full test tube of ammonia, hold a piece of red litmus paper near the mouth of the test tube. When the red litmus turns blue, the basic ammonia has filled the test tube.

DEHYDRATION OF A SALT

A common laboratory experiment calls for finding the formula of a hydrated salt. In this experiment a crucible and its cover (see Figure 13.5) are weighed. A sample of a hydrated salt is added to the crucible and the new mass is found. The crucible is placed on a wire triangle that has been mounted on an iron ring clamped to a ring stand. The crucible cover is placed on the crucible with a slight opening to allow steam to escape. The sample is heated until no more steam is produced. The setup is allowed to cool and the crucible, cover, and salt are reweighed.

The difference in the mass of the setup is the mass of the water. This mass can be converted to find the number of moles of water lost. The difference in the mass of the salt can be found and converted to moles as well. This will

Crucible and Cover

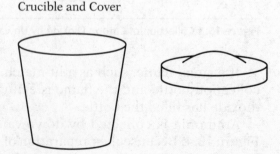

Figure 13.5 A Crucible and Cover

give the formula of the salt: X SALT • Y H_2O. A sample data collection table for the dehydration of copper sulfate pentahydrate looks like this:

Measurements:	W = weight of crucible and cover		Calculations:	
Before heating, find the mass of a sample:				
1. (A)	W + $CuSO_4 \cdot XH_2O$	g	(F) $CuSO_4$	g
2. (B)	– W	g	MM of $CuSO_4$ = 160 g/mole	
3. (E)	$CuSO_4 \cdot XH_2O$	g	$\dfrac{(F)}{160}$ = moles $CuSO_4$	moles
After heating, find the mass of just the salt:				
4. (C)	W + $CuSO_4$	g	(G) XH_2O	g
5. (D)	– W	g	MM of H_2O = 18 g/mol	
6. (F)	$CuSO_4$	g	$\dfrac{(G)}{18}$	mole
Calculate the mass of the water:				
7. (E)	$CuSO_4 \cdot XH_2O$	g	$X = \dfrac{\text{Moles } H_2O}{\text{Moles } CuSO_4}$	mole
8. (F)	– $CuSO_4$	g		
9. (G)	XH_2O	g		

METHODS FOR SEPARATION

There are a number of ways to separate substances that have been combined or mixed together. Three properties of substances that can be called upon to complete a separation are solubility in water, density, and boiling point. Consider a mixture of sand and sodium chloride. Is it feasible to pick out all of the grains of sand with tweezers? Because the NaCl is soluble in water and the sand is not, the sample could be placed into a beaker of water and stirred. This would dissolve the salt but not the sand. The mixture is then filtered through moist filter paper that has been fit to the inside of a funnel. The *filtrate* that passes through the filter paper contains the NaCl. This filtrate is captured and can later be evaporated so that the salt is left behind. The *residue,* or sand in this case, is what is left behind in the filter paper. Because it does not dissolve in water, the residue can be washed from the filter paper. It can then be dried and analyzed.

When two different liquids with different polarities are put together, they will not mix and one of the liquids will float on top of the other. You can take advantage of this by using a separatory funnel to separate the two substances. The separatory funnel (see Figure 13.6) has a stopcock at the bottom of the

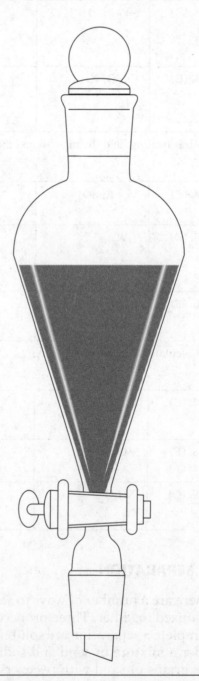

Figure 13.6 Separatory Funnel

apparatus. This allows the liquid with the greater density to be drained through the bottom once the stopcock has been opened (and the stopper top has been removed as well).

To obtain the liquid with the lower density, the liquid is poured out through the top of the separatory funnel.

Distillation separates substances based on their boiling points. A distillation flask is set up with a thermometer and rubber stopper closing off the top of the flask. A condenser is set up with cold water running through it. Finally, a beaker or graduated cylinder is used to collect the distillate as shown in Figure 13.7.

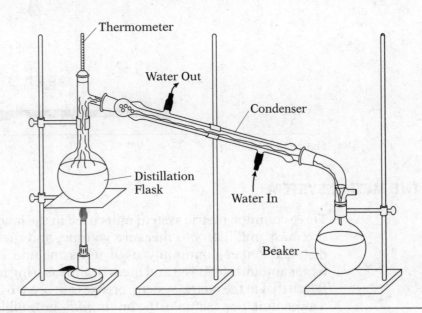

Figure 13.7 Distillation

As the mixture of liquids is heated, one of the liquids will reach its boiling point before the other. The temperature on the thermometer will rise and then stay constant as the first liquid distills. When the temperature starts to increase again, the collection flask is changed and the next substance is captured as it distills.

TITRATION

The titration process was discussed in detail in Chapter 9. Figure 13.8 is a setup showing the two burettes used in the experiment, one for the acid and one for the base. They are clamped to a ring stand via a burette clamp. The titration is performed using a flask because the angled sides of the flask avoid splashing while the flask is being stirred continuously as the drops of acid and base are added. Finally, not shown, is a white piece of paper. Placing the flask on a white piece of paper makes it easier to see the color changes of the indicator.

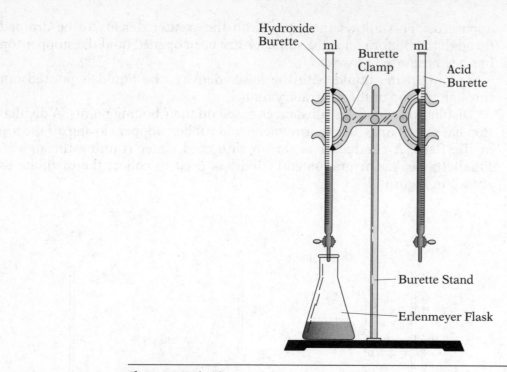

Figure 13.8 Titration

THE METRIC SYSTEM

Three common metric system units used in the laboratory are grams to measure mass, milliliters to measure volume, and meters to measure length or distance. Other commonly used units include kilograms (for measuring larger amounts of mass) and liters (for measuring larger amounts of volume). Recall that in the metric system, prefixes are used to indicate powers of 10 times a given unit. For example, the prefix *milli-* in "milliliter" indicates that a milliliter is equal to 1 liter × 10^{-3} or one one-thousandth of a liter. The prefix *kilo-* in "kilogram" indicates that a kilogram is equal to 1 gram × 10^3 or one thousand grams. The following chart shows the prefixes in units of measure commonly used in the laboratory, the symbol for each prefix, and the power of 10 that each one indicates.

Prefix	Symbol	Power of 10
Kilo-	k	10^3
Centi-	c	10^{-2}
Milli-	m	10^{-3}
Micro-	μ	10^{-6}
Nano-	n	10^{-9}
Pico-	p	10^{-12}

The prefix *micro-*, for example, is useful when determining the parts per million of an ion or substance in drinking water. You have already used the prefix *nano-* when you looked at the spectral lines produced by the return of an electron from the excited state to the ground state. The line spectrum for hydrogen showed that there is an emission with the wavelength of 656 nm (nanometers). This comes out to be 0.000000656 meters! A method for converting from one unit to another is covered in Appendix 1, Mathematical Skills Review.

PERCENT ERROR

Quite often when you perform an experiment, your results come close to, but do not actually match, a particular value that you are looking for. You will then want to find out how far off the mark your results were; that is, the percent error. For example, suppose you know that the accepted value for the boiling point of vanillin is 83°C, but in your experiment you measured it as 85°C. You could say that you were "off by two degrees." Instead, you calculate a percent error by using the formula:

$$\frac{\text{Measured value} - \text{Accepted value}}{\text{Accepted value}} \times 100\%$$

Calculate your percent error for the melting point of vanillin by substituting into the equation:

$$\frac{85°C - 83°C}{83°C} \times 100\%$$

The percent error is 2.41%. In general, a percent error under 5% indicates that the experiment was performed with proper technique and procedure.

ACCURACY VERSUS PRECISION

Anyone who has taken a course in quantitative analysis will tell you about repeating an experiment over and over again to get a large sample of data. It's not uncommon for experiments requiring a titration to be repeated anywhere from 7 to 10 times! Once a number of experiments have been completed, you then want to compare the results to see if they are both accurate and precise. *Accuracy* is the term used to define how close the data have come to the accepted value. *Precision* is the term used to define how closely the data agree with data obtained from other performances of the same experiment. Hopefully your data will be both accurate and precise. Look at the following data a student obtained from a titration experiment involving an acid and base:

Trial 1: 0.178 *M* HCl   Trial 2: 0.181 *M* HCl   Trial 3: 0.177 *M* HCl

Trial 4: 0.178 *M* HCl   Trial 5: 0.179 *M* HCl   Trial 6: 0.180 *M* HCl

If the actual molarity of the acid is 0.125 *M* HCl, you can conclude that while the data are precise, they are far from accurate (you could take an average of the results and calculate a percent error to confirm this). This shows that the student performed the experiment consistently but with a built-in error.

CHAPTER REVIEW QUESTIONS

1. For which two pieces of equipment is it critical to read the meniscus correctly?

   (A) Spatula and Bunsen burner
   (B) Graduated cylinder and stopcock
   (C) Burette and graduated cylinder
   (D) Thistle tube and trough
   (E) Crucible and cover

2. Which is not necessarily needed for a titration?

   (A) An appropriate indicator
   (B) Erlenmeyer flask
   (C) Acid
   (D) Base
   (E) Mortar and pestle

3. A student finds the percent hydration of a salt to be 67% water. The accepted value for percent hydration is 55% water. What is the student's percent error?

   (A) 12%
   (B) 21.8%
   (C) –12%
   (D) 2.41%
   (E) 0.218%

4. To deliver amounts of liquids one would NOT use a

   (A) volumetric pipet
   (B) burette
   (C) graduated cylinder
   (D) evaporating dish
   (E) beaker

5. A student wants to heat up a sample of water to boiling over a certain amount of time to get an idea of how much heat is released by the flame per minute. Which of these items is not needed for this experiment?

   (A) A time-keeping device
   (B) The value of the specific heat of water
   (C) The change in temperature of the water
   (D) The mass of the sample of water
   (E) The volume of water in its gas phase

6. A student took a melting point of an organic compound four times. His work produced the following results: 97°C, 99°C, 100°C, and 97°C. If the real melting point of this compound is 88°C, then his results can be summarized as

   (A) both accurate and precise
   (B) neither accurate nor precise
   (C) accurate but not precise
   (D) precise but not accurate
   (E) extrapolated

Directions: The following question consists of two statements. Determine whether statement I in the leftmost column is true (T) or false (F) and whether statement II in the rightmost column is true (T) or false (F).

	I		II
7.	1,100 grams is equal to 1.100 kilograms	BECAUSE	to convert grams to kilograms one must divide by 100.
8.	The spattering of a salt from a crucible during a salt dehydration is acceptable	BECAUSE	the mass of the salt before and after the dehydration does not play a part in finding the percent hydration of a salt.
9.	It is possible to separate a water-soluble salt from a water-insoluble substance	BECAUSE	the water-soluble substance will be caught in the filter paper, while the insoluble substance passes through filter paper.
10.	When diluting an acid, the acid is added to water	BECAUSE	the dilution of concentrated acid can release heat.

ANSWERS:

1. (C)　　　2. (E)　　　3. (B)　　　4. (D)　　　5. (E)

6. (D)　　　7. (T, F)　　　8. (F, F)　　　9. (T, F)　　　10. (T, T, CE)

PART IV

FOUR FULL-LENGTH PRACTICE TESTS

PRACTICE TEST 1

ANSWER SHEET

This practice test will help measure your current knowledge of chemistry and familiarize you with the structure of the *SAT Chemistry* test. To simulate exam conditions, use the answer sheet below to record your answers and the appendix tables at the back of this book to obtain necessary information.

1. (A) (B) (C) (D) (E)
2. (A) (B) (C) (D) (E)
3. (A) (B) (C) (D) (E)
4. (A) (B) (C) (D) (E)
5. (A) (B) (C) (D) (E)
6. (A) (B) (C) (D) (E)
7. (A) (B) (C) (D) (E)
8. (A) (B) (C) (D) (E)
9. (A) (B) (C) (D) (E)
10. (A) (B) (C) (D) (E)
11. (A) (B) (C) (D) (E)
12. (A) (B) (C) (D) (E)
13. (A) (B) (C) (D) (E)
14. (A) (B) (C) (D) (E)
15. (A) (B) (C) (D) (E)
16. (A) (B) (C) (D) (E)
17. (A) (B) (C) (D) (E)
18. (A) (B) (C) (D) (E)
19. (A) (B) (C) (D) (E)
20. (A) (B) (C) (D) (E)

21. (A) (B) (C) (D) (E)
22. (A) (B) (C) (D) (E)
23. (A) (B) (C) (D) (E)
24. (A) (B) (C) (D) (E)
25. (A) (B) (C) (D) (E)
26. (A) (B) (C) (D) (E)
27. (A) (B) (C) (D) (E)
28. (A) (B) (C) (D) (E)
29. (A) (B) (C) (D) (E)
30. (A) (B) (C) (D) (E)
31. (A) (B) (C) (D) (E)
32. (A) (B) (C) (D) (E)
33. (A) (B) (C) (D) (E)
34. (A) (B) (C) (D) (E)
35. (A) (B) (C) (D) (E)
36. (A) (B) (C) (D) (E)
37. (A) (B) (C) (D) (E)
38. (A) (B) (C) (D) (E)
39. (A) (B) (C) (D) (E)
40. (A) (B) (C) (D) (E)

41. (A) (B) (C) (D) (E)
42. (A) (B) (C) (D) (E)
43. (A) (B) (C) (D) (E)
44. (A) (B) (C) (D) (E)
45. (A) (B) (C) (D) (E)
46. (A) (B) (C) (D) (E)
47. (A) (B) (C) (D) (E)
48. (A) (B) (C) (D) (E)
49. (A) (B) (C) (D) (E)
50. (A) (B) (C) (D) (E)
51. (A) (B) (C) (D) (E)
52. (A) (B) (C) (D) (E)
53. (A) (B) (C) (D) (E)
54. (A) (B) (C) (D) (E)
55. (A) (B) (C) (D) (E)
56. (A) (B) (C) (D) (E)
57. (A) (B) (C) (D) (E)
58. (A) (B) (C) (D) (E)
59. (A) (B) (C) (D) (E)
60. (A) (B) (C) (D) (E)

61. (A) (B) (C) (D) (E)
62. (A) (B) (C) (D) (E)
63. (A) (B) (C) (D) (E)
64. (A) (B) (C) (D) (E)
65. (A) (B) (C) (D) (E)
66. (A) (B) (C) (D) (E)
67. (A) (B) (C) (D) (E)
68. (A) (B) (C) (D) (E)
69. (A) (B) (C) (D) (E)
70. (A) (B) (C) (D) (E)
71. (A) (B) (C) (D) (E)
72. (A) (B) (C) (D) (E)
73. (A) (B) (C) (D) (E)
74. (A) (B) (C) (D) (E)
75. (A) (B) (C) (D) (E)

Chemistry   *Fill in oval CE only if II is correct explanation of I.

	I	II	CE*		I	II	CE*
101.	(T) (F)	(T) (F)	( )	109.	(T) (F)	(T) (F)	( )
102.	(T) (F)	(T) (F)	( )	110.	(T) (F)	(T) (F)	( )
103.	(T) (F)	(T) (F)	( )	111.	(T) (F)	(T) (F)	( )
104.	(T) (F)	(T) (F)	( )	112.	(T) (F)	(T) (F)	( )
105.	(T) (F)	(T) (F)	( )	113.	(T) (F)	(T) (F)	( )
106.	(T) (F)	(T) (F)	( )	114.	(T) (F)	(T) (F)	( )
107.	(T) (F)	(T) (F)	( )	115.	(T) (F)	(T) (F)	( )
108.	(T) (F)	(T) (F)	( )				

GO ON TO THE NEXT PAGE

PRACTICE TEST 1

Time: 60 minutes

<u>Note:</u> Unless otherwise stated, for all statements involving chemical equations and/or solutions, assume that the system is in pure water.

Part A

<u>Directions:</u> Each of the following sets of lettered choices refers to the numbered formulas or statements immediately below it. For each numbered item, choose the one lettered choice that fits it best. Then fill in the corresponding oval on the answer sheet. Each choice in a set may be used once, more than once, or not at all.

<u>Questions 1–4</u>

    (A) Anions
    (B) Cations
    (C) Element
    (D) Isotope
    (E) Atom

1. A positive ion

2. An atom of the same element that differs by the number of neutrons

3. Cannot be broken down chemically

4. Will migrate through a salt bridge to the anode half cell

GO ON TO THE NEXT PAGE

PRACTICE TEST—*Continued*

<u>Questions 5–7</u>

    (A) Calorimeter

    (B) Geiger Counter

    (C) Burette

    (D) Funnel

    (E) Bunsen burner

5. Used to detect radioactivity

6. Used to deliver acids and bases in a titration

7. Can be lined with moist filter paper to catch insoluble solids

<u>Questions 8–10</u>

    (A) Arrhenius acid

    (B) Arrhenius base

    (C) Lewis acid

    (D) Lewis base

    (E) Brønsted-Lowry acid

8. Yields hydroxide ions as the only negative ions in solution

9. Electron pair acceptor

10. Proton donor

<u>Questions 11–14</u>

    (A) Purple solution

    (B) Brown-orange liquid

    (C) Green gas

    (D) Silver-gray liquid

    (E) Yellow-orange when burned in a flame

11. Potassium permanganate

12. Sodium salt

13. Chlorine

14. Mercury

GO ON TO THE NEXT PAGE

PRACTICE TEST—*Continued*

<u>Questions 15–17</u>

(A) E° is positive
(B) ΔS is negative
(C) ΔG is positive
(D) K_{eq} is greater than 1
(E) K_a is very large

15. Indicates a strong acid

16. A reaction is nonspontaneous

17. Less chaos, disorder, and randomness

<u>Questions 18–21</u>

(A) Alkali metals
(B) Alkaline earth metals
(C) Transition metals
(D) Halogens
(E) Noble or inert gases

18. Group 1

19. Group 10

20. Contains elements in the solid, liquid, and gas phases

21. Will form chlorides with the formula of MCl_2

<u>Questions 22–25</u>

(A) 1+
(B) 1–
(C) 0
(D) 2+
(E) 3+

22. Oxidation number of O in H_2O_2

23. Oxidation number of F in HF

24. Oxidation number of O in O_3

25. Oxidation number of calcium in calcium phosphate

GO ON TO THE NEXT PAGE

PRACTICE TEST—*Continued*

PLEASE GO TO THE SPECIAL SECTION AT THE LOWER LEFT-HAND CORNER OF YOUR
ANSWER SHEET LABELED "CHEMISTRY" AND ANSWER QUESTIONS 101–115
ACCORDING TO THE FOLLOWING DIRECTIONS.

Part B

Directions: Each question below consists of two statements. For each question, determine whether statement I in the left-most column is true or false and whether statement II in the rightmost column is true or false. Fill in the corresponding T or F ovals on the answer sheet provided. Fill in the oval labeled "CE" only if statement II correctly explains statement I.

SAMPLE:

EX 1. The nucleus of an atom has a   BECAUSE   the only positive particles
       positive charge   found in the atom's nucleus are
                                                              protons.

SAMPLE ANSWER

	I	II	CE*
EX 1	● (F)	● (F)	●

I		II

101. Methane is defined as a compound   BECAUSE   methane can be broken down chemically.

102. The burning of a piece of paper is a   BECAUSE   once burned, the chemical properties of the paper
      physical change   remain the same.

103. −273 degrees Celsius is also known as   BECAUSE $C = K + 273$.
      absolute zero

104. The relationship between pressure and   BECAUSE   as pressure increases on a gas, the volume of the
      volume is considered to be an inverse   gas will decrease.
      relationship

105. A liquid can boil at different   BECAUSE   the atmospheric (or surrounding) pressure can
      temperatures   vary.

106. Bromine has an atomic mass of 79.9   BECAUSE   about 50% of all bromine atoms are ^{79}Br and the
                                                                 other 50% are ^{81}Br.

107. Excited tungsten atoms will give off light BECAUSE   as the excited electrons return to their ground state
      energy   they emit energy in the form of light.

108. As you go from left to right across the   BECAUSE   as you go from left to right across the periodic
      periodic table the elements tend to   table the elements tend to lose electrons.
      become more metallic in character

GO ON TO THE NEXT PAGE

PRACTICE TEST—*Continued*

109. The bonds found in a molecule of N_2 are nonpolar covalent   BECAUSE   there is an equal sharing of electrons between the nitrogen atoms.

110. The empirical formula of $C_6H_{12}O_6$ is CH_2O BECAUSE   the empirical formula shows the lowest ratio of the elements present in the molecular formula.

111. A solution of NaCl will conduct electricity   BECAUSE   NaCl will not form ions in solution.

112. Increasing the concentration of reactants will cause a reaction to proceed faster   BECAUSE   more reactants lowers the activation energy of a reaction.

113. Cl^{1-} is the conjugate base of HCl   BECAUSE   a conjugate base is formed once a Brønsted-Lowry acid accepts a proton.

114. $F_2 \rightarrow 2F^{1-} + 2e^-$ is a correctly written half reaction   BECAUSE   this half reaction must demonstrate proper conservation of mass and charge.

115. Ethane is considered to be a saturated hydrocarbon   BECAUSE   ethene has a triple bond.

GO ON TO THE NEXT PAGE

PRACTICE TEST—*Continued*

Part C

> Directions: Each of the multiple-choice questions or incomplete sentences below is followed by five answers or comple-
> tions. Select the one answer that is best in each case and then fill in the corresponding oval on the answer sheet provided.

26. Which of the following would not be attracted or deflected while traveling through an electric field?

    I.   Gamma ray
    II. Beta particle
    III. Neutron

    (A) I only
    (B) II only
    (C) I and II only
    (D) I and III only
    (E) I, II, and III

27. Which substance below is resonance stabilized by delocalized pi electrons?

    (A) Benzene
    (B) Hydrochloric acid
    (C) Hydrogen gas
    (D) Methane
    (E) Potassium bromide

28. Which of the following is true about a solution that has $[OH^{1-}] = 1.0 \times 10^{-6}\ M$?

    (A) The pH is 8 and the solution is acidic.
    (B) The $[H^{1+}] = 1.0 \times 10^{-8}\ M$ and the solution is basic.
    (C) The pH is 6 and the solution is acidic.
    (D) The $[H^{1+}] = 1.0 \times 10^{-6}\ M$ and the solution is basic.
    (E) The $[H^{1+}] = 1.0 \times 10^{-14}\ M$ and the solution is neutral.

29. What will be the products of the following double replacement? $(NH_4)_3PO_4 + Ba(NO_3)_2 \rightarrow$

    (A) Ammonium nitrate and barium nitrate
    (B) Barium nitrate and ammonium phosphate
    (C) Barium phosphate and sodium nitrate
    (D) Ammonium nitrate and barium phosphate
    (E) Ammonium nitrite and barium nitrate

30. Which K_a value is that of an acid that is the weakest electrolyte?

    (A) 1.7×10^{-7}
    (B) 2.7×10^{-8}
    (C) 6.6×10^{-10}
    (D) 4.9×10^{-3}
    (E) 5.2×10^{-4}

31. A student performs a titration using 1.00 M NaOH to find the unknown molarity of a solution of HCl. The student records the data as shown below. What is the molarity of the solution of HCl?

Base: final burette reading	21.05 mL
Base: initial burette reading	6.05 mL
mL of base used	
Acid: final burette reading	44.15 mL
Acid: initial burette reading	14.15 mL
mL of acid used	

    (A) 0.75 M
    (B) 0.50 M
    (C) 0.25 M
    (D) 0.10 M
    (E) 2.00 M

32. Which of the following is not a synthetic polymer?

    (A) Polyvinyl chloride
    (B) Plastic
    (C) Polystyrene
    (D) Polyethylene
    (E) Cellulose

GO ON TO THE NEXT PAGE ▶

PRACTICE TEST—*Continued*

33. Which process is represented by the arrow on the following phase diagram? (See Figure 1.)

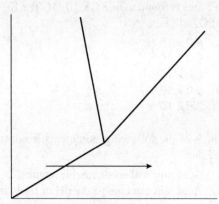

Figure 1

  (A) Evaporation
  (B) Deposition
  (C) Condensation
  (D) Freezing
  (E) Sublimation

34. What is the molar mass of $Ca_3(PO_4)_2$?

  (A) 310 grams/mole
  (B) 154 grams/mole
  (C) 67 grams/mole
  (D) 83 grams/mole
  (E) 115 grams/mole

35. What is the percent composition of oxygen in $C_6H_{12}O_6$ (molar mass = 180)?

  (A) 25%
  (B) 33%
  (C) 40%
  (D) 53%
  (E) 75%

36. The following reaction occurs at STP: $2H_2O(l) \rightarrow 2H_2(g) + O_2(g)$. How many liters of hydrogen gas can be produced by the breakdown of 72 grams of water?

  (A) 5.6 liters
  (B) 11.2 liters
  (C) 22.4 liters
  (D) 44.8 liters
  (E) 89.6 liters

37. What is the mass-action expression for the following reaction at equilibrium?

$$2W(aq) + X(l) \longleftrightarrow 3Y(aq) + 2Z(s)$$

  (A) $\dfrac{[Z]^2[Y]^3}{[W]^2[X]}$

  (B) $\dfrac{[W]^2[X]}{[Z]^2[Y]^3}$

  (C) $\dfrac{[Y]^3}{[W]^2}$

  (D) $\dfrac{[Z][Y]}{[W][X]}$

  (E) $\dfrac{[W][X]}{[Z][Y]}$

38. Which statement best describes the bonding found in formaldehyde, CH_2O?

  (A) The carbon atom is sp hybridized.
  (B) There are three sigma bonds and one pi bond present.
  (C) The bonding gives the molecule a tetrahedral shape.
  (D) The bonds between the atoms are ionic bonds.
  (E) All of the bonds are nonpolar bonds.

39. What is the molarity of a solution that has 29.25 grams of NaCl dissolved to make 1.5 liters of solution?

  (A) 19.5 M
  (B) 3.0 M
  (C) 1.75 M
  (D) 0.33 M
  (E) 1.0 M

40. A sample of a gas at STP contains 3.01×10^{23} molecules and has a mass of 22.0 grams. This gas is most likely

  (A) CO_2
  (B) O_2
  (C) N_2
  (D) CO
  (E) NO

GO ON TO THE NEXT PAGE

PRACTICE TEST—*Continued*

41. What is the value of ΔH for the reaction: $X + 2Y \rightarrow 2Z$?

    $(W + X \rightarrow 2Y \qquad \Delta H = -200 \text{ kcal})$
    $(2W + 3X \rightarrow 2Z + 2Y \qquad \Delta H = -150 \text{ kcal})$

    (A) −550 kcal
    (B) +50 kcal
    (C) −50 kcal
    (D) −350 kcal
    (E) +250 kcal

42. Which fraction would be used to find the new volume of a gas at 760 torr under its new pressure at 900 torr if the temperature is kept constant?

    (A) 900 / 760
    (B) 1.18
    (C) 760 / 900
    (D) 658.7 / 798.7
    (E) 798.7 / 658.7

43. Which of the following can be classified as a precipitation reaction?

    (A) $HCl(aq) + NaOH(aq) \rightarrow$
    (B) $KBr(aq) + NaCl(aq) \rightarrow$
    (C) $AgNO_3(aq) + MgCl_2(aq) \rightarrow$
    (D) $CaCl_2(aq) + KI(aq) \rightarrow$
    (E) $NaNO_3(aq) + HC_2H_3O_2(aq) \rightarrow$

44. Which system at equilibrium will not be influenced by a change in pressure?

    (A) $3O_2(g) \leftarrow\rightarrow 2O_3(g)$
    (B) $N_2(g) + 3H_2(g) \leftarrow\rightarrow 2NH_3(g)$
    (C) $2NO_2(g) \leftarrow\rightarrow N_2O_4(g)$
    (D) $H_2(g) + I_2(g) \leftarrow\rightarrow 2HI(g)$
    (E) $2W(g) + X(g) \leftarrow\rightarrow Y(g) + 2Z(s)$

45. The organic reaction: $C_2H_6 + Cl_2 \rightarrow HCl + C_2H_5Cl$ is best described as

    (A) a substitution reaction
    (B) an addition reaction
    (C) an esterification
    (D) a dehydration synthesis
    (E) a fermentation

46. Enough $CaSO_4(s)$ is dissolved in water at 298 K to produce a saturated solution. The concentration of Ca^{2+} ions is found to be $3.0 \times 10^{-3}M$. The K_{sp} value for $CaSO_4$ will be

    (A) 6.0×10^{-6}
    (B) 9.0×10^{-6}
    (C) 6.0×10^{-3}
    (D) 9.0×10^{-3}
    (E) 3.0×10^{-3}

47. Which of the following statements is not true about acid rain?

    (A) Acid rain will erode marble statues.
    (B) Acid rain can change the pH of lakes and streams.
    (C) Acid rain can be formed from carbon dioxide.
    (D) Acid rain creates holes in the ozone layer.
    (E) Acid rain can be formed from the gases SO_2 and SO_3.

48. Which mole sample of the solids below is best for melting a 500-gram sheet of ice on a sidewalk?

    (A) $NaCl$
    (B) $CaCl_2$
    (C) KBr
    (D) $AgNO_3$
    (E) $NaC_2H_3O_2$

49. Given this reaction that occurs in plants: $6CO_2 + 6H_2O \rightarrow C_6H_{12}O_6 + 6O_2$. If 54 grams of water are consumed by the plant, how many grams $C_6H_{12}O_6$ (molar mass is 180 grams/mole) can be made? Assume an unlimited supply of carbon dioxide for the plant to consume as well.

    (A) 54 grams
    (B) 180 grams
    (C) 540 grams
    (D) 3 grams
    (E) 90 grams

GO ON TO THE NEXT PAGE

PRACTICE TEST—*Continued*

50. Which of the following will not be changed by the addition of a catalyst to a reaction at equilibrium?

    I.   The point of equilibrium
    II. The heat of reaction, ΔH
    III. The potential energy of the products

    (A) I only
    (B) II only
    (C) I and II only
    (D) II and III only
    (E) I, II, and III

51. According to the reaction Pb(s) + S(s) → PbS(s), when 20.7 grams of lead are reacted with 6.4 grams of sulfur

    (A) there will be an excess of 20.7 grams of lead
    (B) the sulfur will be in excess by 3.2 grams
    (C) the lead and sulfur will react completely without any excess reactants
    (D) the sulfur will be the limiting factor in the reaction
    (E) there will be an excess of 10.35 grams of lead

52. Which of the following statements is not part of the kinetic molecular theory?

    (A) The average kinetic energy of gas molecules is proportional to temperature.
    (B) Attractive and repulsive forces are present between gas molecules.
    (C) Collisions between gas molecules are perfectly elastic.
    (D) Gas molecules travel in a continuous, random motion.
    (E) The volume that gas molecules occupy is minimal compared to the volume in which the gas is contained.

53. A student is performing an experiment where a blue salt is being heated to dryness in order to determine the percent of water in the salt. Which pieces of laboratory equipment would be used to help determine this percentage?

    I.   A crucible and cover
    II. Tongs
    III. A triple beam balance

    (A) II only
    (B) III only
    (C) I and III only
    (D) II and III only
    (E) I, II, and III

54. Which of the following is considered to be a dangerous procedure in the laboratory setting?

    (A) Pouring all liquids, especially acids and bases, over the sink
    (B) Wearing goggles
    (C) Pushing glass tubing, thermometers, or glass thistle tubes through a rubber cork
    (D) Pointing the mouth of a test tube that is being heated away from you and others
    (E) Knowing where the fire extinguisher and eye-wash stations are located

55. Given a 4-gram sample of each $H_2(g)$ and $He(g)$, each in separate containers, which of the following statements is true? (Assume STP.)

    (A) The sample of hydrogen gas will occupy 44.8 liters and the sample of helium will contain 6.02×10^{23} molecules.
    (B) The sample of hydrogen gas will occupy 22.4 liters and the sample of helium will contain 3.02×10^{23} molecules.
    (C) The sample of hydrogen gas will occupy 44.8 liters and the sample of helium will contain 1.202×10^{24} molecules.
    (D) The sample of helium will occupy 44.8 liters and the sample of hydrogen gas will contain 6.02×10^{23} molecules.
    (E) None of the above statements is correct.

56. The diagram shows a solid being heated from below its freezing point. Which line segment shows the gas and liquid phases existing at the same time? (See Figure 2.)

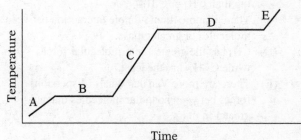

Figure 2

    (A) A
    (B) B
    (C) C
    (D) D
    (E) E

GO ON TO THE NEXT PAGE

PRACTICE TEST—*Continued*

57. Which of the following statements is/are correct regarding molecular geometries?

  I. CH_4 is trigonal pyramidal in shape.
  II. BF_3 is trigonal planar in shape.
  III. XeF_6 is tetrahedral in shape.

  (A) I only
  (B) II only
  (C) III only
  (D) I and III only
  (E) I, II, and III

58. When Uranium-238 undergoes one alpha decay and then one beta decay, the resulting isotope is

  (A) Th-234
  (B) U-234
  (C) Pa-234
  (D) Th-230
  (E) Ra-226

59. Which compound is matched up with its correct name?

  (A) CO—monocarbon monoxide
  (B) CaF_2—calcium difluoride
  (C) CCl_4—carbon tetrachloride
  (D) PCl_3—potassium trichloride
  (E) TiF_4—tin(IV) fluoride

60. Of the statements below, which best explains why CH_4 is a gas at STP, while C_8H_{18} is a liquid and $C_{20}H_{42}$ is a solid?

  (A) $C_{20}H_{42}$ has the greatest ionic interaction between its molecules.
  (B) $C_{20}H_{42}$ has a greater amount of hydrogen bonding than CH_4 or C_8H_{18}.
  (C) There is more dipole-dipole interaction between molecules of greater mass.
  (D) CH_4 has the greatest intermolecular forces while $C_{20}H_{42}$ has the least.
  (E) There are more Van der Waals (dispersion) forces between nonpolar molecules that are greater in mass.

61. Which of the gases listed below would not be collected via water displacement?

  (A) CO_2
  (B) CH_4
  (C) O_2
  (D) NH_3
  (E) H_2

62. Which scientist and discovery are not correctly paired?

  (A) Millikan / neutron
  (B) Rutherford / nucleus
  (C) Charles / relationship between temperature and pressure
  (D) Curie / radioactivity
  (E) Mendeleyev / periodic table

63. Which of the following situations demonstrate(s) an increase in entropy?

  I.   Dissolving a salt into water
  II. Sublimation
  III. Heating up a liquid

  (A) I only
  (B) I and II only
  (C) II and III only
  (D) I and III only
  (E) I, II, and III

64. How many moles of a gas are present in a closed, empty soda bottle that has a volume of 2.0 liters at 22°C and a pressure of 1.05 atm?

  (A) $\dfrac{(1.05\,atm)(2.0\,L)}{(0.0820\,L \cdot atm/K \cdot mol)(22°C)}$

  (B) $\dfrac{(0.0820\,L \cdot atm/K \cdot mol)(295K)}{(1.05\,atm)(2.0\,L)}$

  (C) $\dfrac{(1.05\,atm)(2.0\,L)(295K)}{(0.0820\,L \cdot atm/K \cdot mol)}$

  (D) $\dfrac{(1.05\,atm)(2.0\,L)}{(0.0820\,L \cdot atm/K \cdot mol)(295K)}$

  (E) $\dfrac{(0.0820\,L \cdot atm/K \cdot mol)}{(1.05\,atm)(2.0\,L)(295K)}$

65. Which set of conditions below guarantees that a reaction will be spontaneous?

  (A) $\Delta H(+)$ and $\Delta S(-)$
  (B) $\Delta H(-)$ and $\Delta S(+)$
  (C) $\Delta H(+)$ and $\Delta S(+)$ at a low temperature
  (D) $\Delta H(-)$ and $\Delta S(-)$ at a high temperature
  (E) $\Delta G(+)$

66. How many moles of electrons are transferred in the following reaction? $Ce^{3+} + Pb \rightarrow Ce + Pb^{4+}$

  (A) 14
  (B) 12
  (C) 7
  (D) 24
  (E) 3

GO ON TO THE NEXT PAGE

PRACTICE TEST—*Continued*

67. A gas is confined in the manometer as shown. (See Figure 3.) The stopcock is then opened and the highest level of mercury inside the tube moved to a level that is 80 mm above its lowest level. What is the pressure of the gas?

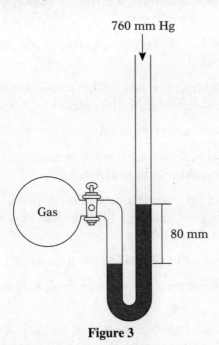

760 mm Hg

Gas

80 mm

Figure 3

(A) 80 mm Hg
(B) 160 mm Hg
(C) 680 mm Hg
(D) 840 mm Hg
(E) The pressure cannot be determined.

68. Which statement best describes the density and rate of effusion of the following gases?

NO_2 C_2H_6 Kr Xe F_2

(A) Fluorine has the lowest density and the lowest rate of effusion.
(B) Xenon has the greatest rate of effusion and the lowest density.
(C) Krypton has the lowest density and greatest rate of effusion.
(D) Ethane has the greatest rate of effusion and the lowest density.
(E) Nitrogen dioxide has the highest density and the greatest rate of effusion.

69. Which indicator is correctly paired up with its proper color if it were added to a base?

I.   Litmus—blue
II. Phenolphthalein—pink
III. Methyl Orange—yellow

(A) I only
(B) II only
(C) III only
(D) I and III only
(E) I, II, and III

70. Which structure below demonstrates a violation of the octet rule? (See Figure 4.)

(A) H—C≡C—H

(B)   O
         ‖
     H—C—H

(C)   OH
        |
    O=S=O
        |
        OH

(D) H—N—H
         |
         H

(E) H—C=C—H
         | |
         H   H

Figure 4

S T O P

IF YOU FINISH BEFORE TIME IS CALLED, GO BACK AND CHECK YOUR WORK.

PRACTICE TEST 1

▮▮▮ ANSWERS AND EXPLANATIONS

1.   B   Cations are ions that are positive in charge.

2.   D   Isotopes are atoms of the same element that have a different mass number because they have a different number of neutrons.

3.   C   By definition, elements cannot be broken down chemically, whole compounds can be broken down chemically.

4.   A   In a voltaic cell, anions will travel through the salt bridge to the anode while cations will travel through the salt bridge to the cathode.

5.   B   Geiger counters are used to detect radioactive emanations.

6.   C   Burettes are used to deliver acids and bases in a titration because of their ability to give precise readings of the acid and base delivered.

7.   D   A funnel can be lined with filter paper so that liquids can be poured through. The liquid that flows through is called the filtrate and the substance left in the filter paper is called the residue.

8.   B   Arrhenius bases yield hydroxide ions as the negative ions in solution.

9.   C   Lewis acids and bases are defined as electron pair acceptors and donors, respectively.

10.   E   While the Arrhenius definition of an acid says that an acid yields hydronium ions as the only positive ions in solution, the Brønsted-Lowry definition says that acids are proton donors.

11.   A   Potassium permanganate is a purple salt that forms a purple solution.

12.   E   Sodium salts, when burned, will turn a flame a bright yellow color.

13.   C   Besides its terrible choking odor, chlorine gas is famous for its green color.

14.   D   Any mercury-filled thermometer will reveal the silver-gray color. Also note that mercury is a metal that is in the liquid state.

15.   E   The larger the K_a value an acid has, the stronger it is as an acid or electrolyte.

16.   C   Gibbs Free Energy tells if a reaction will be spontaneous. When ΔG is positive the reaction is nonspontaneous.

17.   B   Entropy (disorder, chaos, randomness) is designated with the letter "S." If ΔS is negative, there is less entropy.

18.   A   The alkali metals are located in group 1 of the periodic table.

19.   C   The transition elements are located in groups 3 through 12.

PRACTICE TEST—*Continued*

20. D   The halogens, group 17, have elements in the gas (F and Cl), liquid (Br), and solid (I) phases.

21. B   Because they have an oxidation number of 2+ when they form ions, the alkaline earth metals (elements in group 2) will take on two chlorine ions to form chloride salts.

22. B   In peroxides (X_2O_2) the oxidation number of each oxygen atom is 1–.

23. B   When it has reacted, fluorine will have an oxidation number of 1–.

24. C   In a compound made up entirely of a single element, each atom will have an oxidation number of 0.

25. D   Calcium is a group 2 metal and will have an oxidation state of 2+ as described in question 21.

101. T, T, CE   By definition, compounds can be broken down chemically into their elements. Elements cannot be broken down chemically.

102. F, F   Burning a piece of paper is a chemical change because the paper has changed chemically. Tearing the paper would be an example of a physical change.

103. T, F   –273°C is equivalent to 0 K, absolute zero. However, the correct equation is K = C + 273 and not C = K + 273.

104. T, T, CE   In an inverse relationship, such as the one between pressure and volume, as one factor increases in value, the other factor decreases in value.

105. T, T, CE   A liquid boils when the atmospheric (or surrounding) pressure is equal to the vapor pressure of the liquid. Because pressures on a liquid can change, the temperature at which a liquid boils can change as well.

106. T, T, CE   Atomic masses are based upon the average relative abundance. Because about half of all bromine atoms are Br-79 and half are Br-81, you can solve for the atomic mass:

(79)(0.50) = 39.5

(81)(0.50) = 40.5

Total   = 80.0 (or approximately 79.9)

107. T, T, CE   Tungsten makes up the filaments in conventional light bulbs. Electrical energy is used to excite the electrons to a higher energy level. Because a lower energy state is preferred, the electrons move back to a lower energy (more stable) state by giving off the energy gained.

108. F, F   Going left to right across a period, the elements become more nonmetallic in nature and tend to gain electrons.

PRACTICE TEST—*Continued*

109.   T, T,   Nitrogen gas has nonpolar covalent bonds because there is no difference in
        CE   electronegativity between the two nonmetal atoms. No difference in elec-
                 tronegativity also means that the electrons will be shared equally.

110.   T, T,   Empirical formulas show the lowest ratio of the elements present in the
        CE   molecular formula.

111.   T, F   Molten salts and aqueous salts are able to conduct electricity because of the
                 free-moving ions. NaCl will form free-moving ions once molten or dis-
                 solved into solution.

112.   T, F   An increase in concentration increases the frequency of collisions in a reac-
                 tion, causing the reaction to proceed faster. It takes a catalyst, not an
                 increase in concentration, to lower the activation energy.

113.   T, F   A conjugate base is formed from an acid that loses a proton. Brønsted-
                 Lowry acids are proton donors and not acceptors.

114.   F, T   The half reaction in this question shows a charge of 0 on the left but a
                 charge of 4– on the right. Because every half reaction must demonstrate
                 proper conservation of mass and charge, the half reaction in this question
                 is incorrect. It should read $F_2 + 2e^- \rightarrow 2F^{1-}$.

115.   T, F   Alkanes are considered to be saturated because they have all single bonds
                 and the maximum number of hydrogen atoms bonded to the carbon atoms
                 present. Ethene is an alkene and contains a double bond, while alkynes
                 contain a triple bond.

26.   D   Because gamma rays and neutrons have no charge, they will not be
                 attracted or deflected in a charged field.

27.   A   Benzene has alternating single and double bonds between the carbon atoms.
                 These bonds can "rotate" because of the pi electrons that are delocalized.

28.   B   In a solution that has $[OH^{1-}] = 1.0 \times 10^{-6}$ M, the $[H^{1+}]$ must be 1.0×10^{-8} M
                 because $[H^{1+}][OH^{1-}] = 1.0 \times 10^{-14}$. Because the solution has a greater con-
                 centration of hydroxide ions, the solution will be basic.

29.   D   In a double replacement reaction the positive and negative ions "switch
                 partners." Ammonium will bond with the nitrate ion and the barium ion
                 will bond with phosphate.

30.   C   A smaller K_a value is that of a weaker acid and a weaker electrolyte.

31.   B   First subtract to find the volumes of acid and base used. Using the equa-
                 tion: $M_aV_a = M_bV_b$ substitute and solve:

                 (x)(30.0 mL) = (1.00 M)(15.0 mL)
                 x, M_a, will have a value of 0.50 M.

PRACTICE TEST—*Continued*

32. E Cellulose is a polymer that is made naturally by plants. The other choices are polymers, but they are all artificial.

33. E This phase diagram shows a substance changing from the solid phase to the gas phase. This process is called sublimation.

34. A Three atoms of Ca will have a mass of 120 (3 × 40), two atoms of phosphorus will have a mass of 62 (31 × 2), and eight atoms of oxygen will have a mass of 128 (8 × 16). The total mass is 310 grams/mole.

35. D The mass of six oxygen atoms is 96. To find the percent composition, one divides the mass of the atom in question by the total mass of the compound. 96/180 is approximately 0.53 or 53%.

36. E 72 grams of water (molar mass of 18) is 4 moles of water. For every 2 moles of water decomposed, 2 moles of hydrogen gas are produced. So 4 moles of water will form 4 moles of hydrogen gas; 4 moles of a gas at STP will occupy 4 × 22.4 liters, or 89.6 liters.

37. C Remember the rule "Products over reactants, coefficients become powers." Also, solids and liquids are not included in a mass-action equation.

38. B There is a double bond (one sigma and one pi bond) between the C and O and there is a single bond between the C and each of the H atoms (two more sigma bonds).

39. D 29.25 grams of NaCl (molar mass 58.5) is 0.5 moles of NaCl. The equation for molarity is "moles over liters," M = moles/liters. 0.5 moles divided by 1.5 liters is 0.33 M.

40. A 3.01×10^{23} molecules is half a mole of molecules. If half a mole of the gas has a mass of 22.0 grams, then 1 mole of that same gas will have a mass of 44.0 grams. The only gas in this question with the molar mass of 44.0 is carbon dioxide.

41. E Start by reversing and doubling the first reaction. This will put W on both sides of the reaction so that it can cancel with the second equation:

$$4Y \rightarrow 2W + 2X \qquad \Delta H = +400 \text{ kcal}$$

Add on the second equation:

$$2W + 3X \rightarrow 2Z + 2Y \qquad \Delta H = -150 \text{ kcal}$$

Add the two equations together to see the 4Y and 2Y reduce along with the 2X and 3X. The 2W will cancel to get:

$$X + 2Y \rightarrow 2Z \quad \Delta H = +250 \text{ kcal}$$

PRACTICE TEST—*Continued*

42.   C   Boyle's Law states that $P_1V_1 = P_2V_2$. Substituting gives: $(760)V_1 = (900)V_2$. To solve for the new pressure, divide and get:

$$\frac{(760)V_1}{(900)} = V_2$$

43.   C   Salts containing nitrate ions, sodium ions, and halogen ions are soluble. The exceptions for the halides are halide salts containing lead, mercury, or silver. Silver chloride will form in the reaction shown in choice (C).

44.   D   The sum of the coefficients of the products and the sum of the coefficients of the reactants will determine if pressure can shift the equilibrium. The reaction: $H_2(g) + I_2(g) \longleftrightarrow 2HI(g)$ will not shift with a pressure change because the number of moles of gas as reactants is the same as the number of moles of gas as product.

45.   A   This reaction demonstrates the substitution of a hydrogen atom with a chlorine atom.

46.   B   The mass action expression for this slightly soluble salt is $K_{sp} = [Ca^{2+}][SO_4^{2-}] / 1$. For every Ca^{2+} ion formed, one sulfate ion is formed as well. Therefore:

$$K_{sp} = [3.0 \times 10^{-3}M][3.0 \times 10^{-3}M] = 9.0 \times 10^{-6}$$

47.   D   Acid rain can do much damage over time to marine life and to marble structures (made of carbonate rocks). It can be formed from oxides of carbon, nitrogen, and/or sulfur. Acid rain, however, does not attack the ozone layer. CFCs are responsible for destroying the ozone layer.

48.   B   Colligative properties change depending upon the number of moles of particles present in solution. Calcium chloride would do best for melting the ice because it yields 3 moles of ions per mole of solute (Ca^{2+} and $2Cl^{1-}$).

49.   E   54 grams of water (molar mass = 18) is 3 moles of water. Because of the mole ratio:

$$\frac{6 \text{ moles of } H_2O}{3 \text{ moles of } H_2O} = \frac{1 \text{ mole of } C_6H_{12}O_6}{x \text{ moles of } C_6H_{12}O_6}$$

Thus 0.5 moles of $C_6H_{12}O_6$ can be produced. 0.5 moles of $C_6H_{12}O_6$ (molar mass = 180 grams/mole) will have a mass of 90 grams.

50.   E   A catalyst will not change the PE of the reactants, the PE of the products, or the heat of reaction. In addition, a catalyst will not shift or change the point of equilibrium. A catalyst, however, can help the reaction achieve equilibrium faster.

PRACTICE TEST—*Continued*

51.   B   20.7 grams of lead is equal to 0.1 moles of lead. 6.4 grams of sulfur is equal to 0.2 moles of sulfur. For every mole of Pb that reacts, 1 mole of S reacts as well. These amounts dictate that there is twice as much S as is needed. Therefore, 3.2 grams of S will be in excess.

52.   B   Ideally gases will not have attractive or repulsive forces between them.

53.   E   All of these items are needed. The balance is used to weigh the samples before and after heating. Tongs are used to handle hot equipment. Finally, the crucible is what the salt is heated in, while the cover allows water to escape without letting material that may splatter be lost.

54.   C   Glass tubing, thermometers, or glass thistle tubes can break while being pushed through rubber cork. The jagged edges of the glassware can cut skin. It is best to allow the lab specialist to make these adjustments.

55.   A   4 grams of hydrogen gas is equal to 2 moles of this gas while 4 grams of helium is equal to 1 mole of helium. 2 moles of H_2 will occupy 44.8 liters and 1 mole of He will contain 6.02×10^{23} molecules.

56.   D   Boiling allows a phase change from liquid to gas. This part of the diagram is represented by letter D.

57.   B   Methane has a molecular geometry that is tetrahedral, while BF_3 is trigonal planar in shape and XeF_6 is octahedral in shape.

58.   C   This reaction can be demonstrated as follows: $^{238}_{92}U \rightarrow {}^{4}_{2}He + {}^{234}_{90}Th$ and then $^{234}_{90}Th \rightarrow {}^{0}_{-1}e + {}^{234}_{91}Pa$.

59.   C   Covalently bonded substances will use prefixes while ionic compounds will not. Choice A may seem correct, but a covalent compound's name does not start with the prefix "mono-." Calcium fluoride is ionic and will not use prefixes at all. "Tetra-" is correct to indicate "four" in a covalently bonded compound. P is phosphorus and not potassium. Finally, Ti is titanium and not tin, Sn.

60.   E   All of these hydrocarbons are nonpolar and will be affected by Van der Waals (dispersion) forces. As the mass of the nonpolar molecules increases, so do the dispersion forces. This is why the wax (molar mass of 282), $C_{20}H_{42}$, is a solid.

61.   D   Water is a polar substance. Because "like dissolves like," you would not want to collect a polar gas in water, which is also polar. All the other choices are nonpolar substances and can be collected by water displacement.

62.   A   Millikan discovered that the charge of an electron is 1.6×10^{-19} coulombs. The neutron was discovered by Chadwick.

63.   E   All the processes/situations listed are examples of an increase in disorder, randomness, and chaos.

PRACTICE TEST—*Continued*

64.   D   The ideal gas equation, $PV = nRT$ is needed to solve this problem. Solving for moles, $n = PV / RT$. Remember that the value for T must be in Kelvin.

65.   B   Nature prefers a reaction where energy (or heat) has been lowered and entropy has increased. These conditions will guarantee a spontaneous reaction and give ΔG a negative value.

66.   B   First write two half reactions: $3e^- + Ce^{3+} + \rightarrow Ce$ and $Pb \rightarrow Pb^{4+} + 4e^-$. The number of moles of electrons that are lost in the reaction must be the same as the number gained. To accomplish this, multiply the equations by numbers that make the number of moles of electrons equal:

$4(3e^- + Ce^{3+} + \rightarrow Ce)$ becomes $12e^- + 4Ce^{3+} + \rightarrow 4Ce$

$3(Pb \rightarrow Pb^{4+} + 4e^-)$ becomes $3Pb \rightarrow 3Pb^{4+} + 12e^-$

The number of moles of electrons lost is the same as the number gained. The electrons being lost are also the electrons being gained.

67.   D   If the levels of mercury were the same, then the pressure of the gas would be equal to the pressure that is being applied from the opening of the tube. Because the sample of gas moved the mercury to a level that is 80 mm above its lowest level, the pressure of the gas must be greater than the pressure of 760 mm Hg that is being exerted by 80 mm Hg. This means that the pressure of the gas is 840 mm Hg.

68.   D   Ethane has molar mass of 30, the lowest molar mass of any of these gases. Ethane will have the fastest rate of effusion and, because of its low molar mass, the lowest density.

69.   E   All these indicators are paired with their correct color for presence of base.

70.   C   Sulfur can violate the octet rule and form up to six bonds.

PRACTICE TEST 1

▰ SCORE SHEET

Number of questions correct: _____

Less: 0.25 × number of questions wrong: _____

(Remember that omitted questions are not counted as wrong.)

Equals your raw score: _____

Raw Score	Test Score		Raw Score	Test Score		Raw Score	Test Score		Raw Score	Test Score		Raw Score	Test Score
85	800		63	710		41	570		19	440		−3	300
84	800		62	700		40	560		18	430		−4	300
83	800		61	700		39	560		17	430		−5	290
82	800		60	690		38	550		16	420		−6	290
81	800		59	680		37	550		15	420		−7	280
80	800		58	670		36	540		14	410		−8	270
79	790		57	670		35	530		13	400		−9	270
78	790		56	660		34	530		12	400		−10	260
77	790		55	650		33	520		11	390		−11	250
76	780		54	640		32	520		10	390		−12	250
75	780		53	640		31	510		9	380		−13	240
74	770		52	630		30	500		8	370		−14	240
73	760		51	630		29	500		7	360		−15	230
72	760		50	620		28	490		6	360		−16	230
71	750		49	610		27	480		5	350		−17	220
70	740		48	610		26	480		4	350		−18	220
69	740		47	600		25	470		3	340		−19	210
68	730		46	600		24	470		2	330		−20	210
67	730		45	590		23	460		1	330		−21	200
66	720		44	580		22	460		0	320			
65	720		43	580		21	450		−1	320			
64	710		42	570		20	440		−2	310			

Note: This is only a sample scoring scale. Scoring scales differ from exam to exam.

PRACTICE TEST 2

ANSWER SHEET

This practice test will help measure your current knowledge of chemistry and familiarize you with the structure of the *SAT Chemistry* test. To simulate exam conditions, use the answer sheet below to record your answers and the appendix tables at the back of this book to obtain necessary information.

1. Ⓐ Ⓑ Ⓒ Ⓓ Ⓔ	21. Ⓐ Ⓑ Ⓒ Ⓓ Ⓔ	41. Ⓐ Ⓑ Ⓒ Ⓓ Ⓔ	61. Ⓐ Ⓑ Ⓒ Ⓓ Ⓔ
2. Ⓐ Ⓑ Ⓒ Ⓓ Ⓔ	22. Ⓐ Ⓑ Ⓒ Ⓓ Ⓔ	42. Ⓐ Ⓑ Ⓒ Ⓓ Ⓔ	62. Ⓐ Ⓑ Ⓒ Ⓓ Ⓔ
3. Ⓐ Ⓑ Ⓒ Ⓓ Ⓔ	23. Ⓐ Ⓑ Ⓒ Ⓓ Ⓔ	43. Ⓐ Ⓑ Ⓒ Ⓓ Ⓔ	63. Ⓐ Ⓑ Ⓒ Ⓓ Ⓔ
4. Ⓐ Ⓑ Ⓒ Ⓓ Ⓔ	24. Ⓐ Ⓑ Ⓒ Ⓓ Ⓔ	44. Ⓐ Ⓑ Ⓒ Ⓓ Ⓔ	64. Ⓐ Ⓑ Ⓒ Ⓓ Ⓔ
5. Ⓐ Ⓑ Ⓒ Ⓓ Ⓔ	25. Ⓐ Ⓑ Ⓒ Ⓓ Ⓔ	45. Ⓐ Ⓑ Ⓒ Ⓓ Ⓔ	65. Ⓐ Ⓑ Ⓒ Ⓓ Ⓔ
6. Ⓐ Ⓑ Ⓒ Ⓓ Ⓔ	26. Ⓐ Ⓑ Ⓒ Ⓓ Ⓔ	46. Ⓐ Ⓑ Ⓒ Ⓓ Ⓔ	66. Ⓐ Ⓑ Ⓒ Ⓓ Ⓔ
7. Ⓐ Ⓑ Ⓒ Ⓓ Ⓔ	27. Ⓐ Ⓑ Ⓒ Ⓓ Ⓔ	47. Ⓐ Ⓑ Ⓒ Ⓓ Ⓔ	67. Ⓐ Ⓑ Ⓒ Ⓓ Ⓔ
8. Ⓐ Ⓑ Ⓒ Ⓓ Ⓔ	28. Ⓐ Ⓑ Ⓒ Ⓓ Ⓔ	48. Ⓐ Ⓑ Ⓒ Ⓓ Ⓔ	68. Ⓐ Ⓑ Ⓒ Ⓓ Ⓔ
9. Ⓐ Ⓑ Ⓒ Ⓓ Ⓔ	29. Ⓐ Ⓑ Ⓒ Ⓓ Ⓔ	49. Ⓐ Ⓑ Ⓒ Ⓓ Ⓔ	69. Ⓐ Ⓑ Ⓒ Ⓓ Ⓔ
10. Ⓐ Ⓑ Ⓒ Ⓓ Ⓔ	30. Ⓐ Ⓑ Ⓒ Ⓓ Ⓔ	50. Ⓐ Ⓑ Ⓒ Ⓓ Ⓔ	70. Ⓐ Ⓑ Ⓒ Ⓓ Ⓔ
11. Ⓐ Ⓑ Ⓒ Ⓓ Ⓔ	31. Ⓐ Ⓑ Ⓒ Ⓓ Ⓔ	51. Ⓐ Ⓑ Ⓒ Ⓓ Ⓔ	71. Ⓐ Ⓑ Ⓒ Ⓓ Ⓔ
12. Ⓐ Ⓑ Ⓒ Ⓓ Ⓔ	32. Ⓐ Ⓑ Ⓒ Ⓓ Ⓔ	52. Ⓐ Ⓑ Ⓒ Ⓓ Ⓔ	72. Ⓐ Ⓑ Ⓒ Ⓓ Ⓔ
13. Ⓐ Ⓑ Ⓒ Ⓓ Ⓔ	33. Ⓐ Ⓑ Ⓒ Ⓓ Ⓔ	53. Ⓐ Ⓑ Ⓒ Ⓓ Ⓔ	73. Ⓐ Ⓑ Ⓒ Ⓓ Ⓔ
14. Ⓐ Ⓑ Ⓒ Ⓓ Ⓔ	34. Ⓐ Ⓑ Ⓒ Ⓓ Ⓔ	54. Ⓐ Ⓑ Ⓒ Ⓓ Ⓔ	74. Ⓐ Ⓑ Ⓒ Ⓓ Ⓔ
15. Ⓐ Ⓑ Ⓒ Ⓓ Ⓔ	35. Ⓐ Ⓑ Ⓒ Ⓓ Ⓔ	55. Ⓐ Ⓑ Ⓒ Ⓓ Ⓔ	75. Ⓐ Ⓑ Ⓒ Ⓓ Ⓔ
16. Ⓐ Ⓑ Ⓒ Ⓓ Ⓔ	36. Ⓐ Ⓑ Ⓒ Ⓓ Ⓔ	56. Ⓐ Ⓑ Ⓒ Ⓓ Ⓔ	
17. Ⓐ Ⓑ Ⓒ Ⓓ Ⓔ	37. Ⓐ Ⓑ Ⓒ Ⓓ Ⓔ	57. Ⓐ Ⓑ Ⓒ Ⓓ Ⓔ	
18. Ⓐ Ⓑ Ⓒ Ⓓ Ⓔ	38. Ⓐ Ⓑ Ⓒ Ⓓ Ⓔ	58. Ⓐ Ⓑ Ⓒ Ⓓ Ⓔ	
19. Ⓐ Ⓑ Ⓒ Ⓓ Ⓔ	39. Ⓐ Ⓑ Ⓒ Ⓓ Ⓔ	59. Ⓐ Ⓑ Ⓒ Ⓓ Ⓔ	
20. Ⓐ Ⓑ Ⓒ Ⓓ Ⓔ	40. Ⓐ Ⓑ Ⓒ Ⓓ Ⓔ	60. Ⓐ Ⓑ Ⓒ Ⓓ Ⓔ	

Chemistry   *Fill in oval CE only if II is correct explanation of I.

	I	II	CE*		I	II	CE*
101.	Ⓣ Ⓕ	Ⓣ Ⓕ	◯	109.	Ⓣ Ⓕ	Ⓣ Ⓕ	◯
102.	Ⓣ Ⓕ	Ⓣ Ⓕ	◯	110.	Ⓣ Ⓕ	Ⓣ Ⓕ	◯
103.	Ⓣ Ⓕ	Ⓣ Ⓕ	◯	111.	Ⓣ Ⓕ	Ⓣ Ⓕ	◯
104.	Ⓣ Ⓕ	Ⓣ Ⓕ	◯	112.	Ⓣ Ⓕ	Ⓣ Ⓕ	◯
105.	Ⓣ Ⓕ	Ⓣ Ⓕ	◯	113.	Ⓣ Ⓕ	Ⓣ Ⓕ	◯
106.	Ⓣ Ⓕ	Ⓣ Ⓕ	◯	114.	Ⓣ Ⓕ	Ⓣ Ⓕ	◯
107.	Ⓣ Ⓕ	Ⓣ Ⓕ	◯	115.	Ⓣ Ⓕ	Ⓣ Ⓕ	◯
108.	Ⓣ Ⓕ	Ⓣ Ⓕ	◯				

GO ON TO THE NEXT PAGE ➤

PRACTICE TEST 2

Time: 60 minutes

<u>Note:</u> Unless otherwise stated, for all statements involving chemical equations and/or solutions, assume that the system is in pure water.

Part A

<u>Directions:</u> Each of the following sets of lettered choices refers to the numbered formulas or statements immediately below it. For each numbered item, choose the one lettered choice that fits it best. Then fill in the corresponding oval on the answer sheet. Each choice in a set may be used once, more than once, or not at all.

<u>Questions 1–4</u>

(A) Sublimation
(B) Deposition
(C) Vaporization
(D) Condensation
(E) Melting

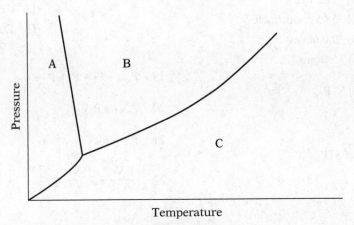

Figure 1

1. Phase A → Phase B

2. Phase C → Phase A

3. Phase A → Phase C

4. Phase B → Phase C

PRACTICE TEST—*Continued*

Questions 5–8

(A) R—COOH
(B) R—CHO
(C) R—CO—R
(D) R—COO—R
(E) R—CO—NH_2

5. Can be neutralized with a base

6. Could be named 2-pentanone

7. Amide functional group

8. Aldehyde functional group

Questions 9–11

(A) 6.02×10^{23} molecules
(B) 11.2 liters
(C) 58.5 grams/mole
(D) 2.0 moles
(E) 5 atoms

9. 88 grams of $CO_2(g)$ at STP

10. 1 molecule of CH_4

11. 32 grams of SO_2 gas at STP

Questions 12–15

(A) Alkali metal
(B) Alkaline earth metal
(C) Transition metal
(D) Halogen
(E) Noble gas

12. Reacts most vigorously with water

13. Is chemically inert

14. Has the highest first ionization energy in its period

15. Forms ions with a 2+ charge

Questions 16–18

(A) Blue
(B) Red
(C) Pink/purple
(D) Colorless
(E) Orange

16. Phenolphthalein in base

17. Litmus in acid

18. Phenolphthalein in acid

Questions 19–22

(A) Boyle's Law
(B) Charles' Law
(C) Ideal Gas Equation
(D) Combined Gas Law
(E) Dalton's Law of Partial Pressures

19. $P_{total} = P_1 + P_2 + P_3 + \cdots$

20. $P_1V_1 = P_2V_2$

21. $PV = nRT$

22. $\dfrac{P_1V_1}{T_1} = \dfrac{P_2V_2}{T_2}$

Questions 23–25

(A) ^4_2He
(B) $^0_{1}e$
(C) γ
(D) ^1_0n
(E) ^1_1H

23. Has a charge of 2+

24. Has the lowest mass

25. Has the greatest mass

GO ON TO THE NEXT PAGE ▶

PRACTICE TEST—*Continued*

PLEASE GO TO THE SPECIAL SECTION AT THE LOWER LEFT-HAND CORNER OF YOUR
ANSWER SHEET LABELED "CHEMISTRY" AND ANSWER QUESTIONS 101–115
ACCORDING TO THE FOLLOWING DIRECTIONS.

Part B

Directions: Each question below consists of two statements. For each question, determine whether statement I in the left-most column is true or false and whether statement II in the rightmost column is true or false. Fill in the corresponding T or F ovals on the answer sheet provided. Fill in the oval labeled "CE" only if statement II correctly explains statement I.

SAMPLE:

EX 1. The nucleus of an atom has a   BECAUSE   the only positive particles
       positive charge   found in the atom's nucleus are
                                                          protons.

SAMPLE ANSWER

	I	II	CE*
EX 1	● F	● F	●

I   II

101. The double and single bonds in benzene   BECAUSE   benzene has delocalized pi electrons that stabilize
      are subject to resonance   its structure.

102. The element with an electron   BECAUSE   the element with an electron configuration of
      configuration of $[He]2s^1$ has a larger $[He]2s^1$ has a greater nuclear charge than fluorine.
      atomic radius than fluorine

103. $1m$ NaCl(aq) will have a higher boiling   BECAUSE   1 mole of NaCl yields 3 moles of ions in solution.
      point than that of $1m$ $CaCl_2$(aq)

104. Neutrons and protons are classified as   BECAUSE   neutrons and protons are both located in the princi-
      nucleons   pal energy levels of the atom.

105. HCl is considered to be an acid   BECAUSE   HCl is a proton acceptor.

106. Powdered zinc reacts faster with acid   BECAUSE   powdered zinc has a greater surface area.
      than a larger piece of zinc

107. NH_3 can best be collected by water   BECAUSE NH_3 is a polar substance.
      displacement

GO ON TO THE NEXT PAGE ▶

PRACTICE TEST—*Continued*

108. At 1 atm, pure water can boil at a temperature less than 273 K

    BECAUSE

    water boils when the vapor pressure of the water is equal to the atmospheric pressure.

109. An exothermic reaction has a negative value for ΔH

    BECAUSE

    in an exothermic reaction the products have less potential energy than the reactants.

110. As pressure on a gas increases, the volume of the gas decreases

    BECAUSE

    pressure and volume have a direct relationship.

111. The addition of H_2 to ethene will form an unsaturated compound called ethane

    BECAUSE

    ethane has as many hydrogen atoms bonded to the carbon atoms as possible.

112. AgCl is insoluble in water

    BECAUSE

    all chlorides are soluble in water except for those of silver, mercury, and lead.

113. ΔS will be positive in value as vaporization occurs

    BECAUSE

    vaporization increases the order of the molecules entering the gas phase.

114. Pure water has a pH of 7

    BECAUSE

    the number of H^{1+} ions is equal to the number OH^{1-} ions.

115. CH_3CH_2—OH and CH_3—O—CH_3 are isomers

    BECAUSE

    CH_3CH_2—OH and CH_3—O—CH_3 have the same molecular formula but different structures.

GO ON TO THE NEXT PAGE ➤

PRACTICE TEST—*Continued*

Part C

Directions: Each of the multiple-choice questions or incomplete sentences below is followed by five answers or completions. Select the one answer that is best in each case and then fill in the corresponding oval on the answer sheet provided.

26. One mole of each of the following substances is dissolved in 1.0 kg of water. Which solution will have the lowest freezing point?

(A) $NaC_2H_3O_2$
(B) $NaCl$
(C) $MgCl_2$
(D) CH_3OH
(E) $C_6H_{12}O_6$

27. Which of these equations is/are properly balanced?

I. $Cl_2 + 2NaBr \rightarrow Br_2 + 2NaCl$
II. $2Na + O_2 \rightarrow Na_2O$
III. $2K + 2H_2O \rightarrow H_2 + 2KOH$

(A) I only
(B) II only
(C) III only
(D) I and III only
(E) I, II, and III

28. Propane and oxygen react according to the equation: $C_3H_8(g) + 5O_2(g) \rightarrow 3CO_2(g) + 4H_2O(g)$. How many grams of water can be produced from the complete combustion of 2.0 moles of $C_3H_8(g)$?

(A) 144.0
(B) 82.0
(C) 8.0
(D) 44.8
(E) 22.4

29. A compound was analyzed and found to be composed of 75% carbon and 25% hydrogen. What is the empirical formula of this compound?

(A) C_2H_4
(B) CH_4
(C) CH_3
(D) CH_2
(E) CH

30. Which compound below has a bent molecular geometry?

(A) H_2SO_4
(B) CH_4
(C) CO_2
(D) H_2S
(E) C_2H_2

31. Of the equipment listed below, which one would require you to read a meniscus?

(A) 100 mL beaker
(B) 500 mL flask
(C) Watch glass
(D) 50 mL buret
(E) Trough

32. Given the following reaction at equilibrium: $Fe^{3+}(aq) + SCN^-(aq) \longleftrightarrow FeSCN^{2+}(aq)$. Which of these would shift the equilibrium to the left?

(A) Adding $FeCl_3$ to the reaction
(B) Adding NH_4SCN to the reaction
(C) Increasing the pressure on the reaction
(D) Adding a catalyst
(E) Adding $FeSCN^{2+}(aq)$ to the reaction

GO ON TO THE NEXT PAGE

PRACTICE TEST—*Continued*

33. Which letter in the boxes below has a value of 7?

Isotope	Number of Protons	Number of Neutrons	Number of Electrons	Mass Number	Atomic Number
^{16}O	A				
^{13}C		B			E
^{23}Na			C		
^{10}B				D	

(A) A
(B) B
(C) C
(D) D
(E) E

34. Each of the elements below is placed in water. Which one will react violently with the water?

(A) Na
(B) Fe
(C) Cu
(D) Au
(E) Ne

35. Which unit is paired incorrectly?

(A) Torr and pressure
(B) Mass and grams
(C) Heat energy and kiloPascals
(D) Volume and milliliter
(E) Temperature and Kelvin

36. Which amount of $Pb(NO_3)_2$, when added to enough water to make 1 liter of solution, will produce a solution with a molarity of 1.0 M?

(A) 144 grams
(B) 331 grams
(C) 317 grams
(D) 0.003 moles
(E) 0.5 moles

37. Enough AgCl(s) is dissolved in water at 298 K to produce a saturated solution. The concentration of Ag^{1+} ions is found to be $1.3 \times 10^{-5}M$. The K_{sp} value for AgCl will be

(A) 2.6×10^{-10}
(B) 1.3×10^{-10}
(C) 1.3×10^{-5}
(D) 1.8×10^{-5}
(E) 1.8×10^{-10}

38. Which statement below is inconsistent with the concept of isotopes?

(A) Each element is composed of atoms.
(B) All atoms of an element are identical.
(C) The atoms of different elements have different chemical and physical properties.
(D) The combining of elements leads to the formation of compounds.
(E) In a compound, the kinds and numbers of atoms are constant.

39. Which sample below has its atoms arranged in a regular, geometric pattern?

(A) $NaC_2H_3O_2(s)$
(B) $H_2O(l)$
(C) Ar(g)
(D) NaCl(aq)
(E) $CH_4(g)$

40. Of the statements below, which holds true for the elements found in Na_2HPO_4?

(A) The total molar mass is 71 grams/mole.
(B) The percent by mass of oxygen is 45%.
(C) The percent by mass of sodium is 16%.
(D) The percent by mass of phosphorus is 44%.
(E) The percent by mass of hydrogen is 13%.

41. Carbon and oxygen react to form carbon dioxide according to the reaction: $C(s) + O_2(g) \rightarrow CO_2(g)$. How much carbon dioxide can be formed from the reaction of 36 grams of carbon with 64 grams of oxygen gas?

(A) 36 grams
(B) 64 grams
(C) 28 grams
(D) 132 grams
(E) 88 grams

GO ON TO THE NEXT PAGE

PRACTICE TEST—*Continued*

42. What is the correct mass-action expression for the reaction $2NO(g) + Cl_2(g) \longleftrightarrow 2NOCl(g)$?

    (A) $\dfrac{[NO][Cl_2]}{[NOCl]}$

    (B) $\dfrac{[NOCl]}{[NO][Cl_2]}$

    (C) $\dfrac{[NO]^2[Cl_2]}{[NOCl]^2}$

    (D) $\dfrac{[NOCl]^2}{[NO]^2[Cl_2]}$

    (E) $\dfrac{[NOCl]^2}{[NO]^2 + [Cl_2]}$

43. Which of the following processes will decrease the rate of a chemical reaction?

    I.   Using highly concentrated reactants
    II. Decreasing the temperature by 25 K
    III. Stirring the reactants

    (A) I only
    (B) II only
    (C) I and III only
    (D) II and III only
    (E) I, II, and III

44. Of the substances below, which is best able to conduct electricity?

    (A) $KBr(l)$
    (B) $NaC_2H_3O_2(s)$
    (C) $C_6H_{12}O_6(aq)$
    (D) $CH_3OH(aq)$
    (E) $NaCl(s)$

45. A voltaic cell is set up and a chemical reaction proceeds spontaneously. Which of the following will not occur in this reaction?

    (A) The electrons will migrate through the wire.
    (B) The cations in the salt bridge will migrate to the anode half cell.
    (C) The cathode will gain mass.
    (D) The anode will lose mass.
    (E) Reduction will occur at the cathode.

46. What is the value for ΔH for the reaction: $D + A + B \rightarrow F$?

    $(A + B \rightarrow C \qquad \Delta H = -390 \text{ kJ})$
    $(D + \frac{1}{2}B \rightarrow E \qquad \Delta H = -280 \text{ kJ})$
    $(F + \frac{1}{2}B \rightarrow C + E \qquad \Delta H = -275 \text{ kcal})$

    (A) -165 kJ
    (B) $+385$ kJ
    (C) -395 kJ
    (D) -945 kJ
    (E) $+400$ kJ

47. The oxidation state of the elements in the choices below will be 1– except for

    (A) F in HF
    (B) Cl in NaCl
    (C) O in H_2O_2
    (D) F in NaF
    (E) H in Na_2HPO_4

48. Which substance is not correctly paired with the type of bonding found between the atoms of that substance?

    (A) CH_4—covalent bonds
    (B) CaO—ionic bonds
    (C) Fe—metallic bonds
    (D) H_3O^{1+}—coordinate covalent bonds
    (E) Cl_2—polar covalent bonds

49. Which electron configuration shows that of an excited atom?

    (A) $1s^2 2s^2 2p^6 3s^1$
    (B) $1s^2 2s^2 2p^6 3s^2 3p^6 3d^1$
    (C) $1s^2 2s^2 2p^4$
    (D) $1s^2 2s^2 2p^6 3s^2 3p^6 4s^2$
    (E) $1s^2 2s^2 2p^6 3s^2 3p^3$

50. Given the chemical reaction $3H_2(g) + N_2(g) \longleftrightarrow 2NH_3(g) + \text{energy}$, the forward reaction can best be described as a(n)

    I.   Synthesis reaction
    II. Phase equilibrium
    III. Exothermic reaction

    (A) II only
    (B) I and II only
    (C) I and III only
    (D) II and III only
    (E) I, II, and III

GO ON TO THE NEXT PAGE

PRACTICE TEST—*Continued*

51. Which of the following is not true regarding conjugates and conjugate pairs?

    (A) HF and F^{1-} are conjugate pairs.
    (B) $NaC_2H_3O_2$ and $C_2H_3O_2^{1-}$ are conjugate pairs.
    (C) CO_3^{2-} is the conjugate base of HCO_3^{1-}.
    (D) NH_4^{1+} is the conjugate acid of NH_3.
    (E) A conjugate pair will differ by an H^{1+} ion.

52. What is the ratio of the rate of effusion of hydrogen gas to that of helium gas?

    (A) 1.41
    (B) 2.00
    (C) 4.00
    (D) 0.50
    (E) 1.00

53. Which substance below will exhibit hydrogen bonding between the molecules of the substance?

    (A) CH_4
    (B) HBr
    (C) HCl
    (D) H_2O
    (E) H_2

54. Given the reaction: $N_2(g) + 3H_2(g) \longleftrightarrow 2NH_3(g) + 22$ kcal, what is the value for ΔH for the reverse reaction when 6 moles of NH_3 are consumed to produce nitrogen gas and hydrogen gas?

    (A) +22 kcal
    (B) +66 kcal
    (C) −22 kcal
    (D) −66 kcal
    (E) +33 kcal

55. A titration is set up so that 40.0 mL of 1.0 *M* NaOH are titrated with 2.0 *M* HCl. If the initial reading of the meniscus of the acid's burette is 3.15 mL what will the final burette reading be?

    (A) 20.00 mL
    (B) 40.00 mL
    (C) 43.15 mL
    (D) 23.15 mL
    (E) 13.15 mL

56. Which of the following best describes the orbital overlap in a molecule of

    Figure 2

    I.   s to s
    II. s to p
    III. sp^2 to sp^2

    (A) I only
    (B) II only
    (C) I and III only
    (D) II and III only
    (E) I, II, and III

57. What will be the new freezing point of the water in a solution of 1 *m* NaCl(aq)?

    (A) −1.86°C
    (B) −0.52°C
    (C) −3.72°C
    (D) 1.86°C
    (E) 3.72°C

58. Which metal will not generate hydrogen gas when placed in HCl(aq)?

    (A) Au
    (B) Mg
    (C) Ca
    (D) Sr
    (E) Zn

59. Which substance is the best oxidizing agent?

    (A) Fe
    (B) O_2
    (C) Na
    (D) Li
    (E) F_2

60. Which substance is not correctly paired with the bonding found between the molecules of that substance?

    (A) NH_3—hydrogen bonding
    (B) F_2—Van der Waals (dispersion) forces
    (C) HCl—dipoles
    (D) CH_4—dipoles
    (E) NaCl(aq)—molecule-ion attraction

GO ON TO THE NEXT PAGE

PRACTICE TEST—*Continued*

61. Which solution is not expected to conduct electricity?

    (A) $NaCl(aq)$
    (B) $C_6H_{12}O_6(aq)$
    (C) $KBr(aq)$
    (D) $HC_2H_3O_2(aq)$
    (E) $NaOH(aq)$

62. Which of the following statements about solubility is correct?

    (A) Gases decrease in solubility with an increase in temperature.
    (B) NaCl is insoluble in water.
    (C) PbI_2 is soluble in water.
    (D) All nitrates are insoluble in water.
    (E) Solubility depends solely upon the amount of solvent used.

63. In 6.20 hours, a 50.0-gram sample of ^{112}Ag decays to 12.5 grams. What is the half-life of ^{112}Ag?

    (A) 1.60 hours
    (B) 3.10 hours
    (C) 6.20 hours
    (D) 12.4 hours
    (E) 18.6 hours

64. The modern periodic table is based upon

    (A) atomic mass of the elements
    (B) number of neutrons in the nucleus
    (C) number of isotopes of an element
    (D) oxidation states
    (E) number of protons in the nucleus

65. The prefix *centi-* means

    (A) one thousand
    (B) one thousandth
    (C) one hundred
    (D) one hundredth
    (E) one millionth

66. What is the pH of a 0.1 M acid solution where the acid has a K_a of 1×10^{-5}?

    (A) 3
    (B) 5
    (C) 6
    (D) 4
    (E) 1

67. Which of the following would you not do in a laboratory setting?

    I.   Pour acids and bases over a sink
    II. Wear goggles
    III. Heat a stoppered test tube

    (A) I only
    (B) II only
    (C) III only
    (D) I and III only
    (E) I, II, and III

68. Which of the following statements is not true regarding the kinetic molecular theory?

    (A) The volume that gas molecules occupy is negligible compared to the volume within which the gas is contained.
    (B) There are no forces present between gas molecules.
    (C) Collisions between gas molecules are perfectly elastic.
    (D) Gas molecules travel in a continuous, random motion.
    (E) The average kinetic energy of gas molecules is inversely proportional to temperature.

69. How many times more basic is a solution with a pH of 10 than a solution with a pH of 8?

    (A) A pH of 10 is two times as basic.
    (B) A pH of 8 is two times as basic.
    (C) A pH of 10 is 2,000 times as basic.
    (D) A pH of 8 is 20 times as basic.
    (E) A pH of 10 is 100 times as basic.

70. Which of the following reactions is not labeled correctly?

    (A) $Fe + Cr^{3+} \rightarrow Fe^{3+} + Cr$ (redox reaction)
    (B) $KBr + H_2O \rightarrow HBr + KOH$ (hydrolysis)
    (C) $CH_4 + 2O_2 \rightarrow CO_2 + 2H_2O$ (combustion)
    (D) $CH_4 + Cl_2 \rightarrow CH_3Cl + HCl$ (addition)
    (E) $CO_2 + H_2O \rightarrow H_2CO_3$ (synthesis)

S T O P

IF YOU FINISH BEFORE TIME IS CALLED, GO BACK AND CHECK YOUR WORK.

PRACTICE TEST 2

ANSWERS AND EXPLANATIONS

1.   E   A is the solid phase and B is the liquid phase. This phase change is called melting.

2.   B   Phase C is the gas phase. When a gas becomes a solid with no apparent liquid phase, the process is called deposition.

3.   A   When a solid changes directly to the gas phase, the process is called sublimation.

4.   C   When a liquid becomes a gas, the process is called vaporization (or evaporation).

5.   A   Acids are neutralized by bases. R—COOH indicates a carboxylic acid.

6.   C   The ending -one indicates a ketone. This corresponds to choice C.

7.   E   Amides contain the element nitrogen bonded to a carbonyl group.

8.   B   Aldehydes are written as R—CHO in shorthand structures.

9.   D   88 grams of carbon dioxide (molar mass = 44) is 2.0 moles of this gas.

10.   E   The only choice that matches with 1 molecule of CH_4 is that 1 molecule of CH_4 contains just 5 atoms.

11.   B   32 grams of SO_2 gas at STP (molar mass = 64) is 0.5 moles of SO_2. 0.5 moles of any gas at STP will occupy 11.2 liters.

12.   A   While both group 1 and group 2 metals are reactive with water, group 1 metals tend to be more reactive than the group 2 metals in the same period.

13.   E   Noble gases have stable octet electron configurations and will not react.

14.   E   Noble gases have stable octet electron configurations and will not allow their electrons to be removed easily, giving them a high ionization energy.

15.   B   Group 2 metals have two valence electrons that will be lost in order for the atom to have a stable octet electron configuration.

16.   C   Phenolphthalein is a pinkish color in base.

17.   B   Litmus will be red in acids and blue in bases.

18.   D   Phenolphthalein is colorless in acid.

19.   E   This choice shows the addition of the partial pressures to add up to the total pressure.

20.   A   Boyle's Law shows the inversely proportional relationship between pressure and volume.

21.   C   The ideal gas equation relates pressure, volume, temperature, and moles.

22.   D   The combined gas law relates pressure, temperature, and volume.

PRACTICE TEST—*Continued*

23.	A	Alpha particles have a charge of 2+ because of the two protons in the particle.
24.	C	The gamma ray is radiation and has no mass at all.
25.	A	The alpha particle has a mass of 4 amu because it consists of 2 protons and 2 neutrons.
101.	T, T, CE	The conjugated systems found in aromatic compounds allow those structures to be stabilized by resonance.
102.	T, F	Lithium has a larger atomic radius than fluorine because fluorine has a greater nuclear charge. This allows fluorine to hold electrons closer to the nucleus.
103.	F, F	Calcium chloride yields 3 moles of ions per 1 mole of salt. This will have a greater impact on boiling and freezing points than is found in sodium chloride.
104.	T, F	The nucleons (protons and neutrons) are located in the nucleus of an atom. An atom's electrons are located in the principal energy levels.
105.	T, F	HCl is an acid because it yields H^{1+} ions and donates a proton. Bases are proton acceptors.
106.	T, T, CE	A powdered substance has a greater surface area and will have a greater reaction rate.
107.	F, T	Ammonia is polar and will dissolve easily in water. This is not an efficient way of collecting ammonia.
108.	F, T	Pure water will be a solid at temperatures below 273 K. Water will boil when the vapor pressure of the water is equal to the surrounding pressure.
109.	T, T, CE	Because the potential energy of the products is less than that of the reactants, when calculating the change in heat the sign for ΔH will be negative.
110.	T, F	As pressure increases on a gas, the volume of the gas decreases. This is an inverse relationship.
111.	F, T	Ethane is a saturated hydrocarbon and cannot bond to any more hydrogen atoms.
112.	T, T, CE	Silver chloride is insoluble in water as are $PbCl_2$ and Hg_2Cl_2.
113.	T, F	Vaporization causes molecules to spread out. This causes more entropy and less order.
114.	T, T, CE	On the pH scale a pH of 7 is neutral. This is because the concentration of H^{1+} ions is equal to the number of OH^- ions.
115.	T, T, CE	Isomers are defined as having the same molecular formula but a different structure. Both compounds have the molecular formula of C_2H_6O, but one is an alcohol and one is an ether.

PRACTICE TEST—*Continued*

26.   C   Magnesium chloride is a soluble salt that will yield 3 moles of ions upon dissolving. Choices A and B will yield only 2 moles of ions. Choices D and E are covalently bonded and will not yield ions at all.

27.   D   Choice I is properly balanced. Choice II should read $2Na + O_2 \rightarrow 2Na_2O$; the coefficient is missing before the sodium oxide. Choice III is properly balanced.

28.   A   If two moles of propane are combusted, then according to the balanced equation, 8 moles of water should be produced.

$$\frac{C_3H_8(g)}{2C_3H_8(g)} = \frac{4H_2O(g)}{8H_2O(g)}$$

8.0 moles of water are then multiplied by the molar mass of water, 18, to get 144 grams of water.

29.   B   First, assume a 100-gram sample. This allows you to change the percent signs to grams and get 75 grams of C and 25 grams of H. Then convert grams to moles to get

75 grams of C divided by 12 gives 6.25 moles of C

25 grams of H divided by 1 gives 25 moles of H

Dividing by the lowest factor of 6.25 gives

C_1H_4 or CH_4.

30.   D   Choices A and B will be tetrahedral in shape while choices C and E will be linear in shape. H_2S has an electron pair geometry that is tetrahedral but with only two atoms of H bonded to the tetrahedron, the geometry of the molecule will be bent.

31.   D   The meniscus of a liquid is the curve of a liquid that is formed due to certain forces that exist. When reading a burette, the meniscus is of utmost importance in determining volume.

32.   E   The addition of a product to a system at equilibrium will shift the equilibrium so that the excess product is consumed. This will produce more reactant.

33.   B   The mass number minus the atomic number gives the number of neutrons in an isotope. The mass number of C-13 is 13 and the atomic number for C is 6. This indicates that there are 7 neutrons in the nucleus of this isotope.

34.   A   Group 1 and group 2 metals are highly reactive. When placed in water they will react to form a base and hydrogen gas.

35.   C   Heat energy is measured in joules or calories. KiloPascals are a unit of pressure.

36.   B   Molarity is moles of solute/liters of solution. To get a 1.0 M solution it takes 1 mole of solute per 1 liter of solution. One mole of lead nitrate (molar mass = 331) weighs 331 grams.

PRACTICE TEST—*Continued*

37.   E   AgCl dissolves according to the equation $AgCl(s) \longleftrightarrow Ag^{1+}(aq) + Cl^{1-}(aq)$.

The $K_{sp} = [Ag^{1+}][Cl^{1-}]$ and for each silver ion produced, one chloride ion is produced. This means that the $K_{sp} = [1.3 \times 10^{-5}M][1.3 \times 10^{-5}M]$ and $K_{sp} = 1.8 \times 10^{-10}$.

38.   B   All atoms of an element are not identical because isotopes have a different number of neutrons.

39.   A   A regular, geometric pattern describes the arrangement of atoms in a solid.

40.   B   The molar mass of this compound is 142 grams/mole. The oxygen atoms make up 64 of the 142 grams. This comes to about 45% of the total mass.

41.   E   36 grams of C indicates that 3 moles of C are being reacted. 64 grams of O_2 indicates that 2 moles of O_2 are being reacted. Because C and O_2 react in a 1:1 ratio, the reaction will be limited by the amount of O_2. The maximum amount of CO_2 that can be produced is 2 moles. 2 moles of CO_2 (molar mass = 44) has a mass of just 88 grams.

42.   D   Remember the mnemonic device, "Products over reactants, coefficients become powers."

43.   B   At lower temperatures the molecules have less kinetic energy and will have a lower rate of reaction. The molecules will collide less frequently and not as effectively.

44.   A   KBr is an ionic compound. In the liquid or aqueous state the ions are mobile and free to carry an electrical current. Choices B and E are ionic compounds but the ions are not mobile. Choices C and D are in solution but they are covalent compounds and will not form ions.

45.   B   Cations in the salt bridge will migrate to the cathode half cell of the voltaic cell.

46.   C   Leave the first reaction as is because A and B need to appear on the left:

$A + B \rightarrow C \qquad \Delta H = -390$ kJ

Leave the second reaction as is because D needs to appear on the left:

$D + \frac{1}{2}B \rightarrow E \qquad \Delta H = -280$ kJ

Switch the third reaction so that B, C, and E can be canceled:

$C + E \rightarrow F + \frac{1}{2}B \qquad \Delta H = +275$ kcal

Add up the equations to get −395 kJ

47.   E   H could have a 1− oxidation state when bonded in a metal hydride like NaH. In Na_2HPO_4 the H has an oxidation state equal to 1+.

48.   E   Because the two chlorine atoms have the same electronegativity, there will be an equal sharing of electrons. This will make the bond nonpolar covalent.

PRACTICE TEST—*Continued*

49.   B   After the 3p orbitals fill up with six electrons, the 4s orbital fills next. After that the 3d orbital is filled.

50.   C   The forward reaction in this case is exothermic because of the release of heat energy. Because one substance is formed from two substances, this reaction is also classified as a synthesis.

51.   B   Conjugate pairs differ by an H^{1+} ion. $NaC_2H_3O_2$ and $C_2H_3O_2^{1-}$ differ by a sodium ion.

52.   A   The rate of hydrogen gas to helium gas can be set up as follows:

$$\frac{r_{H2}}{r_{He}} = \sqrt{\frac{M_{He}}{M_{H2}}}$$

This becomes:

$$\frac{r_{H2}}{r_{He}} = \sqrt{\frac{4}{2}} = \sqrt{2} = 1.41$$

53.   D   Remember the mnemonic device, "I heard about hydrogen bonding on the FON." That is, when hydrogen is bonded to fluorine, oxygen, or nitrogen, the conditions for hydrogen bonding exist.

54.   B   The reverse reaction is an endothermic process, so the sign for ∆H must be positive. Because the amount of the reactants and products has been increased threefold from 2 moles of ammonia to 6 moles of ammonia, the amount of heat in the reaction must increase threefold as well to 66 kcal.

55.   D   In this titration 40.0 mL is the volume of the 1.0 M base. The acid is 2.0 M, but you don't know how much was used. First set up the equation: $M_aV_a = M_bV_b$ and substitute $(2.0\ M)(V_a) = (1.0\ M)(40.0\ mL)$. Solving tells that the volume of acid used is 20.0 mL. Because the initial reading on the burette was 3.15 mL and 20.0 mL of acid was used, the final reading on the burette containing acid is 23.15 mL.

56.   D   In this molecule the carbon atoms are sp^2 hybridized. This means that there is sp^2 bonding between the two carbon atoms. The s orbital from the hydrogen atoms will bond with the p orbitals from the carbon atoms as well. There will be no s to s orbital overlap in this case.

57.   C   1 m NaCl(aq) will act as if it were 2 molal in solution because each mole of NaCl yields 2 moles of ions. The freezing point depression constant for water is 1.86°C/1 m. This means that there is a change of 3.72°C. Because the initial freezing point of water is 0°C, the new freezing point is 3.72°C lower, or –3.72°C.

58.   A   Gold is a precious metal because it rarely reacts. This is why we find gold as a solid nugget and not as a salt that has ions that have undergone a reaction.

PRACTICE TEST—*Continued*

59. E An oxidizing agent is a substance that was reduced. This means that the substance had to gain electrons. With the highest possible electronegativity value of 4.0, fluorine is the best "gainer" of electrons.

60. D Methane, because of its tetrahedral shape, is a nonpolar molecule. Nonpolar molecules will not exhibit dipole attractions.

61. B Glucose is a covalently bonded substance and will not yield the ions needed to carry a current in solution. All the other choices will have mobile ions that can conduct electricity.

62. A Gases generally become less soluble in water with an increase in temperature.

63. B To decay from 50.0 grams to 12.5 grams, this isotope must first undergo one half-life to bring the sample down to 25.0 grams. Decaying from 50.0 grams to 25.0 grams and then to 12.5 grams means that there were two half-lives. Each one would have to take 3.10 hours for there to be a total of 6.20 hours. (See Figure 3.)

$$50.0 \text{ Grams} \xrightarrow{\;3.10 \text{ h}\;} 25.0 \text{ Grams} \xrightarrow{\;3.10 \text{ h}\;} 12.5 \text{ Grams}$$

6.20 Hours Total

Figure 3

64. E The modern periodic table is based upon atomic number, which is defined by the number of protons in the nucleus.

65. D *Centi-* means "one hundredth." For example, a centimeter is one hundredth of a meter.

66. A Start with the expression:

$$Ka = \frac{\left[H^{1+}\right]\left[X^{1-}\right]}{\left[HX\right]} \text{ and then substitute: } 1 \times 10^{-5} = \frac{\left[X\right]\left[X\right]}{\left[0.1 \, M\right]} \text{ and get } 1 \times 10^{-6} = x^2$$

The value of x is (or $[H^{1+}]$) is $1 \times 10^{-3} \, M$. Now that the $[H^{1+}]$ has been obtained, substitute into the equation, $pH = -\log[H^{1+}]$ and get: $pH = -\log[1 \times 10^{-3} \, M]$. The pH will be 3.

67. C Stoppered flasks and test tubes should never be heated. The pressure inside the glassware could forcefully eject the stopper.

68. E As temperature increases, so does the average kinetic energy of the sample. This demonstrates a direct relationship.

69. E The pH is based upon logarithms and powers of 10. The difference between a pH of 10 and a pH of 8 is a 100-fold difference in the amount of OH^{1-} ions.

70. D This organic reaction shows a substitution reaction and not an addition reaction.

PRACTICE TEST 2

SCORE SHEET

Number of questions correct: _____

Less: 0.25 × number of questions wrong: _____

(Remember that omitted questions are not counted as wrong.)

Equals your raw score: _____

Raw Score	Test Score		Raw Score	Test Score		Raw Score	Test Score		Raw Score	Test Score		Raw Score	Test Score
85	800		63	710		41	570		19	440		−3	300
84	800		62	700		40	560		18	430		−4	300
83	800		61	700		39	560		17	430		−5	290
82	800		60	690		38	550		16	420		−6	290
81	800		59	680		37	550		15	420		−7	280
80	800		58	670		36	540		14	410		−8	270
79	790		57	670		35	530		13	400		−9	270
78	790		56	660		34	530		12	400		−10	260
77	790		55	650		33	520		11	390		−11	250
76	780		54	640		32	520		10	390		−12	250
75	780		53	640		31	510		9	380		−13	240
74	770		52	630		30	500		8	370		−14	240
73	760		51	630		29	500		7	360		−15	230
72	760		50	620		28	490		6	360		−16	230
71	750		49	610		27	480		5	350		−17	220
70	740		48	610		26	480		4	350		−18	220
69	740		47	600		25	470		3	340		−19	210
68	730		46	600		24	470		2	330		−20	210
67	730		45	590		23	460		1	330		−21	200
66	720		44	580		22	460		0	320			
65	720		43	580		21	450		−1	320			
64	710		42	570		20	440		−2	310			

Note: This is only a sample scoring scale. Scoring scales differ from exam to exam.

PRACTICE TEST 3

ANSWER SHEET

This practice test will help measure your current knowledge of chemistry and familiarize you with the structure of the *SAT Chemistry* test. To simulate exam conditions, use the answer sheet below to record your answers and the appendix tables at the back of this book to obtain necessary information.

1. (A) (B) (C) (D) (E)	21. (A) (B) (C) (D) (E)	41. (A) (B) (C) (D) (E)	61. (A) (B) (C) (D) (E)
2. (A) (B) (C) (D) (E)	22. (A) (B) (C) (D) (E)	42. (A) (B) (C) (D) (E)	62. (A) (B) (C) (D) (E)
3. (A) (B) (C) (D) (E)	23. (A) (B) (C) (D) (E)	43. (A) (B) (C) (D) (E)	63. (A) (B) (C) (D) (E)
4. (A) (B) (C) (D) (E)	24. (A) (B) (C) (D) (E)	44. (A) (B) (C) (D) (E)	64. (A) (B) (C) (D) (E)
5. (A) (B) (C) (D) (E)	25. (A) (B) (C) (D) (E)	45. (A) (B) (C) (D) (E)	65. (A) (B) (C) (D) (E)
6. (A) (B) (C) (D) (E)	26. (A) (B) (C) (D) (E)	46. (A) (B) (C) (D) (E)	66. (A) (B) (C) (D) (E)
7. (A) (B) (C) (D) (E)	27. (A) (B) (C) (D) (E)	47. (A) (B) (C) (D) (E)	67. (A) (B) (C) (D) (E)
8. (A) (B) (C) (D) (E)	28. (A) (B) (C) (D) (E)	48. (A) (B) (C) (D) (E)	68. (A) (B) (C) (D) (E)
9. (A) (B) (C) (D) (E)	29. (A) (B) (C) (D) (E)	49. (A) (B) (C) (D) (E)	69. (A) (B) (C) (D) (E)
10. (A) (B) (C) (D) (E)	30. (A) (B) (C) (D) (E)	50. (A) (B) (C) (D) (E)	70. (A) (B) (C) (D) (E)
11. (A) (B) (C) (D) (E)	31. (A) (B) (C) (D) (E)	51. (A) (B) (C) (D) (E)	71. (A) (B) (C) (D) (E)
12. (A) (B) (C) (D) (E)	32. (A) (B) (C) (D) (E)	52. (A) (B) (C) (D) (E)	72. (A) (B) (C) (D) (E)
13. (A) (B) (C) (D) (E)	33. (A) (B) (C) (D) (E)	53. (A) (B) (C) (D) (E)	73. (A) (B) (C) (D) (E)
14. (A) (B) (C) (D) (E)	34. (A) (B) (C) (D) (E)	54. (A) (B) (C) (D) (E)	74. (A) (B) (C) (D) (E)
15. (A) (B) (C) (D) (E)	35. (A) (B) (C) (D) (E)	55. (A) (B) (C) (D) (E)	75. (A) (B) (C) (D) (E)
16. (A) (B) (C) (D) (E)	36. (A) (B) (C) (D) (E)	56. (A) (B) (C) (D) (E)	
17. (A) (B) (C) (D) (E)	37. (A) (B) (C) (D) (E)	57. (A) (B) (C) (D) (E)	
18. (A) (B) (C) (D) (E)	38. (A) (B) (C) (D) (E)	58. (A) (B) (C) (D) (E)	
19. (A) (B) (C) (D) (E)	39. (A) (B) (C) (D) (E)	59. (A) (B) (C) (D) (E)	
20. (A) (B) (C) (D) (E)	40. (A) (B) (C) (D) (E)	60. (A) (B) (C) (D) (E)	

Chemistry   *Fill in oval CE only if II is correct explanation of I.

	I	II	CE*		I	II	CE*
101.	(T) (F)	(T) (F)	◯	109.	(T) (F)	(T) (F)	◯
102.	(T) (F)	(T) (F)	◯	110.	(T) (F)	(T) (F)	◯
103.	(T) (F)	(T) (F)	◯	111.	(T) (F)	(T) (F)	◯
104.	(T) (F)	(T) (F)	◯	112.	(T) (F)	(T) (F)	◯
105.	(T) (F)	(T) (F)	◯	113.	(T) (F)	(T) (F)	◯
106.	(T) (F)	(T) (F)	◯	114.	(T) (F)	(T) (F)	◯
107.	(T) (F)	(T) (F)	◯	115.	(T) (F)	(T) (F)	◯
108.	(T) (F)	(T) (F)	◯				

GO ON TO THE NEXT PAGE ➤

PRACTICE TEST 3

Time: 60 minutes

<u>Note:</u> Unless otherwise stated, for all statements involving chemical equations and/or solutions, assume that the system is in pure water.

Part A

<u>Directions:</u> Each of the following sets of lettered choices refers to the numbered formulas or statements immediately below it. For each numbered item, choose the one lettered choice that fits it best. Then fill in the corresponding oval on the answer sheet. Each choice in a set may be used once, more than once, or not at all.

<u>Questions 1–4</u>

(A)

(B)

(C)

(D)

(E)

Figure 1

1. Demonstrates the relationship between pressure (*x*-axis) and volume (*y*-axis) in Boyle's Law

2. Contains a triple point

3. Demonstrates the relationship between temperature (*x*-axis) and volume (*y*-axis) in Charles' Law

4. Shows the relationship between atomic number (*x*-axis) and atomic radius (*y*-axis) for the elements in period 2

GO ON TO THE NEXT PAGE

PRACTICE TEST—*Continued*

Questions 5–8

    (A) Br_2 and Hg
    (B) Cl_2 and F_2
    (C) NH_4^{1+} and H_3O^{1+}
    (D) Fe and Co
    (E) Diamond and graphite

5. These two compounds are in the liquid phase at 293 K.

6. These two compounds have coordinate covalent bonds.

7. These two compounds are allotropes of each other.

8. These two compounds are good oxidizing agents.

Questions 9–11

    (A) R—OH
    (B) R—O—R
    (C) R—NH_2
    (D) R—COO—R
    (E) R—CO—R

9. Ends in *-oate*

10. Ends in *-amine*

11. Ends in *-ol*

Questions 12–15

    (A) $1s^2 2s^2 2p^6 3s^2 3p^6$
    (B) $1s^2 2s^2 2p^6 3s^2 3p^6 4s^2$
    (C) $1s^2 2s^2 2p^6 3s^2 3p^6 4s^1$
    (D) $1s^2$
    (E) $1s^2 2s^2 2p^6 3p^1$

12. The electron configuration for calcium ion

13. The electron configuration for an excited atom

14. The electron configuration for potassium in the ground state

15. The electron configuration for the noble gas with the highest first ionization energy

Questions 16–19

    (A) Sublimation
    (B) Deposition
    (C) Vaporization
    (D) Condensation
    (E) Freezing

16. Solid to gas

17. Gas to solid

18. Liquid to gas

19. Liquid to solid

Questions 20–22

    (A) Nitrogen
    (B) Oxygen
    (C) Chlorine
    (D) Neon
    (E) Beryllium

20. Has 2 valence electrons

21. Has 6 valence electrons

22. Will form an ion with a 3– charge

Questions 23–25

    (A) milli-
    (B) kilo-
    (C) centi
    (D) micro-
    (E) nano-

23. 10^{-9}

24. 10^{-6}

25. 10^3

GO ON TO THE NEXT PAGE

PRACTICE TEST—*Continued*

PLEASE GO TO THE SPECIAL SECTION AT THE LOWER LEFT-HAND CORNER OF YOUR
ANSWER SHEET LABELED "CHEMISTRY" AND ANSWER QUESTIONS 101–115
ACCORDING TO THE FOLLOWING DIRECTIONS.

Part B

Directions: Each question below consists of two statements. For each question, determine whether statement I in the left-most column is true or false and whether statement II in the rightmost column is true or false. Fill in the corresponding T or F ovals on the answer sheet provided. Fill in the oval labeled "CE" only if statement II correctly explains statement I.

SAMPLE:

EX 1. The nucleus of an atom has a   BECAUSE   the only positive particles
       positive charge   found in the atom's nucleus are
                                                          protons.

SAMPLE ANSWER

	I	II	CE*
EX 1	● Ⓕ	● Ⓕ	●

I		II
101. Alpha particles are able to pass through a thin sheet of gold foil	BECAUSE	the atom is mainly empty space.
102. Nitrogen has five valence electrons	BECAUSE	the electron configuration for nitrogen is $1s^2 2s^2 2p^6$.
103. A molecule of ethyne is linear	BECAUSE	the carbon atoms in ethyne are sp hybridized.
104. KNO_3 will not dissolve in water	BECAUSE	all chlorides are soluble in water.
105. CCl_4 is a polar molecule	BECAUSE	the dipole arrows for CCl_4 show counterbalance and symmetry.
106. According to the equation $M_1V_1 = M_2V_2$, as the volume increases the molarity decreases	BECAUSE	as water is added to a solution the solution is diluted.
107. HCl is an Arrhenius acid	BECAUSE	HCl will yield hydronium ions as the only positive ions in solution.

GO ON TO THE NEXT PAGE

PRACTICE TEST—*Continued*

108. Adding more reactants will speed up a   BECAUSE   the reactants will collide less frequently.
      reaction

109. $Al^{3+} + 3e^- \rightarrow Al$ is a correctly balanced   BECAUSE $Al^{3+} + 3e^- \rightarrow Al$ correctly demonstrates conserva-
      oxidation reaction   tion of mass and conservation of charge.

110. ^4_2He is the correct symbol for an   BECAUSE   an alpha particle is a helium-3 nucleus.
      alpha particle

111. Fluorine has the highest value for   BECAUSE   fluorine has the greatest attraction for electrons.
      electronegativity

112. The number 5,007 has three significant   BECAUSE   zeros between non-zero digits are significant.
      figures

113. DNA is a polymer   BECAUSE   DNA has many smaller units bonded to create
                                                                    longer chains.

114. Radiation and radioisotopes can have   BECAUSE   radioisotopes and radiation can be used for radio
      beneficial uses   dating, radiotracers, and food preservation.

115. A $1m$ NaCl(aq) solution will freeze at a   BECAUSE   as a solute is added to a solvent, the boiling point
      temperature below 273 K   increases while the freezing point decreases.

GO ON TO THE NEXT PAGE ➡

PRACTICE TEST—*Continued*

Part C

> Directions: Each of the multiple-choice questions or incomplete sentences below is followed by five answers or completions. Select the one answer that is best in each case and then fill in the corresponding oval on the answer sheet provided.

26. When chlorine gas and hydrogen gas react to form hydrogen chloride, what will be the change of enthalpy of the reaction? (Bond dissociation energies can be found in Appendix 4 at the back of the book.)

   (A) +245 kJ/mol
   (B) +185 kJ/mol
   (C) −185 kJ/mol
   (D) −1105 kJ/mol
   (E) +1105 kJ/mol

27. How much heat is required to raise the temperature of 85 grams of water from 280 K to 342 K?

   (A) 5,270 J
   (B) 355 J
   (C) 259 J
   (D) 151 J
   (E) 22,029 J

28. Which of the following is not part of the Atomic Theory?

   (A) Compounds are made up of combinations of atoms.
   (B) All atoms of a given element are alike.
   (C) All matter is composed of atoms.
   (D) A chemical reaction involves the rearrangement of atoms.
   (E) The atom is mainly empty space.

29. Which of the following compounds will have an atom with a molecular geometry that is described as trigonal planar with respect to other atoms present?

   I. BF_3
   II. $CH_2=CH_2$
   III. Cyclopropane

   (A) I only
   (B) II only
   (C) III only
   (D) I and II only
   (E) I, II, and III

30. Which of the following transmutations demonstrate(s) beta decay?

   I.   Bi-212 → Po-212
   II.   Pb-212 → Bi-212
   III. Ra-228 → Ac-228

   (A) I only
   (B) II only
   (C) II and III only
   (D) I and II only
   (E) I, II, and III

31. A liquid will boil when

   (A) the liquid is hot
   (B) a salt has been added to the liquid
   (C) the vapor pressure of the liquid is equal to the surrounding pressure
   (D) the vapor pressure is reduced
   (E) the surrounding pressure is increased

32. Which sample has atoms that are arranged in a regular geometric pattern?

   (A) KCl(l)
   (B) $NaC_2H_3O_2$(s)
   (C) Fe(l)
   (D) NaCl(aq)
   (E) HCl(aq)

33. Which aqueous solution has a molarity of 1.0 *M*?

   (A) 73 grams of HCl dissolved to make 2.0 liters of solution
   (B) 360 grams of $C_6H_{12}O_6$ dissolved to make 1.5 liters of solution
   (C) 94 grams of K_2O dissolved to make 0.75 liters of solution
   (D) 24 grams of LiOH dissolved to make 1.25 liters of solution
   (E) 40 grams of HF dissolved to make 2.50 liters of solution

GO ON TO THE NEXT PAGE

PRACTICE TEST—*Continued*

34. Which double replacement reaction forms an insoluble precipitate?

    (A) $HCl(aq) + KOH(aq) \rightarrow$
    (B) $KNO_3(aq) + Na_2SO_4(aq) \rightarrow$
    (C) $NaCl(aq) + CaCl_2(aq) \rightarrow$
    (D) $AgNO_3(aq) + KCl(aq) \rightarrow$
    (E) $KBr(aq) + H_2O(aq) \rightarrow$

35. Of the following solutions, which one is expected to be the weakest electrolyte?

    (A) $HCl(aq)$
    (B) $HF(aq)$
    (C) $NaOH(aq)$
    (D) $KI(aq)$
    (E) $HClO_4(aq)$

36. Which of the following indicate(s) a basic solution?

    I.   Litmus paper turns blue.
    II. Phenolphthalein turns pink.
    III. Hydronium ion concentration is greater than hydroxide ion concentration.

    (A) I only
    (B) II only
    (C) III only
    (D) I and II only
    (E) I, II, and III

37. Which of the following half reactions is correctly balanced?

    (A) $MnO_4{}^{1-} \rightarrow Mn^{2+} + 4H_2O$
    (B) $Cu + 2Ag^{1+} \rightarrow 2Ag + Cu^{2+}$
    (C) $H_2 + OH^{1-} \rightarrow 2H_2O$
    (D) $Pb^{2+} + 2e^- \rightarrow Pb$
    (E) $2F^{1-} + 2e^- \rightarrow F_2$

38. The quantity "one mole" will not be equal to

    (A) 22.4 L of $H_2(g)$ at STP
    (B) 6.02×10^{23} carbon atoms
    (C) 64 grams of $SO_2(g)$
    (D) 36 grams of H_2O
    (E) 207 grams of Pb

39. Which statement below is false regarding empirical formulas?

    (A) The empirical formula for butyne is C_2H_3.
    (B) The empirical formula for ammonia is NH_3.
    (C) The empirical formula of CH_2O is $C_6H_{12}O_6$.
    (D) Ionic compounds are written as empirical formulas.
    (E) The empirical and molecular formulas for methane are the same.

40. The percent composition by mass of oxygen in $BaSO_4$ is

    (A) 233.4%
    (B) 66.7%
    (C) 27.4%
    (D) 58.7%
    (E) 13.7%

41. How many grams of Fe_2O_3 can be formed from the rusting of 446 grams of Fe according to the reaction: $4Fe + 3O_2 \rightarrow 2Fe_2O_3$? (Assume an unlimited amount of oxygen gas.)

    (A) 320 grams
    (B) 223 grams
    (C) 159 grams
    (D) 480 grams
    (E) 640 grams

42. Sodium and chlorine react according to the following reaction: $2Na + Cl_2 \rightarrow 2NaCl$. If the reaction starts with 5.0 moles of Na and 3.0 moles of Cl_2 then which statement below is true?

    (A) Cl_2 is the excess reagent and 5.0 moles of NaCl will be produced.
    (B) Na is the excess reagent and 2.5 moles of NaCl will be produced.
    (C) There will be an excess of 2.0 moles of Na.
    (D) Na is the limiting reagent and 2.0 moles of NaCl will be produced.
    (E) Cl is the excess reagent and 2.0 moles of NaCl will be produced.

GO ON TO THE NEXT PAGE ➤

PRACTICE TEST—*Continued*

43. The equilibrium constant expression for the reaction: $2A(g) + B(g) \longleftrightarrow 3C(s) + 2D(g)$ is written as

(A) $K_{eq} = \dfrac{[A]^2[B]}{[C]^3[D]^2}$

(B) $K_{eq} = \dfrac{[C]^3[D]^2}{[A]^2[B]^2}$

(C) $K_{eq} = \dfrac{[D]^2}{[A]^2[B]}$

(D) $K_{eq} = \dfrac{[A]^2[B]}{[D]^2}$

(E) $K_{eq} = \dfrac{[C][D]}{[A][B]}$

44. Given the reaction: $3H_2(g) + N_2(g) \longleftrightarrow 2NH_3(g) +$ heat energy. Which of the following would drive the equilibrium in the direction opposite to that of the other four choices?

(A) Remove ammonia from the reaction.
(B) Increase the temperature of the system.
(C) Increase the pressure on the system.
(D) Add nitrogen gas.
(E) Add hydrogen gas.

45. Which of the following demonstrate(s) $\Delta S(-)$?

I. Raking up leaves
II. Boiling a liquid
III. Emptying a box of confetti onto the floor

(A) I only
(B) II only
(C) I and II only
(D) I and III only
(E) II and III only

46. In which of the following pieces of glassware does a meniscus become of importance?

(A) Watchglass
(B) Burette
(C) Beaker
(D) Flask
(E) Funnel

47. If the pressure on a gas is doubled, the volume of the gas will be

(A) doubled
(B) the same
(C) halved
(D) quartered
(E) quadrupled

48. Which of the following statements about gas collection is false?

(A) Carbon dioxide can be collected by an upward displacement of air.
(B) Ammonia can be tested for by placing red litmus paper at the mouth of the collection glassware.
(C) Ammonia can be collected by water displacement.
(D) Hydrogen gas can be collected by water displacement.
(E) Carbon dioxide can be tested for with a lit match.

49. In the diagram shown below, which letter represents the potential energy of the products minus the potential energy of the reactants? (See Figure 2.)

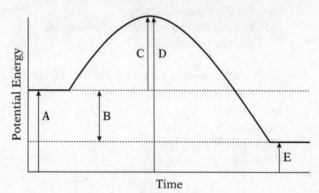

Figure 2

(A) B
(B) D
(C) A
(D) E
(E) C

GO ON TO THE NEXT PAGE

PRACTICE TEST—*Continued*

50. What is the heat of reaction for $A + B \rightarrow F$?

    $(A + B \rightarrow 2C \qquad \Delta H = +150$ kcal)
    $(C \rightarrow 2D + 2E \qquad \Delta H = -450$ kcal)
    $(F \rightarrow 4D + 4E \qquad \Delta H = +725$ kcal)

    (A)   −1475 kcal
    (B)   +25 kcal
    (C)   −1025 kcal
    (D)   +325 kcal
    (E)   +300 kcal

51. Molten KBr is allowed to undergo the process of electrolysis. Which reaction occurs at the anode?

    (A) $K^{1+} + 1e^- \rightarrow K(s)$
    (B) $2Br^{1-} \rightarrow Br_2 + 2e^-$
    (C) $K(s) \rightarrow K^{1+} + 1e^-$
    (D) $Br_2 \rightarrow 2Br^{1-} + 2e^-$
    (E) $Br_2 + 2e^- \rightarrow 2Br^{1-}$

52. Which will not happen when sodium sulfate is added to a saturated solution of $PbSO_4$ that is at equilibrium? $[PbSO_4(s) \longleftrightarrow Pb^{2+}(aq) + SO_4^{2-}(aq)]$

    (A)   The solubility of the lead sulfate will decrease.
    (B)   The concentration of lead ions will decrease.
    (C)   The reaction will shift to the left.
    (D)   The K_{sp} value will change.
    (E)   The equilibrium will shift to consume the increase in sulfate ions.

53. What is the voltage of the voltaic cell $Cu / Cu^{2+} // Zn / Zn^{2+}$ at 298 K if $[Zn^{2+}] = 0.2$ M and $[Cu^{2+}] = 4.0$ M?

    (A)   +1.10 V
    (B)   −1.10 V
    (C)   +1.07 V
    (D)   +1.14 V
    (E)   −1.07 V

54. Which of the following molecules has polar bonds but is a nonpolar molecule?

    (A) H_2
    (B) H_2O
    (C) NH_3
    (D)   NaCl
    (E) CO_2

55. A titration is set up so that 35.0 mL of 1.0 M NaOH are titrated with 1.5 M HCl. How many milliliters of acid are needed to completely titrate this amount of base?

    (A)   15.00 mL
    (B)   35.00 mL
    (C)   23.33 mL
    (D)   58.33 mL
    (E)   20.00 mL

56. Which statement is inconsistent with the concept of isotopes of the same element?

    (A)   Isotopes have the same number of protons.
    (B)   Isotopes have the same atomic number.
    (C)   Isotopes differ in mass number.
    (D)   Isotopes differ in number of neutrons present.
    (E)   Isotopes differ in their nuclear charge.

57. Which of the following pairs of substances can be broken down chemically?

    (A)   Ammonia and iron
    (B)   Helium and argon
    (C)   Methane and water
    (D)   Potassium and lithium
    (E)   Water and carbon

58. What is the volume of 2.3 moles of an ideal gas at 300 K and a pressure of 1.1 atmospheres?

    (A) $\dfrac{(2.3 \text{ moles})(0.0820 \text{ L·atm/mol·K})(300 \text{ K})}{(1.1 \text{ atm})}$

    (B) $\dfrac{(1.1 \text{ atm})}{(2.3 \text{ moles})(0.0820 \text{ L·atm/mol·K})(300 \text{ K})}$

    (C) $\dfrac{(2.3 \text{ moles})(0.0820 \text{ L·atm/mol·K})}{(300 \text{ K})(1.1 \text{ atm})}$

    (D) $\dfrac{(300 \text{ K})(0.0820 \text{ L·atm/mol·K})}{(2.3 \text{ moles})(1.1 \text{ atm})}$

    (E) $\dfrac{(2.3 \text{ moles})(1.1 \text{ atm})(300 \text{ K})}{(0.0820 \text{ L·atm/mol·K})}$

GO ON TO THE NEXT PAGE

PRACTICE TEST—*Continued*

59. Substance X has three common isotopes: X-48, X-49, and X-51. If the relative abundances of these isotopes are 42%, 38%, and 20%, respectively, what is the atomic mass of substance X?

    (A) 49.33
    (B) 48.62
    (C) 50.67
    (D) 48.98
    (E) 49.67

60. Which choice below would affect the rate of reaction in the opposite way from the other four?

    (A) Cool the reaction down.
    (B) Add a catalyst.
    (C) Decrease the pressure.
    (D) Use larger pieces of solid reactants.
    (E) Decrease the concentration of the reactants.

61. One mole of an ideal gas at STP has its temperature changed to 15°C and its pressure changed to 700 torr. What is the new volume of this gas?

    (A) $\dfrac{(760 \text{ torr})(22.4 \text{ L})(288 \text{ K})}{(273 \text{ K})(700 \text{ torr})}$

    (B) $\dfrac{(273 \text{ K})(700 \text{ torr})}{(760 \text{ torr})(22.4 \text{ L})(288 \text{ K})}$

    (C) $\dfrac{(760 \text{ torr})(22.4 \text{ L})(273 \text{ K})}{(288 \text{ K})(700 \text{ torr})}$

    (D) $\dfrac{(700 \text{ torr})(22.4 \text{ L})(287 \text{ K})}{(273 \text{ K})(760 \text{ torr})}$

    (E) $\dfrac{(760 \text{ torr})(1.0 \text{ L})(288 \text{ K})}{(273 \text{ K})(700 \text{ torr})}$

62. Which reaction will not occur spontaneously?

    (A) $Au^{3+} + 3e^- \rightarrow Au$
    (B) $Mg + 2H^{1+} \rightarrow Mg^{2+} + H_2$
    (C) $F_2 + 2e^- \rightarrow 2F^{1-}$
    (D) $Li^{1+} + 1e^- \rightarrow Li$
    (E) $2Na + Cl_2 + 2e^- \rightarrow 2NaCl$

63. As you go from left to right across a period on the periodic table there in a decrease in

    (A) first ionization energy
    (B) nuclear charge
    (C) electronegativity
    (D) the ability to gain electrons
    (E) metallic character

64. Which of the following statements is false?

    (A) H_2 has just one sigma bond.
    (B) HCl has just one sigma bond.
    (C) H—C≡C—H has four pi bonds and three sigma bonds.
    (D) CH_2=CH_2 has five sigma bonds and one pi bond.
    (E) H_2O has two sigma bonds and two lone pairs.

65. What is the correct formula for iron(III) sulfate?

    (A) $FeSO_4$
    (B) $Fe_2(SO_4)_3$
    (C) $Fe(SO_4)_3$
    (D) Fe_3SO_4
    (E) $Fe_3(SO_4)_2$

66. A solution has a pH of 6.0. What is the concentration of OH^{1-} ions in solution?

    (A) $6.0 \times 10^{-14} M$
    (B) $1.0 \times 10^{-6} M$
    (C) $1.0 \times 10^{-14} M$
    (D) $6.0 \times 10^{-8} M$
    (E) $1.0 \times 10^{-8} M$

67. Which of the following statements about bonding is correct?

    (A) Van der Waals forces exist between polar molecules.
    (B) Dipoles are the result of the equal sharing of electrons.
    (C) Cu(s) is a network solid.
    (D) Hydrogen bonds exist between the molecules of HCl.
    (E) NaCl(aq) has attraction between the molecules and the ions.

GO ON TO THE NEXT PAGE

PRACTICE TEST—*Continued*

68. A radioactive substance decays from 100 grams to 6.25 grams in 100 days. What is the half-life of this radioactive substance?

    (A) 25 days
    (B) 6.25 days
    (C) 12.5 days
    (D) 100 days
    (E) 50 days

69. Which choice or choices demonstrate amphoterism?

    I. $HCl + H_2O \rightarrow H_3O^{1+} + Cl^{1-}$ and
         $H_2O + NH_3 \rightarrow OH^{1-} + NH_4^{1+}$
    II. $HS^{1-} + HCl \rightarrow Cl^{1-} + H_2S$ and
         $HS^{1-} + NH_3 \rightarrow NH_4^{1+} + S^{2-}$
    III. $HCl + NaOH \rightarrow NaCl + H_2O$ and
         $NaCl + H_2O \rightarrow HCl + NaOH$

    (A) I only
    (B) II only
    (C) III only
    (D) I and II only
    (E) I and III only

70. Which statement below is incorrect regarding balanced equations?

    (A) $C + O_2 \rightarrow CO_2$ is balanced and is a synthesis reaction.
    (B) $CaCO_3 \rightarrow CaO + CO_2$ is balanced and is a decomposition reaction.
    (C) $Na + Cl_2 \rightarrow NaCl$ is not balanced but demonstrates a synthesis reaction.
    (D) $KI + Pb(NO_3)_2 \rightarrow PbI_2 + KNO_3$ is balanced and is a single replacement reaction.
    (E) $2H_2O \rightarrow 2H_2 + O_2$ is balanced and demonstrates a redox reaction.

S T O P

IF YOU FINISH BEFORE TIME IS CALLED, GO BACK AND CHECK YOUR WORK.

PRACTICE TEST 3

ANSWERS AND EXPLANATIONS

1.   E   As the pressure increases on a gas, the volume of the gas will decrease. The graph that demonstrates Boyle's Law is curved as indicated by choice E.

2.   C   This graph is a phase diagram, which contains a triple point.

3.   A   As temperature increases, so does the volume of a gas.

4.   D   Moving from lithium to fluorine in period 2, the atomic radius decreases.

5.   A   Bromine and mercury are liquids at room temperature.

6.   C   Both the ammonium and hydronium ions have bonded a proton by donating two electrons to the bond that has formed with the proton.

7.   E   Allotropes are different forms of the same element. Both diamond and graphite are made of carbon.

8.   B   Better oxidizing agents are easily reduced (gain electrons). With their higher electronegativities, both chlorine and fluorine are good oxidizing agents.

9.   D   The ending *-oate* is that of an ester, R—COO—R.

10.   C   The functional group R—NH_2 is that of an amine.

11.   A   Alcohols have an —OH group attached to them and end in *-ol*.

12.   A   A calcium atom has 20 electrons, $1s^2 2s^2 2p^6 3s^2 3p^6 4s^2$. A calcium ion will have 18 electrons because calcium forms an ion with a charge of 2+, $1s^2 2s^2 2p^6 3s^2 3p^6$.

13.   E   This choice shows that the 3s sublevel has been skipped over.

14.   C   Potassium has 19 electrons and the configuration of $1s^2 2s^2 2p^6 3s^2 3p^6 4s^1$.

15.   D   The noble gas with the most stable configuration and highest first ionization energy is helium.

16.   A   The changing of a solid to a gas without a liquid phase is called sublimation.

17.   B   Deposition is the changing of a gas to a solid without a liquid phase.

18.   C   Vaporization of a liquid will cause the liquid to enter the gas phase.

19.   E   Freezing a liquid will turn the liquid into a solid.

20.   E   Beryllium has two valence electrons, $1s^2 2s^2$.

21.   B   Oxygen has six valence electrons, $1s^2 2s^2 2p^4$.

22.   A   Nitrogen has five valence electrons. In an effort to make a complete octet, nitrogen will gain three electrons and have an ion with a charge of 3–.

23.   E 10^{-9}, a billionth, is the value that is represented by the prefix *nano-*.

24.   D 10^{-6}, a millionth, is the value that is represented by the prefix *micro-*.

PRACTICE TEST—*Continued*

25.	B	10^3, a thousand, is the value that is represented by the prefix *kilo-*.
101.	T, T, CE	Rutherford's gold foil experiment showed that alpha particles can pass through a sheet of gold foil, proving that the atom is mainly empty space.
102.	T, F	Nitrogen does have five valance electrons, but the electron configuration shown is that of the nitrogen ion when it has gained three electrons.
103.	T, T, CE	Ethyne has two carbon atoms that have a triple bond between them. The triple bond demands an sp hybridization and the resulting linear geometry.
104.	F, F	All nitrates are soluble in water, but the chlorides of silver, mercury, and lead are not.
105.	F, T	The bonds in carbon tetrachloride are polar but because the molecule is symmetrical, the resulting molecules will be nonpolar.
106.	T, T, CE	Concentration and volume have an inverse relationship because, as water is added to a solution with a given concentration, the concentration will decrease.
107.	T, T, CE	HCl is matched up perfectly with the definition of an Arrhenius acid in this problem.
108.	T, F	Adding more reactants to a reaction will speed up the reaction by increasing the frequency of collisions.
109.	F, T	Although the half reaction does demonstrate conservation of mass and charge, the half reaction is a reduction and not an oxidation.
110.	T, F	Looking at the mass number for the alpha particle shows that there are four nucleons present. An alpha particle is a helium-4 nucleus.
111.	T, T, CE	Electronegativity is an atom's ability to attract electrons. With a value of 4.0, fluorine has the highest electronegativity and attraction for electrons.
112.	F, T	While 5,007 has three different digits, there are four significant figures present because both zeros between the nonzero digits are significant.
113.	T, T, CE	DNA is a polymer made up of many individual nucleic acids.
114.	T, T, CE	Even though radiation and radioactive substances can have negative effects on humans, if used correctly and in the right amounts, they can also be beneficial to humans as well.
115.	T, T, CE	Adding a solute to a solvent causes the colligative properties of the solvent to change. This includes the elevation of the boiling point and the depression of the freezing point.

PRACTICE TEST—*Continued*

26.   C   This reaction calls for one H to H bond to be broken, one Cl to Cl bond to be broken, and two H to Cl bonds to be formed. This is set up as follows:

$[1(H—H) + 1(Cl—Cl)] – [2(H—Cl)]$. Substitution gives:

$[435 + 240] – [2(430)] = 675 – 860 = –185$ kJ.

27.   E   The amount of heat absorbed by water can be calculated using the equation $q = mc\Delta T$. Substituting gives (85 grams)(4.18J/g K)(62 K) = 22,028.6 J.

28.   E   All the statements are included in the Atomic Theory except for the empty space concept of the atom. This was concluded by Rutherford in his gold foil experiment.

29.   D   The sp^2 hybridization of the two carbon atoms in ethane and the six valence electrons preferred by boron, gives rise to a trigonal planar molecular geometry. Cyclopropane has all single bonds in its molecule and will have sp^3 hybridized carbon atoms.

30.   E   The three equations are $^{212}_{83}Bi \rightarrow {}^{212}_{84}Po + {}^{0}_{-1}e$, $^{212}_{82}Pb \rightarrow {}^{212}_{83}Bi + {}^{0}_{-1}e$, and $^{228}_{88}Ra \rightarrow {}^{228}_{89}Ac + {}^{0}_{-1}e$. All three equations show conservation of mass because the mass numbers add up, and they all show conservation of charge because the nuclear charges (atomic numbers) add up as well.

31.   C   When the vapor pressure of a liquid is equal to the atmospheric pressure that surrounds the liquid, the liquid will boil.

32.   B   Solids have their atoms set in a fixed position. This allows for a regular geometric pattern such as the lattice in NaCl(s) to be formed.

33.   A   The equation for calculating the molarity of a solution is moles of solute / total liters of solution. Because 73 grams of HCl (molar mass is 36.5) is 2 moles of HCl dissolved to make 2.0 liters of solution total, the solution will be 1.0 *M*.

34.   D   When a double replacement reaction occurs between $AgNO_3$(aq) and KCl(aq), the products formed are AgCl and KNO_3. AgCl is insoluble in water and forms a white precipitate.

35.   B   HF is a weak acid and will also act like a weak electrolyte. The other substances are either strong acids, strong bases, or water-soluble salts.

36.   D   Litmus will be blue and phenolphthalein will be pink in a basic solution. The concentration of hydroxide ions should also be greater than the concentration of hydronium ions.

37.   D   The reaction $Pb^{2+} + 2e^- \rightarrow Pb$ shows conservation of mass and charge. Notice that choice B is not a half reaction; it is full redox reaction that is properly balanced.

38.   D   36 grams of water (molar mass is 18) will be equal to 2 moles of water.

PRACTICE TEST—*Continued*

39.   C   All of these statements are correct except for choice C, which should read, "The empirical formula of $C_6H_{12}O_6$ is CH_2O."

40.   C   The total mass of this compound is 233.4. Because the oxygen makes up 64 of the 233.4, the oxygen is $64/233 \times 100\% = 27.4\%$.

41.   E   446 grams of Fe (atomic mass is 55.8) is 8 moles of iron. If 8 moles of Fe react, then according to the balanced equation, 4 moles of Fe_2O_3 should be produced:

$$\frac{4\ Fe}{8\ Fe} = \frac{2\ Fe_2O_3}{x\ moles\ Fe_2O_3}$$

4 moles of Fe_2O_3 (molar mass is 160) is 640 grams.

42.   A   5.0 moles of sodium would require only 2.5 moles of chlorine gas because twice as much sodium is consumed in the reaction as chlorine (as seen in the balanced equation). This means that sodium is the limiting reagent and chlorine is in excess. If all 5.0 moles of the sodium are used up, then 5.0 moles of NaCl will be produced because of their 1:1 ratio in the balanced equation.

43.   C   Remember that the equilibrium expression calls for "products over reactants, coefficients become powers." Solids and liquids are not included in the expressions. This means that the correct expression is

$$K_{eq} = \frac{[D]^2}{[A]^2[B]}$$

44.   B   Increasing the temperature of a system at equilibrium will shift the equilibrium away from the side that has the heat energy. In this case the equilibrium will shift to the left. In all the other cases the equilibrium will shift to the right.

45.   A   A negative sign for entropy means more order (or less chaos and less disorder). Raking up leaves is a more orderly state.

46.   B   The burette is the most precise piece of glassware listed. Because a burette is narrow, the meniscus will be prominent and of importance.

47.   C   Because pressure and volume are inversely proportional, an increase in pressure will decrease the volume. Use Boyle's Law and substitute with "mock" values to get: $1P_1 1V_1 = 2P_2 V_2$. Solving for the new volume V_2, you find that $1P_1$ has been divided by 2:

$$\frac{1P_1\ 1V_1}{2P_2} = V_2$$

The volume has been halved.

PRACTICE TEST—*Continued*

48.   C   Ammonia is a polar molecule and will dissolve in a polar substance like water. This means that water displacement is a poor method for the collection of ammonia.

49.   A   The potential energy of the products minus the potential energy of the reactants is the heat of reaction. This is designated by the letter "B" on the diagram.

50.   A   Because A and B are on the left in the overall reaction you have:

$$A + B \rightarrow 2C \qquad \Delta H = +150 \text{ kcal}$$

To cancel out the 2C that appears on the right in the step above, double the step so that it reads:

$$2C \rightarrow 4D + 4E \qquad \Delta H = -900 \text{ kcal}$$

Finally, you need to have F on the right as dictated in the overall equation. You also need to cancel out the D and E formed in the last step:

$$4D + 4E \rightarrow F \qquad \Delta H = -725 \text{ kcal}$$

Add up the heats of reaction from the three steps to get: $\Delta H = -1475$ kcal.

51.   B   Remember the mnemonic device, "An Ox," and that oxidation occurs at the anode. This means that the half reaction at the anode will show a loss of electrons. The negatively charged bromine ions will serve this purpose. The half reaction must also be written correctly as well.

52.   D   The solubility product constant will change only if there is a change in temperature.

53.   D   To calculate the new electrode potential for nonstandard condition, use the Nernst equation:

$$E = E° \frac{-2.30 \text{ RT}}{nF} (\log Q). \text{ Substitution gives:}$$

$$E = +1.10 \text{ V} \frac{-0.059 \text{ V}}{2} (\log [0.2]/[4.0]) \text{ and becomes}$$

$$E = +1.10 \text{ V} \frac{-0.059 \text{ V}}{2} (-1.30)$$

Solving gives a new potential of about 1.14 volts.

54.   E   Water, ammonia, and carbon dioxide all have polar bonds. Because it is a symmetrical molecule, carbon dioxide will have a counterbalance of the dipole forces and be a nonpolar molecule.

55.   C   To calculate the amount of acid needed use the titration formula: $M_a V_a = M_b V_b$. Substitution gives: $(1.5 M)(V_a) = (1.0 M)(35.0 \text{ mL})$. Solve to find that the volume of acid is 23.33 mL.

PRACTICE TEST—*Continued*

56.   E   Because isotopes are the same element with different mass numbers, they will have the same number of protons and the same nuclear charge.

57.   C   Compounds can be broken down chemically while elements cannot. Methane and water are examples of compounds.

58.   A   This problem requires the use of the ideal gas equation, $PV = nRT$. Rearranging to solve for V gives:

$$V = nRT/P \text{ or } \frac{(2.3 \text{ moles})(0.0820 \text{ L·atm/mol·K})(300 \text{ K})}{(1.1 \text{ atm})}.$$

59.   D   Multiply the mass numbers by their abundances:

$(48)(0.42) = 20.16$

$(49)(0.38) = 18.62$

$(51)(0.20) = 10.20$

Total comes to 48.98.

60.   B   All the factors mentioned will decrease the rate of reaction except for adding a catalyst, which will increase the rate of reaction.

61.   A   One mole of the gas occupies 22.4 liters initially at 273 K and 760 torr. Using the combined gas law and solving for the new volume gives: $V_2 = P_1V_1T_2 / T_1P_2$. The new temperature is 288 K (using $K = C + 273$) and substitution gives:

$$\frac{(760 \text{ torr})(22.4 \text{ L})(288 \text{ K})}{(273 \text{ K})(700 \text{ torr})}$$

Even though the question does not ask for the new volume, you can still predict that the volume will be more than 22.4 liters. This is because the pressure decreased and the temperature increased, both suggesting an increase in the volume.

62.   D   Lithium is an active metal and will most likely lose electrons. This choice shows the lithium ion gaining an electron to form solid lithium, a reaction that has an electrode potential that is negative and will not occur spontaneously.

63.   E   Going from left to right across period 2, the elements become more non-metallic and there is a decrease in the metallic character.

64.   C   Because there is a triple bond between the carbon atoms, there are two pi bonds and one sigma bond. Add to this the two sigma bonds between the carbon and hydrogen atoms and the total is two pi bonds and three sigma bonds.

PRACTICE TEST—*Continued*

65.   B   The roman numeral III means that the iron ion has a charge of 3+. Sulfate has a charge of 2–. Using the crisscross method and using parentheses for the polyatomic ion, the correct formula is $Fe_2(SO_4)_3$.

66.   E   Because the pH is 6, the $[H^{1+}]$ is 1.0×10^{-6} M. Using the equation $1.0 \times 10^{-14} = [H^{1+}][OH^{1-}]$ and substituting, $1.0 \times 10^{-14} = [1.0 \times 10^{-6}$ $M][OH^{1-}]$, you find that the hydroxide ion concentration will be 1.0×10^{-8} M.

67.   E   When ions are dissolved in water to make a solution, there is an attraction between the polar water molecules and the charged ions. This is called the molecule-ion attraction.

68.   A   This substance had undergone four half-life periods as shown here:

100 grams $\rightarrow$ 50 grams $\rightarrow$ 25 grams $\rightarrow$ 12.5 grams $\rightarrow$ 6.25 grams

There were four half-life periods totaling 100 days. This means that each half-life period is 25 days.

69.   D   In reaction I the water acts like an acid and a base. In reaction II the HS^{1-} ion acts like an acid and a base. Reaction III shows neutralization and hydrolysis, but not amphoterism.

70.   D   This reaction is not balanced. It should be: $2KI + Pb(NO_3)_2 \rightarrow PbI_2 + 2KNO_3$. Also, this reaction is classified as a double replacement.

PRACTICE TEST 3

SCORE SHEET

Number of questions correct: _____

Less: 0.25 × number of questions wrong: _____

(Remember that omitted questions are not counted as wrong.)

Equals your raw score: _____

Raw Score	Test Score		Raw Score	Test Score		Raw Score	Test Score		Raw Score	Test Score		Raw Score	Test Score
85	800		63	710		41	570		19	440		−3	300
84	800		62	700		40	560		18	430		−4	300
83	800		61	700		39	560		17	430		−5	290
82	800		60	690		38	550		16	420		−6	290
81	800		59	680		37	550		15	420		−7	280
80	800		58	670		36	540		14	410		−8	270
79	790		57	670		35	530		13	400		−9	270
78	790		56	660		34	530		12	400		−10	260
77	790		55	650		33	520		11	390		−11	250
76	780		54	640		32	520		10	390		−12	250
75	780		53	640		31	510		9	380		−13	240
74	770		52	630		30	500		8	370		−14	240
73	760		51	630		29	500		7	360		−15	230
72	760		50	620		28	490		6	360		−16	230
71	750		49	610		27	480		5	350		−17	220
70	740		48	610		26	480		4	350		−18	220
69	740		47	600		25	470		3	340		−19	210
68	730		46	600		24	470		2	330		−20	210
67	730		45	590		23	460		1	330		−21	200
66	720		44	580		22	460		0	320			
65	720		43	580		21	450		−1	320			
64	710		42	570		20	440		−2	310			

Note: This is only a sample scoring scale. Scoring scales differ from exam to exam.

PRACTICE TEST 4

▬▬ ANSWER SHEET

This practice test will help measure your current knowledge of chemistry and familiarize you with the structure of the *SAT Chemistry* test. To simulate exam conditions, use the answer sheet below to record your answers and the appendix tables at the back of this book to obtain necessary information.

1. Ⓐ Ⓑ Ⓒ Ⓓ Ⓔ	21. Ⓐ Ⓑ Ⓒ Ⓓ Ⓔ	41. Ⓐ Ⓑ Ⓒ Ⓓ Ⓔ	61. Ⓐ Ⓑ Ⓒ Ⓓ Ⓔ
2. Ⓐ Ⓑ Ⓒ Ⓓ Ⓔ	22. Ⓐ Ⓑ Ⓒ Ⓓ Ⓔ	42. Ⓐ Ⓑ Ⓒ Ⓓ Ⓔ	62. Ⓐ Ⓑ Ⓒ Ⓓ Ⓔ
3. Ⓐ Ⓑ Ⓒ Ⓓ Ⓔ	23. Ⓐ Ⓑ Ⓒ Ⓓ Ⓔ	43. Ⓐ Ⓑ Ⓒ Ⓓ Ⓔ	63. Ⓐ Ⓑ Ⓒ Ⓓ Ⓔ
4. Ⓐ Ⓑ Ⓒ Ⓓ Ⓔ	24. Ⓐ Ⓑ Ⓒ Ⓓ Ⓔ	44. Ⓐ Ⓑ Ⓒ Ⓓ Ⓔ	64. Ⓐ Ⓑ Ⓒ Ⓓ Ⓔ
5. Ⓐ Ⓑ Ⓒ Ⓓ Ⓔ	25. Ⓐ Ⓑ Ⓒ Ⓓ Ⓔ	45. Ⓐ Ⓑ Ⓒ Ⓓ Ⓔ	65. Ⓐ Ⓑ Ⓒ Ⓓ Ⓔ
6. Ⓐ Ⓑ Ⓒ Ⓓ Ⓔ	26. Ⓐ Ⓑ Ⓒ Ⓓ Ⓔ	46. Ⓐ Ⓑ Ⓒ Ⓓ Ⓔ	66. Ⓐ Ⓑ Ⓒ Ⓓ Ⓔ
7. Ⓐ Ⓑ Ⓒ Ⓓ Ⓔ	27. Ⓐ Ⓑ Ⓒ Ⓓ Ⓔ	47. Ⓐ Ⓑ Ⓒ Ⓓ Ⓔ	67. Ⓐ Ⓑ Ⓒ Ⓓ Ⓔ
8. Ⓐ Ⓑ Ⓒ Ⓓ Ⓔ	28. Ⓐ Ⓑ Ⓒ Ⓓ Ⓔ	48. Ⓐ Ⓑ Ⓒ Ⓓ Ⓔ	68. Ⓐ Ⓑ Ⓒ Ⓓ Ⓔ
9. Ⓐ Ⓑ Ⓒ Ⓓ Ⓔ	29. Ⓐ Ⓑ Ⓒ Ⓓ Ⓔ	49. Ⓐ Ⓑ Ⓒ Ⓓ Ⓔ	69. Ⓐ Ⓑ Ⓒ Ⓓ Ⓔ
10. Ⓐ Ⓑ Ⓒ Ⓓ Ⓔ	30. Ⓐ Ⓑ Ⓒ Ⓓ Ⓔ	50. Ⓐ Ⓑ Ⓒ Ⓓ Ⓔ	70. Ⓐ Ⓑ Ⓒ Ⓓ Ⓔ
11. Ⓐ Ⓑ Ⓒ Ⓓ Ⓔ	31. Ⓐ Ⓑ Ⓒ Ⓓ Ⓔ	51. Ⓐ Ⓑ Ⓒ Ⓓ Ⓔ	71. Ⓐ Ⓑ Ⓒ Ⓓ Ⓔ
12. Ⓐ Ⓑ Ⓒ Ⓓ Ⓔ	32. Ⓐ Ⓑ Ⓒ Ⓓ Ⓔ	52. Ⓐ Ⓑ Ⓒ Ⓓ Ⓔ	72. Ⓐ Ⓑ Ⓒ Ⓓ Ⓔ
13. Ⓐ Ⓑ Ⓒ Ⓓ Ⓔ	33. Ⓐ Ⓑ Ⓒ Ⓓ Ⓔ	53. Ⓐ Ⓑ Ⓒ Ⓓ Ⓔ	73. Ⓐ Ⓑ Ⓒ Ⓓ Ⓔ
14. Ⓐ Ⓑ Ⓒ Ⓓ Ⓔ	34. Ⓐ Ⓑ Ⓒ Ⓓ Ⓔ	54. Ⓐ Ⓑ Ⓒ Ⓓ Ⓔ	74. Ⓐ Ⓑ Ⓒ Ⓓ Ⓔ
15. Ⓐ Ⓑ Ⓒ Ⓓ Ⓔ	35. Ⓐ Ⓑ Ⓒ Ⓓ Ⓔ	55. Ⓐ Ⓑ Ⓒ Ⓓ Ⓔ	75. Ⓐ Ⓑ Ⓒ Ⓓ Ⓔ
16. Ⓐ Ⓑ Ⓒ Ⓓ Ⓔ	36. Ⓐ Ⓑ Ⓒ Ⓓ Ⓔ	56. Ⓐ Ⓑ Ⓒ Ⓓ Ⓔ	
17. Ⓐ Ⓑ Ⓒ Ⓓ Ⓔ	37. Ⓐ Ⓑ Ⓒ Ⓓ Ⓔ	57. Ⓐ Ⓑ Ⓒ Ⓓ Ⓔ	
18. Ⓐ Ⓑ Ⓒ Ⓓ Ⓔ	38. Ⓐ Ⓑ Ⓒ Ⓓ Ⓔ	58. Ⓐ Ⓑ Ⓒ Ⓓ Ⓔ	
19. Ⓐ Ⓑ Ⓒ Ⓓ Ⓔ	39. Ⓐ Ⓑ Ⓒ Ⓓ Ⓔ	59. Ⓐ Ⓑ Ⓒ Ⓓ Ⓔ	
20. Ⓐ Ⓑ Ⓒ Ⓓ Ⓔ	40. Ⓐ Ⓑ Ⓒ Ⓓ Ⓔ	60. Ⓐ Ⓑ Ⓒ Ⓓ Ⓔ	

Chemistry   *Fill in oval CE only if II is correct explanation of I.

	I	II	CE*		I	II	CE*
101.	Ⓣ Ⓕ	Ⓣ Ⓕ	◯	109.	Ⓣ Ⓕ	Ⓣ Ⓕ	◯
102.	Ⓣ Ⓕ	Ⓣ Ⓕ	◯	110.	Ⓣ Ⓕ	Ⓣ Ⓕ	◯
103.	Ⓣ Ⓕ	Ⓣ Ⓕ	◯	111.	Ⓣ Ⓕ	Ⓣ Ⓕ	◯
104.	Ⓣ Ⓕ	Ⓣ Ⓕ	◯	112.	Ⓣ Ⓕ	Ⓣ Ⓕ	◯
105.	Ⓣ Ⓕ	Ⓣ Ⓕ	◯	113.	Ⓣ Ⓕ	Ⓣ Ⓕ	◯
106.	Ⓣ Ⓕ	Ⓣ Ⓕ	◯	114.	Ⓣ Ⓕ	Ⓣ Ⓕ	◯
107.	Ⓣ Ⓕ	Ⓣ Ⓕ	◯	115.	Ⓣ Ⓕ	Ⓣ Ⓕ	◯
108.	Ⓣ Ⓕ	Ⓣ Ⓕ	◯				

GO ON TO THE NEXT PAGE ➤

PRACTICE TEST 4

Time: 60 minutes

<u>Note:</u> Unless otherwise stated, for all statements involving chemical equations and/or solutions, assume that the system is in pure water.

Part A

<u>Directions:</u> Each of the following sets of lettered choices refers to the numbered formulas or statements immediately below it. For each numbered item, choose the one lettered choice that fits it best. Then fill in the corresponding oval on the answer sheet. Each choice in a set may be used once, more than once, or not at all.

<u>Questions 1–4</u>

    (A) The point of equilibrium

    (B) The triple point

    (C) The freezing point

    (D) The point where reactants first form products

    (E) The boiling point

1. A specific temperature and pressure where solid, liquid, and gas phases exist simultaneously

2. Can be shifted by adding more reactants

3. Vapor pressure of a liquid is equal to the pressure of the surroundings

4. The activated complex

PRACTICE TEST—*Continued*

Questions 5–8

    (A) Red

    (B) Purple

    (C) Orange

    (D) Green

    (E) Blue

5. Copper(II) sulfate solution

6. Chlorine gas

7. $KMnO_4$ solution

8. Bromine liquid

Questions 9–11

    (A) Voltaic cell

    (B) Electrolytic cell

    (C) Geiger Counter

    (D) pH meter

    (E) Calorimeter

9. Requires an external current to make a redox reaction spontaneous

10. Requires a salt bridge

11. Detects radioactive particles

Questions 12–15

    (A) Halogens

    (B) Alkali metals

    (C) Alkaline earth metals

    (D) Noble gases

    (E) Lanthanides

12. Valence electrons are located in the f orbitals

13. Need to lose one electron to form a stable octet

14. Will have the highest first ionization energies

15. Contain elements in the solid, liquid, and gas phases at STP

Questions 16–19

    (A) 9.03×10^{23} molecules

    (B) 44.8 liters

    (C) 3.5 moles

    (D) 6.0 grams

    (E) 3.01×10^{23} atoms

16. 0.25 moles of O_2 at STP

17. 3.0 moles of H_2 at STP

18. 56 grams of N_2 at STP

19. 96.0 grams of SO_2 at STP

Questions 20–22

    (A) Water

    (B) Hydrogen bromide

    (C) Iron

    (D) Argon

    (E) Sodium chloride

20. Hydrogen bonding

21. Highly polar

22. Dispersion forces

Questions 23–25

    (A) Alpha particle

    (B) Beta particle

    (C) Gamma radiation

    (D) Positron

    (E) Deuteron

23. Po-218 → At-218 + X

24. Tc-99 → Tc-99 + X

25. Ne-19 → F-19 + X

GO ON TO THE NEXT PAGE

PRACTICE TEST—*Continued*

PLEASE GO TO THE SPECIAL SECTION AT THE LOWER LEFT-HAND CORNER OF YOUR
ANSWER SHEET LABELED "CHEMISTRY" AND ANSWER QUESTIONS 101–115
ACCORDING TO THE FOLLOWING DIRECTIONS.

Part B

Directions: Each question below consists of two statements. For each question, determine whether statement I in the left-most column is true or false and whether statement II in the rightmost column is true or false. Fill in the corresponding T or F ovals on the answer sheet provided. Fill in the oval labeled "CE" only if statement II correctly explains statement I.

SAMPLE:

EX 1. The nucleus of an atom has a positive charge   BECAUSE   the only positive particles found in the atom's nucleus are protons.

SAMPLE ANSWER

	I	II	CE*
EX 1	● (F)	● (F)	●

	I		II
101.	An element's nuclear charge is equal to the number of protons in the nucleus	BECAUSE	the only charged particles in the nucleus are neutrons.
102.	A reaction will be spontaneous if ΔH is negative and ΔS is positive	BECAUSE	ΔG will be negative when there is a decrease in enthalpy and an increase in entropy.
103.	Cl^{1-} is the conjugate base of HCl	BECAUSE	a conjugate base is formed when an acid gains a proton.
104.	An electrolytic cell makes a nonspontaneous redox reaction occur	BECAUSE	an electrolytic cell uses an external current to drive a redox reaction.
105.	The maximum number of electrons allowed in the third principal energy level is 18	BECAUSE	the maximum number of electrons allowed in a principal energy level is dictated by the equation $2n^2$.
106.	3,000 kilograms is equal to 3 grams	BECAUSE	the prefix *kilo-* means "one thousandth."
107.	An increase in temperature will cause a gas to expand	BECAUSE	temperature and volume have a direct relationship.

GO ON TO THE NEXT PAGE →

PRACTICE TEST—*Continued*

108. A catalyst will change the heat of reaction   BECAUSE   a catalyst will lower the potential energy of the activated complex in a reaction.

109. Helium will have fewer dispersion forces between its atoms than the other noble gases   BECAUSE   as the mass of nonpolar atoms and molecules increases, dispersion forces increase.

110. Nitrogen gas will have a greater rate of effusion than oxygen gas   BECAUSE   lighter, less dense gases travel faster than heavier, more dense gases.

111. Propane can be decomposed chemically   BECAUSE   propane is a compound that is made up of simpler elements.

112. A mixture of two different liquids can be separated via distillation   BECAUSE   different liquids have different boiling points.

113. Isotopes have different atomic numbers   BECAUSE   isotopes must have different numbers of electrons.

114. Butene can be converted into butane   BECAUSE   the addition reaction of hydrogen gas to an alkene will form an alkane.

115. NaCl is a basic salt   BECAUSE   hydrolysis of NaCl reveals the formation of NaOH and HCl.

GO ON TO THE NEXT PAGE

PRACTICE TEST—*Continued*

Part C

Directions: Each of the multiple-choice questions or incomplete sentences below is followed by five answers or completions. Select the one answer that is best in each case and then fill in the corresponding oval on the answer sheet provided.

26. When 58 grams of water is heated from 275 K to 365 K, the water

    (A) absorbs 21,820 joules of heat
    (B) absorbs 377 joules of heat
    (C) releases 5,220 joules of heat
    (D) absorbs 242 joules of heat
    (E) releases 90 joules of heat

27. Which of the following are uses for radiation and radioactivity that are of benefit to us?

    I.   Nuclear waste
    II. Radioisotopes
    III. Excess exposure

    (A) I only
    (B) II only
    (C) III only
    (D) I and II only
    (E) I and III only

28. Which of the following statements is not part of the kinetic molecular theory?

    (A) The average kinetic energy of gas molecules is directly proportional to temperature.
    (B) Attractive and repulsive forces are present between gas molecules.
    (C) Collisions between gas molecules are perfectly elastic.
    (D) Gas molecules travel in a continuous, random motion.
    (E) The volume that gas molecules occupy is minimal compared to the volume within which the gas is contained.

29. The following redox reaction occurs in an acidic solution: $Ce^{4+} + Bi \rightarrow Ce^{3+} + BiO^{1+}$. What is the coefficient before the Ce^{4+} when the equation is fully balanced?

    (A) 1
    (B) 2
    (C) 3
    (D) 6
    (E) 9

30. Which statement regarding significant figures is false?

    (A) Zeros can be significant.
    (B) When multiplying, the answer is determined by the number of significant figures.
    (C) When adding, the answer is determined by the number of decimal places.
    (D) When dividing, the answer is determined by the number of decimal places.
    (E) The number 50,004 has five significant figures.

31. Which statement below best describes the molecule in question?

    (A) Water has a bent molecular geometry and one lone pair of electrons.
    (B) Ammonia has a trigonal pyramidal molecular geometry and two lone pairs of electrons.
    (C) Methane has a trigonal planar molecular geometry.
    (D) Carbon dioxide is linear because it has one single bond and one triple bond.
    (E) The carbon atoms in ethane are sp^3 hybridized.

GO ON TO THE NEXT PAGE

PRACTICE TEST—*Continued*

32. A compound was analyzed and found to be 12.1% C, 71.7% Cl, and 16.2% O. What is the empirical formula for this compound?

   (A) C_2OCl
   (B) $COCl$
   (C) CO_2Cl_2
   (D) C_2O_2Cl
   (E) CCl_2O

33. Which statement is true about the percent composition by mass in $C_6H_{12}O_6$?

   (A) Carbon is 6.7% by mass.
   (B) Oxygen is 53.3% by mass.
   (C) Hydrogen is 12% by mass.
   (D) Carbon is 72% by mass.
   (E) Carbon is 20% by mass.

34. Which process would have a positive value for the change in entropy?

   I.   The expansion of the universe
   II. The condensation of a liquid
   III. A food fight in a school cafeteria

   (A) I only
   (B) II only
   (C) III only
   (D) II and III only
   (E) I and III only

35. Of the gases below, which would react with rain water to produce acid rain?

   I.   CFCs
   II. Methane
   III. Carbon dioxide

   (A) I only
   (B) II only
   (C) III only
   (D) I and III only
   (E) I, II, and III

36. A sample of gas is trapped in a manometer and the stopcock is opened. (See Figure 1.) The level of mercury moves to a new height as can be seen in the diagram. If the pressure of the gas inside the manometer is 815 torr, what is the atmospheric pressure in this case?

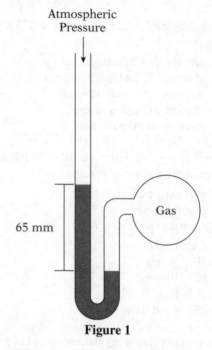

Figure 1

   (A) 760 torr
   (B) 740 torr
   (C) 750 torr
   (D) 815 torr
   (E) 880 torr

37. Which aqueous solution is expected to have the highest boiling point?

   (A) 1.5 m $FeCl_2$
   (B) 3.0 m CH_3OH
   (C) 2.5 m $C_6H_{12}O_6$
   (D) 2.5 m $NaCl$
   (E) 1.0 m $CaCl_2$

38. Which K_a value is that of a better electrolyte?

   (A) 1.0×10^{-2}
   (B) 2.0×10^{-12}
   (C) 5.0×10^{-7}
   (D) 3.0×10^{-4}
   (E) 1.0×10^{-6}

GO ON TO THE NEXT PAGE

39. The following substances were all dissolved in 100 grams of water at 290 K to produce saturated solutions. If the solution is heated to 310 K, which substance will have a decrease in its solubility?

    (A) NaCl
    (B) KI
    (C) $CaCl_2$
    (D) HCl
    (E) KNO_3

40. Methane undergoes a combustion reaction according to the reaction $CH_4(g) + 2O_2(g) \rightarrow CO_2(g) + 2H_2O(l)$. How many grams of methane gas were burned if 67.2 liters of carbon dioxide gas are produced in the reaction? (Assume STP.)

    (A) 16 grams
    (B) 48 grams
    (C) 3 grams
    (D) 132 grams
    (E) 22.4 grams

41. A closed system contains the following reaction at STP: $Cl_2(g) + 2NO_2(g) \longleftrightarrow 2NO_2Cl(g)$. What is the equilibrium constant expression for this reaction?

    (A) $K_{eq} = \dfrac{[NO_2Cl]^2}{[Cl_2][NO_2]}$

    (B) $K_{eq} = \dfrac{[Cl_2][NO_2]^2}{[NO_2Cl]}$

    (C) $K_{eq} = \dfrac{[Cl_2][NO_2]^2}{[NO_2Cl]^2}$

    (D) $K_{eq} = \dfrac{[NO_2Cl]^2}{[Cl_2][NO_2]^2}$

    (E) $K_{eq} = \dfrac{[NO_2Cl]^2}{[Cl_2]^2[NO_2]^2}$

42. At a particular temperature, the equilibrium concentrations of the substances in question 41 are as follows:

    $[NO_2Cl] = 0.5\ M$ $[Cl_2] = 0.3\ M$ $[NO_2] = 0.2\ M$

    What is the value of the equilibrium constant for this reaction?

    (A) 2.1
    (B) 0.48
    (C) 0.0357
    (D) 20.83
    (E) 208.83

43. Which Lewis structure below has been drawn incorrectly in Figure 2?

    (A) H:H

    (B) H:C:::N:

    (C) H:Ö:
        H

    (D) :N:::N:

    (E) F:B̈:F
        F

    Figure 2

44. Which reaction below demonstrates the Lewis definition of acids and bases?

    (A) $HCl + NaOH \rightarrow HOH + NaCl$
    (B) $H_2O + NH_3 \rightarrow OH^{1-} + NH_4^{1+}$
    (C) $NH_3 + BF_3 \rightarrow NH_3BF_3$
    (D) $HI + KOH \rightarrow H_2O + KI$
    (E) $H^+ + OH^{1-} \rightarrow H_2O$

45. Which sample is a homogeneous mixture?

    (A) KI(aq)
    (B) Fe(s)
    (C) $CO_2(g)$
    (D) $NH_3(l)$
    (E) NaCl(s)

GO ON TO THE NEXT PAGE

PRACTICE TEST—*Continued*

46. Which pair below represents isomers of the same compound?

   (A) $CH_3CH_2CH_2OH$ and $HOCH_2CH_2CH_3$
   (B) $CH_3CH_2CH_3$ and $CH_3CH_2CH_2CH_3$
   (C) $CH_3CH(Cl)CH_3$ and $CH_3CH_2CH_2Cl$
   (D) CH_3COCH_3 and $CH_3CH_2CH_2CHO$
   (E) $ClCH_2CH_2Br$ and $BrCH_2CH_2Cl$

47. Which would you never do in a laboratory setting?

   I.   Eat and drink in the laboratory
   II.   Push a thermometer through a rubber stopper
   III.   Remove your goggles to take a better look at a reaction

   (A) I only
   (B) II only
   (C) III only
   (D) I and III only
   (E) I, II, and III

48. How many pi bonds are there in a molecule of $N{\equiv}C{-}CH_2{-}CH_2{-}CO{-}NH{-}CH{=}CH_2$?

   (A) 7
   (B) 4
   (C) 12
   (D) 10
   (E) 5

49. When the equation: $C_2H_6 + O_2 \rightarrow CO_2 + H_2O$ is completely balanced using the lowest whole number coefficients, the sum of the coefficients will be

   (A) 4
   (B) 9.5
   (C) 19
   (D) 15.5
   (E) 11

50. From the heats of reaction of these individual reactions:

   $A + B \rightarrow 2C$ $\Delta H = -500$ kJ,
   $D + 2B \rightarrow E$ $\Delta H = -700$ kJ,
   $2D + 2A \rightarrow F$ $\Delta H = +50$ kJ

   Find the heat of reaction for $F + 6B \rightarrow 2E + 4C$.

   (A) +450 kJ
   (B) −1,100 kJ
   (C) +2,350 kJ
   (D) −350 kJ
   (E) −2,450 kJ

51. Which solutions have a concentration of 1.0 *M*?

   I.   74 grams of calcium hydroxide dissolved to make 1 liter of solution
   II.   74.5 grams of potassium chloride dissolved to make 1 liter of solution
   III.   87 grams of lithium bromide dissolved to make 1 liter solution

   (A) I only
   (B) III only
   (C) I and III only
   (D) II and III only
   (E) I, II, and III

52. According to the reaction $3H_2 + N_2 \rightarrow 2NH_3$, how many grams of hydrogen gas and nitrogen gas are needed to make exactly 68 grams of ammonia?

   (A) 2 grams of hydrogen gas and 28 grams of nitrogen gas
   (B) 3 grams of hydrogen gas and 1 gram of nitrogen gas
   (C) 12 grams of hydrogen gas and 56 grams of nitrogen gas
   (D) 102 grams of hydrogen gas and 34 grams of nitrogen gas
   (E) 6 grams of hydrogen gas and 2 grams of nitrogen gas

53. Which compound is not paired with its correct name?

   (A) $FeCl_2$ / iron(II) chloride
   (B) K_2O / potassium oxide
   (C) NO_2 / nitrogen dioxide
   (D) PCl_3 / potassium trichloride
   (E) NH_4Cl / ammonium chloride

GO ON TO THE NEXT PAGE

PRACTICE TEST—*Continued*

54. How many grams of HI can be made from 6 grams of H_2 and 800 grams of I_2 when hydrogen gas and diatomic iodine react according to the equation: $H_2 + I_2 \rightarrow 2HI$?

   (A) 800 grams of HI can be made with 38 grams of iodine in excess.
   (B) 768 grams of HI can be made with 6 grams of hydrogen in excess.
   (C) 768 grams of HI can be made with 38 grams of iodine in excess.
   (D) 2286 grams of HI can be made with no excess reactants.
   (E) 806 grams of HI can be made with no excess reactants.

55. 500 mL of a 0.2 M solution has 200 mL of water added to it. What is the new molarity of this solution?

   (A) 0.50 M
   (B) 0.28 M
   (C) 0.70 M
   (D) 0.14 M
   (E) 0.40 M

56. Which mixture is correctly paired with a method for separation of the mixture?

   (A) Oil and water—filter paper
   (B) Salt water—distillation
   (C) Sand and water—separatory funnel
   (D) Sand and sugar—tweezers
   (E) Sugar water—filter paper

57. Which reaction between ions does not form a precipitate?

   (A) $Ag^{1+} + Cl^{1-}$
   (B) $Pb^{2+} + 2I^{1-}$
   (C) $Ca^{2+} + CO_3^{2-}$
   (D) $Hg^{2+} + 2Br^{1-}$
   (E) $Na^{1+} + OH^{1-}$

58. Which will happen when sodium sulfate is added to a saturated solution of $CaSO_4$ that is at equilibrium? $[CaSO_4(s) \longleftrightarrow Ca^{2+}(aq) + SO_4^{2-}(aq)]$

   (A) The solubility of the calcium sulfate will decrease.
   (B) The concentration of calcium ions will increase.
   (C) The reaction will shift to the right.
   (D) The K_{sp} value will change.
   (E) The equilibrium will shift to consume the decrease in sulfate ions.

59. Given the reaction $2A(g) + B(g) + Heat \longleftrightarrow 3C(g) + D(g)$, what could be done to the reaction to shift the equilibrium so that more D is made?

   (A) Increase the concentration of D.
   (B) Increase the concentration of C.
   (C) Increase the temperature.
   (D) Increase the pressure.
   (E) Remove B from the reaction.

60. A 16-gram sample of water at 273 K is cooled so that it becomes a completely solid ice cube at 273 K. How much heat was released by the sample of water to form this ice cube?

   (A)   16 J
   (B) 4,368 J
   (C) 18,258 J
   (D)   350 J
   (E) 5,334 J

61. Sublimation is the process by which a solid becomes a gas without having a liquid phase. Which of these can sublime?
   I.   Iodine
   II. Naphthalene
   III. Carbon dioxide

   (A) I only
   (B) II only
   (C) III only
   (D) I and III only
   (E) I, II, and III

GO ON TO THE NEXT PAGE

PRACTICE TEST—*Continued*

62. Which of the following will decrease the rate of a reaction?

    (A) Using powdered solids instead of whole pieces
    (B) Selecting ionic reactants that have been dissolved in water
    (C) Decreasing the temperature
    (D) Increasing the pressure
    (E) Adding a catalyst

63. Three gases are mixed in a sealed container. The container has 0.3 moles of gas A, 0.4 moles of gas B, and 0.3 moles of gas C. The total pressure of the gases is 660 torr. What is true about the partial pressures of the gases?

    (A) The partial pressure of gas A is 264 torr.
    (B) The partial pressure of gas B is 396 torr.
    (C) The partial pressure of gas C is 220 torr.
    (D) The partial pressures of gases A and C are each 198 torr.
    (E) The partial pressure of gas B is 660 torr.

Questions 64 and 65 refer to the voltaic cell below in Figure 2:

64. What is the half reaction that occurs at the cathode?

    (A) $Al \rightarrow Al^{3+} + 3e^-$
    (B) $Ni^{2+} + 2e^- \rightarrow Ni$
    (C) $Ni \rightarrow Ni^{2+} + 2e^-$
    (D) $2Al^{3+} + 6e^- \rightarrow 2Al$
    (E) $Al^{3+} + 3e^- \rightarrow Al$

65. Which statement is true about the setup above?

    (A) The electrode potential for this cell is 1.40 V.
    (B) The electrode potential for this cell is 2.54 V.
    (C) Electrons will be carried by the salt bridge.
    (D) Ions will be carried through the wire.
    (E) The reaction is nonspontaneous.

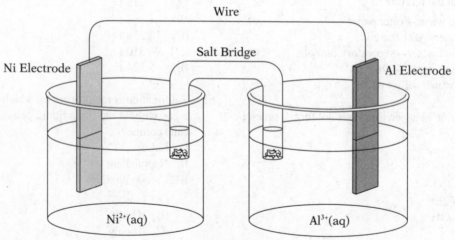

Figure 2

PRACTICE TEST—*Continued*

66. Over a number of years the average pH of a stream changes from a pH of 6.9 to a pH of 5.9 due to acid rain. Which statement is true about the pH of the stream?

    (A) The pH of the stream now is one time more acidic than it was years ago.
    (B) The stream now has 10 times more hydroxide ions than it did years ago.
    (C) The pH of the stream is now 10 times more acidic than it was years ago.
    (D) The stream is more basic now than it was years ago.
    (E) The concentration of hydronium ion in the stream has decreased over the years.

67. An alkaline earth metal, element M, reacts with oxygen. What is going to be the general formula for the compound formed?

    (A) M_2O
    (B) MO
    (C) MO_2
    (D) M_2O_3
    (E) M_3O_2

68. Which functional group below does not contain a carbonyl group?

    (A) Aldehydes
    (B) Ketones
    (C) Esters
    (D) Ethers
    (E) Carboxylic acids

69. Using the bond dissociation energies found in the appendix, calculate the change in the heat of reaction for $2H_2 + O_2 \rightarrow 2H_2O$.

    (A) −118 kJ
    (B) +118 kJ
    (C) −91 kJ
    (D) −1,042 kJ
    (E) −833 kJ

70. Equilibrium

    (A) is defined as equal concentrations of reactants and products
    (B) is defined as equal rates for forward and reverse reactions
    (C) can be shifted by adding a catalyst
    (D) can exist for chemical changes but not physical changes
    (E) must always favor the formation of products

STOP

IF YOU FINISH BEFORE TIME IS CALLED, GO BACK AND CHECK YOUR WORK.

PRACTICE TEST 4

ANSWERS AND EXPLANATIONS

1. B   The triple point on a phase diagram tells the temperature and pressure needed for a solid, liquid, and gas to exist at the same time.

2. A   If a system is at equilibrium, adding more reactants will shift the equilibrium to form more products.

3. E   The boiling point of a liquid is the temperature in which the vapor pressure of a liquid is equal to the atmospheric pressure.

4. D   The activated complex on the potential energy diagram is where the first appearance of products occurs.

5. E   Copper sulfate is a blue salt that forms blue solutions.

6. D   Chlorine is a green gas.

7. B   Permanganate ion will form a dark purple solution.

8. C   Bromine is an orange liquid at STP.

9. B   An electrolytic cell uses an externally applied current to drive a nonspontaneous reaction.

10. A   The salt bridge in the voltaic cell allows ions to migrate from one half cell to another.

11. C   The Geiger counter detects radioactive emanations.

12. E   The lanthanides and actinides are the elements whose valence electrons are located in the f orbitals.

13. B   The alkali metals have a valence electron configuration of s^1 and need to lose just one electron to form a stable, complete, outermost principal energy level.

14. D   Because the noble gases have very stable electron configurations, it takes much energy to remove an electron from the stable, complete, outermost principal energy level.

15. A   Fluorine and chlorine are gases at STP, while bromine is a liquid and iodine is a solid.

16. E   0.25 moles of O_2 will have 0.50 moles of oxygen atoms. 0.50 moles of atoms is the same as 3.01×10^{23} atoms.

17. D   3.0 moles of hydrogen gas (molar mass is 2) will have a mass of 6.0 grams.

18. B   56 grams of nitrogen gas at STP is 2 moles of nitrogen gas. This sample will occupy 44.8 liters.

19. A   96.0 grams of sulfur dioxide (molar mass is 64) is the same as 1.5 moles of sulfur dioxide. This will equal 9.03×10^{23} molecules of sulfur dioxide.

PRACTICE TEST—*Continued*

20. A Water is the only substance from the list that will exhibit hydrogen bonds. Remember the mnemonic device "FON."

21. A Water is the only substance from the list that is a polar molecule.

22. D Argon is monatomic and nonpolar. The noble gases will exhibit dispersion (or Van der Waals) forces between their atoms.

23. B Filling in the atomic numbers shows the full reaction: $^{218}_{84}Po \rightarrow ^{218}_{85}At + X$. X has no mass and a negative one charge, $^{0}_{-1}X$. This makes X a beta particle.

24. C Tc-99 did not change in mass or in atomic number. It must have given off something without mass or charge. Energy was released. Gamma radiation is the only energy that is listed among the choices.

25. D Filling in the atomic numbers shows the reaction: $^{19}_{10}Ne \rightarrow ^{19}_{9}F + ^{0}_{+1}X$. X is a particle with zero mass and a 1+ charge, a positron.

101. T, F The number of protons defines the atomic number and the nuclear charge for an element. Neutrons do not have any charge.

102. T, T, CE According to the Gibbs Free Energy equation, $\Delta G = \Delta H - T\Delta S$, the favored decrease in enthalpy (ΔH is –) and the favored increase in entropy (ΔS is +); ΔG will have a negative value.

103. T, F Conjugate pairs differ by a proton. When an acid loses a proton, it forms a conjugate base.

104. T, T, CE The electrolytic cell uses an external power supply to drive a redox reaction that normally has an electrode potential that is negative in value.

105. T, T, CE The expression $2n^2$, where n is the principal energy level number, is used to determine the maximum number of electrons that can be held in a principal energy level. Remember to square the number first, then multiply by 2.

106. F, F *Kilo-* means "one thousand." 3,000 grams would be equivalent to 3 kilograms.

107. T, T, CE According to Charles' Law, as the temperature of a gas increases, the volume of the gas will increase as well. This is a direct relationship.

108. F, T While a catalyst can lower the potential energy of the activated complex and activation energy in a reaction, a catalyst will not change the heat of a reaction.

109. T, T, CE Dispersion forces increase between nonpolar molecules as the mass increases. Helium is the lightest of the noble gases and has the fewest dispersion forces between its atoms.

110. T, T, CE Nitrogen is lighter and less dense than oxygen given equal conditions. This means that the rate of effusion for nitrogen gas will be greater.

111. T, T, CE Propane is a compound made up of elements. Compounds can be decomposed in a chemical reaction, whereas elements cannot.

PRACTICE TEST—*Continued*

112. T, T, CE   A distillation uses heat to separate a mixture of liquids based upon their different boiling points.

113. F, F   Isotopes are the same element with a different number of neutrons and a different mass number.

114. T, T, CE   The addition reaction allows a diatomic molecule to be added to the double and triple bonds of organic and other compounds.

115. F, T   When NaCl undergoes hydrolysis, water is added and the original acid and base are formed, NaOH and HCl. Because these are both strong, they will neutralize each other. This means that NaCl is a neutral salt.

26.   A   Use the equation q = mcΔT to find the amount of heat absorbed by the sample of water. Substitution and solving gives: (58 grams)(4.18 J/g K)(90) = 21,820 joules.

27.   B   Radioisotopes (and radiotracers) can be used to help diagnose problems in certain organs in our bodies. The use of radioisotopes is beneficial if the right dosage is used correctly.

28.   B   Ideally, gas molecules should not have any attractive or repulsive forces between the molecules.

29.   C   There are a number of steps required to balance this redox reaction:

(i) Separate the two half reactions:

$Ce^{4+} \rightarrow Ce^{3+}$ and $Bi \rightarrow BiO^{1+}$

(ii) Add water to balance the oxygen atoms:

$H_2O + Bi \rightarrow BiO^{1+}$

(iii) Add H^{1+} ions to balance the hydrogen atoms:

$H_2O + Bi \rightarrow BiO^{1+} + 2H^{1+}$

(iv) Add electrons to balance the charges:

$1e^- + Ce^{4+} \rightarrow Ce^{3+}$ and $H_2O + Bi \rightarrow BiO^{1+} + 2H^{1+} + 3e^-$

(v) Use the distributive property to balance the number of electrons:

$3(1e^- + Ce^{4+} \rightarrow Ce^{3+})$ becomes $3e^- + 3Ce^{4+} \rightarrow 3Ce^{3+}$

(vi) Add the two reactions together and cancel the common substances:

$3Ce^{4+} + H_2O + Bi \rightarrow 3Ce^{3+} + BiO^{1+} + 2H^{1+}$

The coefficient before the cesium ions is 3.

30.   D   When multiplying and dividing, the final answer contains the same number of significant figures as the number with the fewest significant figures.

31.   E   Ethane is a hydrocarbon with all single bonds. This means that the two carbon atoms will both be sp^3 hybridized.

PRACTICE TEST—*Continued*

32.   E   There are three steps in completing this problem:

(i) Change the percent sign to grams (assume a 100-gram sample):

12.1 grams of C, 71.7 grams Cl, and 16.2 grams of O.

(ii) Convert the grams of each element to moles:

1 mole of C, 2 moles of Cl, and 1 mole of O.

(iii) Divide by the lowest number of moles:

In this case it is the number "1." The empirical formula will be CCl_2O.

33.   B   The total mass of this compound is 180. The oxygen makes up 96 of the 180 which is about 53%.

34.   E   The positive value for the change in entropy means that there will be more disorder. An expanding universe and a food fight are sure signs of more disorder.

35.   C   Carbon dioxide and water can react to produce carbonic acid. CFCs are responsible for causing a hole in the ozone layer and methane is a greenhouse gas that can trap heat on earth.

36.   C   Because the gas pressure inside the manometer is pushing the level of mercury higher and toward the opening of the tube, the pressure inside the manometer must be greater than the atmospheric pressure. The mercury rose to a level that is 65 mm above the height of the bulb in the manometer. This means that the atmospheric pressure is 65 mm lower than the pressure of the gas. 815 – 65 = 750 torr.

37.   D   2.5 molal NaCl will be, in effect, 5.0 molal because 1 mole of sodium chloride yields 2 moles of ions. This is the highest concentration of any of the choices and will have the greatest effect on the boiling and freezing points of water.

38.   A   A stronger acid or base will also be a better electrolyte because more ions will be released into solution. The greatest value listed is 1.0×10^{-2}.

39.   D   Gases, like HCl, will experience a decrease in solubility as the temperature of the solution they are dissolved in increases. The solids will all have an increase in solubility as the temperature increases.

40.   B   67.2 liters of carbon dioxide equates to 3 moles of carbon dioxide. Because carbon dioxide and methane are in a 1:1 ratio, for each mole of carbon dioxide produced, 1 mole of methane reacted. This means that 3 moles of methane were burned. Because the molar mass of methane is 16, 3 moles of methane would weigh 48 grams.

41.   D   Remember for equilibrium constants, "Products over reactants, coefficients become powers." This is demonstrated by choice D.

PRACTICE TEST—*Continued*

42. D Substitute into the expression

$$K_{eq} = \frac{[NO_2Cl]^2}{[Cl_2][NO_2]^2} \quad \text{to get:} \quad K_{eq} = \frac{[0.5]^2}{[0.3][0.2]^2}$$

Solving gives an answer of 20.83 indicating that the product was favored in this reaction.

43. E Boron will not form an octet, as shown correctly in the other four choices. Boron prefers six electrons in its outermost principal energy level.

44. C The Lewis definition of acids states that acids are electron pair acceptors while bases are electron pair donors. Choices A, D, and E show the Arrhenius definition whereas choice B shows the Brønsted-Lowry definition.

45. A By definition, an aqueous solution must be homogeneous. The KI(aq) tells that there is a homogeneous solution of water and KI.

46. C Isomers have the same molecular formula but a different structure. This also means that isomers will have different names. The isomers in this question are 2-chloropropane and 1-chlorpropane.

47. E Eating, drinking, and removing one's goggles are all unsafe in the laboratory setting. Pushing a glass thermometer through a rubber stopper is also dangerous and should be done by a trained laboratory specialist.

48. B There are 4 pi bonds in this molecule. Two of them are between the N and C. One is between the two carbon atoms that have a double bond. The last pi bond is between the C and the O. When C and O are written as shown in this problem, it means that a double bond is present. A double bond has 1 pi bond.

49. C When balancing a reaction, leave the simplest substance for last. In this case it will be the oxygen gas. Balancing the carbon atoms and hydrogen atoms gives $C_2H_6 + O_2 \rightarrow 2CO_2 + 3H_2O$. Balancing the oxygen atoms gives $C_2H_6 + 3.5O_2 \rightarrow 2CO_2 + 3H_2O$. To get whole number coefficients, multiply the entire equation by 2: $2C_2H_6 + 7O_2 \rightarrow 4CO_2 + 6H_2O$.

50. E First, double the reaction for A, B and C so that 4C ends up on the product side:

$2A + 2B \rightarrow 4C \ \Delta H = -1,000 \text{ kJ}$

Next, double the second reaction so that 2E can be produced:

$2D + 4B \rightarrow 2E \ \Delta H = -1400 \text{ kJ}$

Finally, switch the last reaction so that *F* appears on the reactant side:

$F \rightarrow 2D + 2A \ \Delta H = -50 \text{ kJ}$

Adding up the three steps shows that the heat of reaction is −2,450 kJ.

51. E All three masses given are equivalent to 1 mole of the compounds in question. Because the 1 mole samples are all dissolved to make 1 liter of solution, each solution is 1 molar.

PRACTICE TEST—*Continued*

52.　　C　　68 grams of ammonia (molar mass is 17) is 4 moles of ammonia. Setting up a proportion, you see that 6 moles of hydrogen gas and 2 moles of nitrogen gas are needed: $3H_2 + N_2 \rightarrow 2NH_3$ is doubled and becomes $6H_2 + 2N_2 \rightarrow 4NH_3$. 6 moles of hydrogen gas (molar mass is 2) is 12 grams of hydrogen gas. 2 moles of nitrogen gas (molar mass is 28) is 56 grams of nitrogen gas.

53.　　D　　Although it does use the prefix "tri-" correctly for a covalent compound, PCl_3 is phosphorus trichloride.

54.　　C　　6 grams of hydrogen gas (molar mass is 2) is 3 moles of hydrogen gas. This means that 3 moles of iodine will react too because the hydrogen and iodine are in a 1:1 ratio. 3 moles of iodine (molar mass is 127) are needed, so it turns out that 762 grams of iodine are needed. This puts 38 grams of iodine in excess. Because the reactants have been tripled in quantity, the amount of HI made from the balanced equation will also triple so that the equation reads: $3H_2 + 3I_2 \rightarrow 6HI$. 6 moles of HI (molar mass is 128) is 768 grams of HI.

55.　　D　　The initial volume, V_1, is 500 mL and the initial molarity, M_1, is 0.2 M. The new volume, V_2, is 700 mL because 200 mL of water were added to the original 500 mL. Using the equation $M_1V_1 = M_2V_2$ substitute and find that $(0.2)(500) = (M_2)(700)$. Solving for M_2 you see that the concentration has decreased as it should when diluted: M_2 is 0.14 M.

56.　　B　　A distillation can boil the solution and drive off the water from the solution. The water vapor will then enter a condenser where it is cooled and turns back into a liquid. This separates the water from the salt that will be left behind.

57.　　E　　Halides of lead, mercury, and silver will form precipitates, so choices A, B, and D are all precipitates. Calcium carbonate is also a solid. Sodium hydroxide is soluble in water.

58.　　A　　Because there were already sulfate ions in solution and more sulfate ions were added, sulfate ion is called the common ion. Adding sodium sulfate to the solution increases the concentration of sulfate ion in solution driving the reverse reaction. This is called the common ion effect and more of the solid calcium sulfate will be made. If the solid is being formed that means that it is not dissolving and the solubility has decreased.

59.　　C　　An increase in temperature will increase the amount of heat, which is one of the reactants. Because a reactant was added, more products will be made.

60.　　E　　This calculation requires using the heat of fusion of water. The equation is $q = H_f m$. Substitution gives $q = (333.6 \text{ J/g})(16 \text{ grams}) = 5,334$ joules of heat.

61.　　E　　All three substances can sublime. Naphthalene is the substance that is used to make mothballs. Solid carbon dioxide is called dry ice. Iodine is a purple solid that can sublime as well.

62.　　C　　A decrease in temperature will cause the molecules to move with less kinetic energy. This means that the collisions will occur less frequently and will not be as effective.

PRACTICE TEST—*Continued*

63.   D   An equal number of moles of a gas will contain an equal number of gas molecules. An equal number of gas molecules will contribute to the pressure equally. The partial pressures of these gases are:

A is 30% of the mixture so (0.3)(660) = 198 torr.

B is 40% of the mixture so (0.4)(660) = 264 torr.

C is 30% of the mixture so (0.3)(660) = 198 torr.

64.   B   Looking at the two reactions as oxidations shows that $Al \rightarrow Al^{3+} + 3e^-$ has a potential of +1.66 V while $Ni \rightarrow Ni^{2+} + 2e^-$ has a potential of +0.26 V. This indicates that the loss of electrons from Al is more of a spontaneous process than the loss of electrons from Ni. Al has a more positive electrode potential and will lose electrons to the Ni. Al will lose electrons and be the anode. The Ni electrode will gain electrons and be the cathode. At the cathode there is a gain of electrons for the ions that are in the solution where the cathode is located. The electrons will react with the Ni^{2+} ions according to the reaction $Ni^{2+} + 2e^- \rightarrow Ni$.

65.   A   From above we have the two half reactions:

$Al \rightarrow Al^{3+} + 3e^-$ +1.66 V

$Ni^{2+} + 2e^- \rightarrow Ni$ −0.26 V

The total is 1.40 V. Also remember that the electrode potential is never multiplied by any coefficients in a balanced equation.

66.   C   Remember that pH is based upon logarithms and base-10. Because the pH changed by a value of 1.0, it has changed by a power of 10. Because the pH value dropped, the stream has become more acidic.

67.   B   An alkaline earth metal is found in group 2 of the periodic table. This means that it will have two valance electrons and form an ion with a charge of 2+. When it reacts with oxygen's ionic charge of 2−, the two ions will combine in a 1:1 ratio.

68.   D   A carbonyl group is characterized by a C==O. Ethers have an oxygen atom but the oxygen has only single bonds, R—O—R.

69.   E   Set up this problem:

[2(H—H) + 1(O—O)] − [4(H—O)]. Use the reference tables and substitute to get:

[2(435) + 1(145)] − [4(462)] =

[870+145] − [1,848] =

[1,015] − [1,848] = −833 kJ.

70.   B   A reversible reaction that has equal forward and reverse rates is a reaction that has achieved equilibrium.

PRACTICE TEST 4

▒ SCORE SHEET

Number of questions correct: _____

Less: 0.25 × number of questions wrong: _____

(Remember that omitted questions are not counted as wrong.)

Equals your raw score: _____

Raw Score	Test Score		Raw Score	Test Score		Raw Score	Test Score		Raw Score	Test Score		Raw Score	Test Score
85	800		63	710		41	570		19	440		−3	300
84	800		62	700		40	560		18	430		−4	300
83	800		61	700		39	560		17	430		−5	290
82	800		60	690		38	550		16	420		−6	290
81	800		59	680		37	550		15	420		−7	280
80	800		58	670		36	540		14	410		−8	270
79	790		57	670		35	530		13	400		−9	270
78	790		56	660		34	530		12	400		−10	260
77	790		55	650		33	520		11	390		−11	250
76	780		54	640		32	520		10	390		−12	250
75	780		53	640		31	510		9	380		−13	240
74	770		52	630		30	500		8	370		−14	240
73	760		51	630		29	500		7	360		−15	230
72	760		50	620		28	490		6	360		−16	230
71	750		49	610		27	480		5	350		−17	220
70	740		48	610		26	480		4	350		−18	220
69	740		47	600		25	470		3	340		−19	210
68	730		46	600		24	470		2	330		−20	210
67	730		45	590		23	460		1	330		−21	200
66	720		44	580		22	460		0	320			
65	720		43	580		21	450		−1	320			
64	710		42	570		20	440		−2	310			

Note: This is only a sample scoring scale. Scoring scales differ from exam to exam.

APPENDIXES

APPENDIX 1
MATHEMATICAL SKILLS REVIEW

IN THIS SECTION YOU WILL LEARN ABOUT...

Converting Units
Scientific Notation and Exponents
Logarithms
Significant Figures

CONVERTING UNITS

It is not difficult to convert units provided that you know the meaning and value of the prefixes that are used in the metric system. There are two methods for converting units: dimensional analysis and setting up a proportion. Even though setting up a proportion may seem easier because you have been doing it since elementary school, dimensional analysis can be just as easy and effective with a little bit of practice. Look at both methods as they are presented here, and then choose the one that works best for you. Remember that the SAT II exam is an all multiple-choice exam. So the method you use for solving problems is not going to be checked.

Consider the example of eggs and the term "dozen." The first "burning question" will be, "How many eggs are there in two dozen eggs?" Off the top of your head you can easily say that there are 24 eggs in two dozen eggs—easy to do when you have been doing it all your life. If you were to set up a proportion you could say that if one dozen has 12 eggs, then two dozen has x number of eggs:

$$\frac{1 \text{ dozen}}{2 \text{ dozen}} = \frac{12 \text{ eggs}}{x}$$

Cross-multiplying and dividing gives 24 = 1x and x equals 24 eggs upon solving. To use dimensional analysis, you would start with the unit and value given, and then multiply it by the conversion factor. The conversion factor has a numerator and a denominator. Place the unit that you want to cancel out in the denominator and the unit you are converting to in the numerator:

$$2 \text{ dozen eggs} \frac{(12 \text{ eggs})}{(1 \text{ dozen eggs})} = 24 \text{ eggs}$$

In effect you have multiplied the "2 dozen" by the number 1 because 12 eggs and 1 dozen are exactly the same thing. The units "dozen eggs" cancel and the unit "eggs" remains.

Now convert 3.5 kilograms to grams. By proportion you could set up the following:

$$\frac{1 \text{ kilogram}}{3.5 \text{ kilograms}} = \frac{1,000 \text{ grams}}{x \text{ grams}} \qquad \text{or you can set up a dimensional analysis:}$$

$$3.5 \text{ kilograms} \frac{(1,000 \text{ grams})}{(1 \text{ kilogram})} = 3,500 \text{ grams}$$

Sometimes you might be required to set up multiple conversions. Such is the case when finding the number of hours in a week:

$$(1 \text{ week}) \frac{(7 \text{ days})(24 \text{ hours})}{(1 \text{ week})(1 \text{ day})} = 168 \text{ hours}$$

Here the units "week" and "day" canceled out. Simply multiply straight across to arrive at the answer.

This last example uses a dimensional analysis to convert between moles, mass, and volume. How many liters will 150 grams of $SO_2(g)$ occupy at STP? Remember the importance of keeping careful track of the numbers, units, and substance in a problem such as this one. Start by converting to moles and then to volume:

$$150 \text{ grams } SO_2 \ \frac{(1 \text{ mole } SO_2)}{(64 \text{ grams } SO_2)} \frac{(22.4 \text{ L } SO_2)}{(1 \text{ mole } SO_2)} = 52.5 \text{ liters}$$

Next, multiply 150 by 22.4 and then divide that by 64. The final answer is 52.5 liters.

SCIENTIFIC NOTATION AND EXPONENTS

Earlier in this book you were presented with Avogadro's number: the mole or 6.02×10^{23}. Surely you agree that writing this number in scientific notation is easier than writing 602 followed by 21 more zeros! The first rule of scientific notation is that there must be "no naked decimals"; that is, there must be a digit to the left and to the right of the decimal. The next thing to do is to interpret the exponents properly. Exponents that are positive in value tell you to move the decimal to the right by "so many" places; negative exponents tell you to move the decimal to the left. For example, the number 5.0×10^3 translates into 5,000 while the number 5.0×10^{-3} translates into 0.005.

Multiplying exponents is done by adding the exponents that are to be multiplied. For example $(5.0 \times 10^3)(1.5 \times 10^3) = 7.5 \times 10^6$. To divide exponents, subtract the exponent in the denominator from the exponent in the numerator. For example,

$$\frac{6.0}{2.0} \times \frac{10^8}{10^5} = 3.0 \times 10^3$$

Sometimes the exponents are negative and you must remember the rules for subtracting numbers that are negative. One such rule is as follows: when subtracting from a negative number, change the minus sign to a plus sign and then change the sign of the number after the minus sign. For example, in the division problem $8.0 \times 10^{-8} / 2.0 \times 10^{-3}$ you know that the answer will be 4.0×10 "to the something." That something is: $-8 - (-3)$ which becomes $-8 + (+3) = -5$. The answer is 4.0×10^{-5}.

LOGARITHMS

Logarithms are based on powers of 10. The general formula for the base-10 logarithm is $10^x = y$ or $\log y = x$. For example, because 10^2 equals 100, the log 100 is equal to 2. In Chapter 9 you saw how to calculate pH from the concentration of hydronium ion in solution. The equation used was $pH = -\log[H^{1+}]$. To find the pH, first type in the concentration of the hydronium ion, hit the [LOG] key on the calculator and then the [+/-] key to negate the calculation. If the $[H^{1+}]$ were $2.5 \times 10^{-4}\ M$ in solution, you would first type 2.5×10^{-4} into the calculator. Hitting the [LOG] key you get -4.60. After hitting the [+/-] key the final answer is 4.60.

What do you do if you are given a pH and asked to find the $[H^{1+}]$? What is the $[H^{1+}]$ in a solution with a pH of 3.50? In this case you have to work things "backward." First, negate the pH, -3.50. Then press the [SHIFT] or "second function" button on the calculator, and then press the [LOG] key to perform an "antilog." The result is $3.16 \times 10^{-4}\ M$.

One last word about logarithms and "taking the log" of a number: recall from high school mathematics that $\log xy = \log x + \log y$ and that $\log x/y = \log x - \log y$. Earlier in this book you saw the equation $[H^{1+}][OH^{1-}] = 1 \times 10^{-14}$. If you apply $\log xy = \log x + \log y$ to the equation for the autoionization of water, you arrive at a new equation, $pH + pOH = 14$.

SIGNIFICANT FIGURES

Take a calculator and perform the calculation $5.31 \div 8.568$. How many decimal places does your calculator provide? If you used a calculator provided with your computer you probably got an answer on the order of 0.6197478991596638655462184873949 6! Given that many of us are money conscious, most people would give an answer that goes to two decimals as on a price tag and say that the answer is 0.62. The rules for significant figures tell a different story. There is absolutely no need for 32 decimal places! The rules for significant figures tell first, how to identify which figures are worthy of attention and, second, how to "treat" numbers in a calculation.

The rules for significant figures and examples of each rule are as follows:

- All nonzero digits are significant. Examples are 1, 2, 3, etc. . . . .
- Zeros between nonzero digits are significant. These zeros are holding places between the other digits. An example is 403. Here all three digits are significant. The "4" tell you the number of hundreds, the "0" tells you the number of tens, and the "3" tells you the number of ones.
- Initial zeros are not significant. For example, in 0.0203 the first two zeros can be replaced with scientific notation and the number can be written as 2.03×10^{-2}. This example has only three significant figures.

- Final zeros are not significant if no decimal is present. For example, in 6,700 only the six and the seven are significant. The final two zeros can be replaced with scientific notation and the number can be written as 6.7×10^3.
- Final zeros are significant if a decimal is present. For example, 200. and 2,500. have three and four significant figures, respectively.
- Final zeros located after a decimal are significant. For example, the numbers 2.50 and 3.7770 have three and five significant figures, respectively.

Now that you can identify the figures that are significant and "have meaning," you can perform the four basic mathematical functions. When adding and subtracting numbers, the answer has the same number of decimal places as the number in the calculation with the fewest number of decimal places. The calculation is rounded off to the proper number of decimal places.

Example:

34.34

2.6   ← has the fewest number of decimal places (one decimal place)

+ 1.341

38.281 becomes 38.3 (has one decimal place and was rounded off)

When multiplying and dividing numbers, the answer has the same number of significant figures as the number in the calculation with the fewest number of significant figures. Once again, round off the final answer.

Example:

6.24   ← has three significant figures

× 4.3   ← has two significant figures

26.832 becomes 27!

In this case the answer must have two significant figures. Rounding off, 26.832 becomes 27—no decimals at all!

Returning to the original problem, when you divide 5.31 by 8.568, you get 0.61974789915966386554621848739496 on your calculator. However, because you used the mathematical operation division, your answer must have just three significant figures. Rounding off 0.6197 becomes 0.620.

APPENDIX 2

EQUATIONS AND SYMBOLS

Density	$D = m/V$	C: Degrees Celsius
Density of Gases	D_{gas} = molar mass/22.4 L	c: Specific Heat
Change in Heat	$\Delta H = PEP - PER$	D: Density
Boyle's Law	$P_1V_1 = P_2V_2$	E: Calculated Electrode Potential
Charles' Law	$\dfrac{V_1}{T_1} = \dfrac{V_2}{T_2}$	E°: Standard Electrode Potential
Temperature in Kelvin	$K = C + 273$	$\mathcal{F}$: Faraday
Combined Gas Law	$\dfrac{P_1V_1}{T_1} = \dfrac{P_2V_2}{T_2}$	G: Gibbs Free Energy
Dalton's Law of Partial Pressures	$P_{total} = P_{gas1} + P_{gas2} + P_{gas3}$	H: Heat of Reaction
Graham's Law of Effusion	$\dfrac{r_1}{r_2} = \sqrt{\dfrac{M_2}{M_1}}$	H_f: Heat of Fusion H_v: Heat of Vaporization
Ideal Gas Law Equation	$PV = nRT$	K: Kelvin
Maximum Number of Electrons in a PEL	$2n^2$	K_{eq}: Equilibrium Constant
Percent Composition	$\dfrac{\text{Total mass of element}}{\text{Molar mass of compound}} \times 100\%$	K_w: Equilibrium Constant for Water
Molarity	$M = \dfrac{\text{Moles of solute}}{\text{Liters of solution}}$	m: Mass *m:* Molality
Molality	$m = \dfrac{\text{Moles of particles}}{\text{Kilograms of solvent}}$	M: Molar Mass *M:* Molarity
Dilution	$M_1V_1 = M_2V_2$	n: Moles
General Equation for Equilibrium Constant	$K_{eq} = \dfrac{[C]^c[D]^d}{[A]^a[B]^b}$	n: PEL Number
Titration	$M_aV_a = M_bV_b$	P: Pressure
Equilibrium Constant for Water	$K_w = [H^{1+}][OH^{1-}]$	PE: Potential Energy
Heat	$q = mc\Delta T$	Q: Concentration of Products Divided by Concentration of Reactants
Heat of Fusion	$q = H_f m$	

(continued)

Heat of Vaporization	$q = H_v m$	q: Heat
Gibbs Free Energy	$\Delta G = \Delta H - T\Delta S$	r: Rate of Effusion
Nernst	$E = E° - \dfrac{2.30\,RT}{n\mathcal{F}}\,(\log Q)$	S: Entropy
		T: Temperature
Percent Composition	$\dfrac{\text{Measured value } - \text{ Accepted value}}{\text{Accepted value}} \times 100\%$	V: Volume

APPENDIX 3

PERIODIC TABLE OF THE ELEMENTS

Periodic Table of the Elements

1	2	3	4	5	6	7	8	9	10	11	12	13	14	15	16	17	18
1 H 1.0079																	2 He 4.0026
3 Li 6.941	4 Be 9.0122											5 B 10.81	6 C 12.011	7 N 14.007	8 O 15.999	9 F 18.998	10 Ne 20.179
11 Na 22.989	12 Mg 24.305											13 Al 26.981	14 Si 28.086	15 P 30.974	16 S 32.06	17 Cl 35.453	18 Ar 39.948
19 K 39.098	20 Ca 40.08	21 Sc 44.956	22 Ti 47.88	23 V 50.941	24 Cr 51.996	25 Mn 54.938	26 Fe 55.847	27 Co 58.933	28 Ni 58.69	29 Cu 63.546	30 Zn 65.38	31 Ga 59.72	32 Ge 72.59	33 As 74.922	34 Se 78.96	35 Br 79.904	36 Kr 83.80
37 Rb 85.468	38 Sr 87.62	39 Y 88.906	40 Zr 91.22	41 Nb 92.905	42 Mo 95.94	43 Tc (98)	44 Ru 101.07	45 Rh 102.91	46 Pd 106.42	47 Ag 107.87	48 Cd 112.41	49 In 114.82	50 Sn 118.69	51 Sb 121.75	52 Te 127.60	53 I 126.90	54 Xe 131.29
55 Cs 132.91	56 Ba 137.33	57 * La 138.90	72 Hf 178.49	73 Ta 180.95	74 W 183.85	75 Re 186.21	76 Os 190.2	77 Ir 192.22	78 Pt 195.08	79 Au 196.97	80 Hg 200.59	81 Tl 204.38	82 Pb 207.2	83 Bi 208.98	84 Po (209)	85 At (210)	86 Rn (222)
87 Fr (223)	88 Ra 226.0	89 # Ac 227.03	104 Rf (261)	105 Db (262)	106 Sg (263)	107 Bh (262)	108 Hs (265)	109 Mt (266)	110 Uun (269)	111 Uuu (272)	112 Uub (277)						

* Lanthanides

58 Ce 140.12	59 Pr 140.91	60 Nd 144.24	61 Pm (145)	62 Sm 150.36	63 Eu 151.96	64 Gd 157.25	65 Tb 158.92	66 Dy 162.50	67 Ho 164.93	68 Er 167.26	69 Tm 168.93	70 Yb 173.04	71 Lu 174.97

Actinides

90 Th 232.03	91 Pa 231.03	92 U 238.03	93 Np 237.05	94 Pu (244)	95 Am (243)	96 Cm (247)	97 Bk (247)	98 Cf (251)	99 Es (254)	100 Fm (257)	101 Md (257)	102 No (255)	103 Lr (256)

APPENDIX 4

REFERENCE TABLES

Physical Constants for Water

Normal freezing point	0°C or 273 K
Normal boiling point	100°C or 373 K
Freezing point depression	1.86°C / 1 m
Boiling point elevation	0.52°C / 1 m
Autoionization constant of water at 298 K	$K_w = 1.0 \times 10^{-14}$
Specific heat	4.18 J/g K
Heat of fusion	333.6 J/g
Heat of vaporization	2259 J/g

Electronegativity Values

Aluminum	1.6
Bromine	3.0
Calcium	1.0
Carbon	2.6
Chlorine	3.2
Fluorine	4.0
Hydrogen	2.2
Iodine	2.7
Lithium	1.0
Magnesium	1.3
Nitrogen	3.0
Oxygen	3.5
Phosphorus	2.2
Potassium	0.8
Sodium	0.9
Sulfur	2.6

Solubility Product Constants at 298 K

Lead iodide—PbI_2	7.1×10^{-9}
Lead sulfate—$PbSO_4$	1.6×10^{-8}
Magnesium hydroxide—$Mg(OH)_2$	1.8×10^{-11}
Silver chloride—$AgCl$	1.8×10^{-10}

K_{a1} Constants for Weak Acids at 298 K

Acetic, $HC_2H_3O_2$	1.8×10^{-5}
Chlorous, $HClO_2$	1.2×10^{-2}
Hydrofluoric, HF	6.8×10^{-4}
Hydrogen Sulfide, H_2S	5.7×10^{-8}
Hypochlorous, HClO	3.0×10^{-8}
Phosphoric, H_3PO_4	7.5×10^{-3}
Sulfurous, H_2SO_3	1.7×10^{-2}

Bond Dissociation Energies in kJ/mol

C—C	349
C—Cl	329
C—H	412
C=O	798
Cl—Cl	240
H—Cl	430
H—H	435
N—H	390
N—N	163
N≡N	941
O—H	462
O—O	145

Select Polyatomic Ions

Ammonium	NH_4^{1+}
Carbonate	CO_3^{2-}
Chlorate	ClO_3^{1-}
Chlorite	ClO_2^{1-}
Chromate	CrO_4^{2-}
Cyanide	CN^{1-}
Dichromate	$Cr_2O_7^{2-}$
Hydronium	H_3O^{1+}
Hydroxide	OH^{1-}
Nitrate	NO_3^{1-}
Nitrite	NO_3^{1-}
Permanganate	MnO_4^{1-}
Phosphate	PO_4^{3-}
Sulfate	SO_4^{2-}
Sulfite	SO_3^{2-}

Standard Electrode Potentials for Elements on the Activity Series

Nonmetals

$F_2 + 2e^- \rightarrow 2F^{1-}$	+2.87 V
$Cl_2 + 2e^- \rightarrow 2Cl^{1-}$	+1.51 V
$Br_2 + 2e^- \rightarrow 2Br^{1-}$	+1.06 V
$I_2 + 2e^- \rightarrow 2I^{1-}$	+0.54 V

Metals

$Li^{1+} + 1e^- \rightarrow Li$	−3.05 V
$K^{1+} + 1e^- \rightarrow K$	−2.93 V
$Na^{1+} + 1e^- \rightarrow Na$	−2.71 V
$Mg^{2+} + 2e^- \rightarrow Mg$	−2.37 V
$Al^{3+} + 3e^- \rightarrow Al$	−1.66 V
$Zn^{2+} + 2e^- \rightarrow Zn$	−0.76 V
$Cr^{3+} + 3e^- \rightarrow Cr$	−0.74 V
$Fe^{2+} + 2e^- \rightarrow Fe$	−0.45 V
$Co^{2+} + 2e^- \rightarrow Co$	−0.28 V
$Ni^{2+} + 2e^- \rightarrow Ni$	−0.26 V
$Sn^{2+} + 2e^- \rightarrow Sn$	−0.14 V
$Pb^{2+} + 2e^- \rightarrow Pb$	−0.13 V
$2H^{1+} + 2e^- \rightarrow H_2$	**0.00 V***
$Cu^{2+} + 2e^- \rightarrow Cu$	+0.34 V
$Ag^{1+} + 1e^- \rightarrow Ag$	+0.80 V
$Au^{3+} + 3e^- \rightarrow Au$	+1.50 V

*Denotes arbitrary standard.

APPENDIX 5
GLOSSARY

Absolute Zero	The lowest achievable temperature of 0 Kelvin or –273°C.
Accuracy	How close data come to the accepted or "real" value.
Actinides	Elements with the atomic numbers 90 through 103.
Activation Energy	The energy needed to start a reaction.
Alcohols	Organic compounds that have the function group R—OH.
Alkali Metals	Group 1 metals of the periodic table.
Alkaline Earth Metals	Group 2 metals of the periodic table.
Alkanes	Saturated hydrocarbons that contain all single bonds.
Alkenes	Unsaturated hydrocarbons that have a double bond between two carbons.
Alkyl Halides	Class of organic compounds in which a halogen is bonded to the organic molecule.
Alkynes	Unsaturated hydrocarbons that have a triple bond between two carbons.
Allotropes	Different substances in the same phase formed from the same elements.
Alpha Particles	Particles containing two protons and two neutrons. These particles are identical to helium-4 nuclei. The symbols are ^4_2He or α.
Amides	Organic compounds that have the function group R—CO—NH$_2$.
Amines	Organic compounds that have the function group R—NH$_2$.
Amphoteric	Describes a substance that can act as either an acid or a base.
Anion	A negatively charged ion.
Anode	Electrode where oxidation occurs.
Artificial Transmutation	A nuclear reaction in which an isotope is being bombarded with a particle to trigger the transmutation.
Atom	Composed of protons, neutrons, and electrons, an atom is a particle that defines an element.
Atomic Mass	The atomic mass takes into account all the masses of the isotopes of an atom and their relative abundance.
Atomic Number	Number of protons located in the nucleus of an atom. Can also be defined as the nuclear charge of an atom.

Atomic Radius — The distance from the atom's nucleus to the outermost electron of that atom.

Atomic Theory — Theory of the atom as stated by John Dalton: All matter is composed of atoms; all atoms of a given atom are alike; compounds are made up of atoms combining in fixed proportions; a chemical reaction involves the rearrangement of atoms; and atoms are neither created nor destroyed in a chemical reaction.

Avogadro's Number — One mole, or 6.02×10^{23}.

Beta Particle — An electron that is ejected from the atom's nucleus.

Binary Compounds — Compounds that have only two different elements present.

Boiling Point — The point at which the vapor pressure of a liquid is equal to the surrounding/atmospheric pressure.

Boyle's Law — A gas law stating that at constant temperature, pressure and volume have an inverse relationship.

Buffer — A solution that is resistant to changes in pH.

Calorie — A measure of heat energy; 1 calorie is equal to 4.18 joules.

Carbonyl Group — Part of an organic compound characterized by the double bond between a carbon atom and an oxygen atom, C=O.

Cathode — Electrode that is the site of reduction.

Cation — An ion with a positive charge.

Celsius — A measure of temperature in which the freezing point of water is 0°C and the boiling point of water is 100°C.

Chain Reaction — Reaction in which one event causes multiple events to occur until all materials have been consumed.

Charles' Law — A gas law stating that at constant pressure, temperature and volume are directly proportional.

Chemical Formulas — An expression of the composition of a compound by a combination of symbols and figures that show which elements are present and how much of each element is in a compound.

Chemical Properties — Properties that are observed with regard to how a substance reacts with other substances.

Coefficient — Numerical indication of the quantity of a substance in an equation.

Colligative Properties — The properties of a solvent that depend on the concentration of dissolved particles present.

Combined Gas Law — A gas law that combines the laws of Charles and Boyle.

Common Ion Effect — A decrease in the solubility of a salt due to the shift in equilibrium when an ion is added to the solution.

Compound — Two or more elements combined with definite proportions.

Conjugate Acid — The acid formed when a Brønsted-Lowry base gains a proton.

Conjugate Base	The base formed when a Brønsted-Lowry acid loses a proton.
Conjugate Pair	An acid or base that differs only in the presence or absence of a proton.
Conservation of Charge	The sum of the charges of the reactants will be equal to the sum of the products.
Coordinate Covalent Bond	A covalent bond in which one atom donates both electrons.
Covalent Bond	A bond formed when two nonmetal atoms share electrons in order to satisfy their need to have a full outermost principal energy level.
Dalton's Law of Partial Pressures	A law stating that the combined pressure of a combination of gases is equal to the sum of the individual pressures of the gases.
Decay Series	Series of decays an isotope will undergo until a stable isotope is formed.
Decomposition	The process by which one compound breaks down into many substances.
Density	Mass per unit of volume.
Deposition	Changing from the gas phase to the liquid phase without any apparent solid phase in between.
Dipole	The condition in which a molecule has a "buildup" of negative charge on one side and a positive charge on another side.
Dispersion Forces	Weak forces existing between nonpopular molecules. Also known as Van der Waals forces.
Double Bond	A covalent bond that involves the sharing of two pairs of electrons.
Double Replacement	Reaction in which two elements exchange anions and cations to form the products.
Ductile	Has the ability to be rolled into thin wires.
Electrode Potentials	Voltage of a given oxidation or reduction half reaction.
Electrodes	Sites for oxidation and reduction.
Electrolysis	A reaction in which electricity is used to make a nonspontaneous reaction occur.
Electrolyte	A solute that creates ions in solution that can carry an electrical current.
Electrolytic Cell	A device that requires an outside source of current to make a non-spontaneous reaction occur.
Electron	A negatively charged particle that orbits the nucleus of an atom in the principal energy levels.
Electronegativity	A measure of an atom's ability to attract electrons.
Electroplating	Coating a substance with a metal.
Element	A substance that is unable to be broken down chemically.
Empirical Formula	Shows the lowest ratio of all the elements of a compound to each other.
End Point	Point of a titration where the indicator changes color.

Endothermic   When more energy is absorbed than released in a chemical reaction.

Energy   The ability to do work.

Enthalpy   The heat absorbed or released in a chemical reaction. Also known as the heat of reaction.

Entropy   Used to describe chaos, randomness, and disorder.

Equilibrium   A state of balance between two opposing reactions that are occurring at the same rate.

Ethers   Organic compounds that have the functional group R—O—R.

Excess Reagent   The compound that does not completely react in a chemical reaction.

Excited State   The movement of electrons to a higher energy level once energy has been added to an atom.

Exothermic   Describing a chemical reaction in which more energy is released than absorbed.

Families   The vertical columns on the periodic table.

Faraday   The charge of one mole of electrons. A charge of approximately 96,500 coulombs.

Filtrate   The aqueous portion of a sample that has been poured through filter paper.

Fission   The splitting of larger nuclei into smaller ones, causing a release of nuclear energy.

Freezing   The process by which particles of the liquid phase enter the solid phase.

Functional Groups   Particular arrangement of atoms in organic compounds.

Fusion   The joining of smaller nuclei to form a larger one, causing a release of nuclear energy.

Gamma Rays   High-energy electromagnetic radiation emitted from the nucleus of a radioactive atom.

Gas   A phase of matter characterized as having no definite volume or shape and having molecules spaced far apart.

Geiger Counter   A device used to detect and measure the activity of radioactive particles.

Gibbs Free Energy   Equation used to determine if a reaction will be spontaneous.

Graham's Law   At the same temperature and pressure, gases effuse at a rate inversely proportional to the square roots of their molecular masses.

Ground State   When the electrons are in their lowest energy state.

Group   Vertical column on the periodic table.

Half Cell   Part of a voltaic cell where oxidation or reduction can occur.

Half-Life   The amount of time it takes for half a radioactive substance to decay.

Half Reactions   Two separate reactions that show the oxidation and reduction reactions separately.

Halogens   Elements found in group 17 of the periodic table.

Heat of Reaction	The heat absorbed or released in a chemical reaction. Also known as *enthalpy.*
Hess's Law	The sum of the heats of reaction of the steps in a reaction is equal to the overall heat of reaction.
Heterogeneous	Describing a mixture that is not the same throughout.
Homogeneous	Describing a mixture that is the same throughout.
Hund's Rule	Electrons will fill an orbital singly to the maximum extent possible before pairing up.
Hybridization	The promotion of an electron to a higher energy level so that the atom can bond to another atom.
Hydrocarbon	An organic compound that consists of only the elements hydrogen and carbon.
Hydrogen Bonding	A weak force that comes about when hydrogen is bonded to fluorine, oxygen, or nitrogen.
Hydrolysis	The addition of water to a salt to form the acid and base from which the salt was made.
Ideal Gas Law	A law that states that an ideal gas obeys the equation $PV = nRT$.
Indictors	Substances that change color to indicate if a substance is acidic or basic.
Intermolecular Bond	A bond that exists between molecules.
Intramolecular Bond	A bond that exists between atoms.
Ion	An atom that has gained or lost electrons.
Ionic Bonds	Very strong bonds that are formed between a cation and an anion.
Ionization Energy	The energy needed to remove an electron from an atom to form an ion.
Isomers	Compounds with the same molecular formula but different structures.
Isotopes	Atoms that have the same atomic number but a different mass number due to having a different number of neutrons.
Joule	A measure of heat energy. 4.18 Joules is equal to 1.0 calories.
Kelvin	The Kelvin scale is based upon the lowest temperature that can be achieved, 0 K (absolute zero) or –273°C.
Kinetic Energy	Energy that is in motion.
Kinetic Molecular Theory	Set of rules that are assumed to govern the motion of molecules.
Lanthanides	Elements with the atomic numbers 58 through 71.
Lattice	Regular structure among the atoms in a solid.
Law of Conservation of Mass	The law stating that mass cannot be created or destroyed in a chemical reaction.

Le Châtelier's Principle — When a stress or change in conditions is applied to a system at equilibrium, the point of equilibrium will shift in such a manner as to relieve the applied stress.

Lewis Structure — A drawing of the structure of a compound in which the arrangement of the valence electrons is represented by the use of dots.

Limiting Reagent — Substance that is completely used up in a chemical reaction.

Line Spectrum — Specific wavelengths of light emitted from an atom when the electrons return to the ground state from the excited state.

Liquids — Have a definite volume, take the shape of the container they are placed in, and have touching molecules.

Litmus — Indicator that turns red in acid and blue in base.

Malleable — Has the ability to be hammered into thin sheets.

Mass — Measure of the quantity of particles in an object.

Mass Action Equation — An equation written that shows the product of the concentrations of the products raised to the power of their coefficients divided by the product of the concentrations of the reactants raised to the power of their coefficients is equal to a constant.

Mass Defect — The amount of mass of the particles involved in the nuclear reaction that is converted to energy.

Mass Number — The total number of nucleons (protons and neutrons) found in an atom.

Matter — Anything that has mass and takes up space.

Melting — Particles of the solid phase entering the liquid phase.

Melting Point — The temperature at which the particles of the solid phase enter the liquid phase.

Meniscus — The curvature of a liquid that is the result of the adhesive forces between the liquid's molecules and the walls of a glass container.

Metallic Bond — A bond in which the electrons are free to move among the metal atoms.

Metalloids — Elements that exhibit some of the properties of metals and nonmetals.

Metals — Elements that are characterized by the ability to conduct heat and electricity, have a shiny luster, and lose electrons in a chemical reaction.

Mixtures — The result of combinations of elements and/or compounds.

Molality — Way of expressing concentration. Ratio of moles of solute to kilograms of solvent.

Molar Volume — Volume (22.4 liters) that one mole of a gas will occupy at STP.

Molarity — Way of expressing concentration. The ratio of moles of solute to total liters of solution.

Mole — A unit of Avogadro's number. A mole of particles is equal to 6.02×10^{23} of those particles.

Mole Ratio — The ratio of the number of moles of one substance to the moles of another substance as dictated by the balanced equation.

Molecular Formulas	Indicate the total number of atoms of each element that are present in a covalently bonded molecule.
Molecule-Ion Attraction	Attraction between charged ions and polar molecules in a solution.
Natural Transmutation	Transmutation that does not need to be triggered by a particle bombarding the isotope.
Network Solid	Nonmetal atoms bonding to each other in a covalent fashion to form a continuous network.
Neutralization	The process in which an acid and a base react to form salt and water.
Neutron	A particle with no charge that is found in the nucleus of an atom.
Noble (Inert) Gases	Gases found in group 18 of the periodic table.
Nonmetals	Elements that are characterized by being poor conductors of heat and electricity, being soft and brittle, and tending to gain electrons to form anions.
Nonpolar Covalent Bond	A covalent bond in which the electrons are shared and distributed equally.
Nucleons	Particles found in the nucleus of an atom (protons and neutrons).
Octet Rule	An atom will desire eight electrons in its outermost principal energy level to maximize its stability.
Orbital	Region around the atom where electrons are most likely to be found.
Organic Chemistry	Study of carbon and carbon-containing compounds.
Oxidation	A loss of electrons.
Oxidation Number	The charge on an ion or the charge that an atom "feels."
Oxidizing Agent	The reducing substance that causes the oxidation of other substances.
Pauli Exclusion Principle	A rule that states there cannot be more than two electrons in an atomic orbital. It also states that no two electrons can have the same four quantum numbers.
Percent Composition	Ratio of the total mass of an element in a compound to the total mass of the compound.
Period	Horizontal row on the periodic table.
pH	Negative logarithm of the hydrogen ion concentration of a solution.
Phenolphthalein	Indicator that is colorless in acid, and pink (or purple, magenta) in base.
Physical Properties	Observable and measurable properties of a substance.
Pi Bond	The second or third bond that is formed between hybridized atoms that have orbitals which overlap.
Polar Covalent Bond	A covalent bond that involves electrons not being shared equally.
Polyatomic Ions	Ions that have many atoms in them.
Positron	A particle that has the same mass as an electron but a charge of 1+.
Potential Energy	Energy that is stored.

Precipitate   An insoluble substance that separates from, and forms in, a solution.

Precision   How closely results from the same experiment agree with one another.

Pressure   The measurement of the ratio of the force exerted on an area.

Products   The results of a chemical reaction.

Proton   A particle found in the nucleus of an atom with a positive charge.

Quarks   Subatomic particles that make up protons and neutrons. Quarks have charges of +2/3 or –1/3.

Rate   Change in concentration over time.

Reactant   A substance used at the start of an equation.

Redox   Another term for oxidation and reduction.

Reducing Agent   The oxidized substance causing the reduction of other substances.

Reduction   A gain of electrons.

Residue   The solids that are trapped by filter paper.

Reversible Reaction   A reaction in which products formed further react to form the original reactants.

Rows   Horizontal rows on the periodic table.

Salt Bridge   An apparatus that allows ions to migrate from one half cell to another.

Saturated   Describing a solution in which a dissolved solute and an undissolved solute are in equilibrium.

Semimetals   Elements that exhibit some of the properties of metals and nonmetals.

Sigma Bond   A bond that arises from the overlap of two s orbitals or from the overlap of one s and one p orbital.

Single Replacement   Reaction where one element replaces another element.

Solids   Substances that have definite shape and volume. The atoms are in a rigid, fixed, regular geometric pattern.

Solubility   The ability of a substance to dissolve in another substance.

Solubility Product Constant The equilibrium constant of a slightly soluble salt.

Solute   A substance that is dissolved into a solvent.

Solution   A homogenous mixture of a solute and a solvent.

Solvent   A substance that a solute is dissolved into.

Spectators   Substances that do not take part in a reaction.

Spontaneous   A process that occurs without added external energy or without additional intervention.

Standard Pressure   Pressure characterized by pressures equal to 760 mm Hg, 760 torr, 101.3 kPa, or 1.0 atm.

Standard Temperature and Pressure	A common standard of conditions, defined as 0°C and 1 atm (273 K and 760 torr).
Stock Method	Method for naming compounds where a roman numeral is used to indicate the amount of positive charge on the cation.
Stoichiometry	The branch of chemistry that deals with the amounts of products produced from certain amounts of reactants.
Sublimation	Changing from the solid phase to the gas phase without any apparent liquid phase in-between.
Substance	A variety of matter with identical properties and composition.
Supersaturated	When a solution contains more solute than a saturated solution would at a given temperature.
Symbol	Letter(s) designation for an element.
Synthesis	When many substances come together to form one compound.
Temperature	Average kinetic energy possessed by a sample.
Titration	The process by which acids and bases can be measured out in exact quantities so that they neutralize each other exactly and without any excess.
Transition Metals	Metals found in groups 3 through 10 of the periodic table.
Transmutation	Formation of a new element when an element undergoes nuclear disintegration.
Triple Bond	A covalent bond that involves the sharing of three pairs of electrons.
Triple Point	A specific point in temperature and pressure at which solid, liquid, and gas exist at the same time.
Unsaturated	Describing a solution that contains less solute than a saturated solution would at a given temperature.
Valence Electrons	The electrons that are located in the outermost principal energy level.
Van der Waals Forces	Weak forces existing between nonpolar molecules. Also known as *dispersion forces*.
Vapor Pressure	Pressure exerted by the vapor of a liquid as the molecules of the liquid evaporate.
Vaporization	Process by which a liquid enters the gas phase.
Voltaic Cell	A setup that allows a redox reaction to occur spontaneously so that the electrons can be used to do work.
Volume	The space an object occupies.

Here's what people are saying about *Step into the Bible*

"This book is a children's Christian classic. I am happy to recommend it to all parents and urge them to take advantage of its unique approach. I know of no book on the market like it."

Billy Graham

"I was raised on Foster's *First Steps for Little Feet*. When our five children were growing up, I found it an invaluable help. You can start with a three-year-old and take him gradually through the Scriptures. I urged my daughter, Ruth, to re-do and bring it up to date. She has done this beautifully — an undertaking I am sure Foster himself would approve and be grateful for."

Ruth Bell Graham

"Ruth Graham has put together an absolutely wonderful book for families. The questions and Memory Verses are a perfect way to help your children learn to love God. When parents mentor their children, the legacy of faith continues from generation to generation."

Jim Burns, Ph.D.
President, HomeWord
Author of *Confident Parenting*

"As a mom of five children, this book will be on the top of my list of tools to help me teach my children biblical principles in a relevant and applicable way. Thank you, Ruth, for updating this classic for families to use today. I give this book my highest recommendation."

Lysa TerKeurst
President of Proverbs 31 Ministries,
speaker and author

"How many times as parents have we longed for the luxury of raising our children in the gentle wisdom of 'the good old days.' *Step into the Bible* provides the heritage of timeless truths with delightful illustrations and photographs that captivate a child's attention and heart."

Lisa Whelchel
Bestselling author of *Creative Correction*,
The Facts of Life and Other Lessons My Father Taught Me,
and *Taking Care of the "Me" in Mommy*

Step
INTO THE
Bible

100 Family Devotions to
Help Grow Your Child's Faith

Ruth Graham

ZONDERVAN®

ZONDERVAN.com/
AUTHORTRACKER
follow your favorite authors

16 17 18 19 /DSC/ 10 9 8 7 6 5 4

Contents

In loving memory of my mother, who taught me the Scriptures with
First Steps for Little Feet Along Gospel Paths by Charles Foster.

May the little people in your life learn about the
Bible in this revised and updated edition of Foster's original book.

~ R.G.

Introduction

Four Generations

Step into the Bible is based on the book *First Steps for Little Feet Along Gospel Paths* by Charles Foster. First published in the 1800s, this book taught the Bible to four generations of Graham children. It has stood the test of time with updates and re-releases over the past 150-plus years.

Many years ago, when my grandparents were medical missionaries to China, they chose *First Steps for Little Feet Along Gospel Paths* to teach their children biblical truths. They saw that this book built precept upon precept in teaching children the great principles and facts of the Christian faith.

In turn, my mother used this book in our preschool and elementary training. She not only read it to us and quizzed us with the questions following each story, she also taught us this was a book to be loved. It gave us an early appreciation for the Bible and told us of the historical examples to follow as we grew up.

It was a memorable day when I received my own copy of *First Steps for Little Feet Along Gospel Paths*. I cherished the little book. I could not read yet, but I remember gently leafing through the book, looking at the pictures, and anticipating the day when I could read it myself.

When my own children were ready for family prayer time, we proceeded with gusto. I went to the store and bought all the new Bible storybooks, books on how to have exciting family devotions, and idea books. For the first few days, things went fairly well, but eventually the plan fizzled.

Finally, I remembered *First Steps for Little Feet Along Gospel Paths* on my bookshelf. My children had been delighted to see my name scrawled in pencil in the front. They were excited about a book their mommy read when she was little and even their Tai Tai (what they called my mother; pronounced Teddy, meaning "old lady" in Chinese) read when she was a little girl in China.

My children enjoyed the short stories. The family devotion time took us a total of about ten minutes. My three-and-a-half-year-old daughter thought it was great and did very well. And even though my two-year-old son didn't answer the questions, he was learning from his sister.

Updating

As we progressed through the book, it became increasingly clear that the book needed updating. The language of the 1800s sounded strange to us. The concept of the book was still excellent—it was a classic. So I began to edit words and phrases as I went along.

On a visit home to my parents' home in North Carolina one Easter in the late 1970s, I told my mother that the book needed to be rewritten and suggested that she do it. But she turned around and said that she thought I should do it. As I continued to edit, rewrite, and add some original material, I became aware of the value of this book as a tool in teaching children. Mr. Foster's concept of building upon a foundation, precept upon precept, was strong. And when I double-checked a fact in Scripture, I was impressed with the accuracy of the book. It was an exciting project for me.

Thus came the publication of the 1980 edition, called *First Steps in the Bible*.

Over 25 years after the 1980 edition released, I once again revised stories, added new ones, and removed weak stories. You have the newest edition of the classic book in your hands: *Step into the Bible*. The title represents the interactive features that encourage families to develop their faith during devotional time.

A Unique Book

As you go through this book with your children, you will notice that it is far from being a surface Bible storybook. Through Bible stories, I introduce the basic doctrines, obedience to God, original sin, Christ's atoning death, and eternal life. Children understand these principles and readily accept them. It is easy for adults to make things too complicated. Children want to believe and have such open hearts toward God.

Over 300 full-color illustrations and photographs will capture your children's imaginations. Also, throughout the book there are Parent Notes to help you guide your children into an understanding of the principles set forth.

Every story has an appropriate Memory Verse. I feel it is important for children to begin memorizing verses early to become a lifelong habit. In addition, every tenth devotional in the book is designated as a "resting" period. These devotions contain only Scripture. Use these resting steps as an opportunity to immerse yourself and your family in the Word. Questions follow each story, which help both children and adults learn about themselves, their faith, and God.

Devotions for the Family

When I was growing up, we had family prayers right after breakfast. I found that a good time for my children was right after dinner. Things didn't seem quite so rushed, and we were all together and refreshed by a good meal. It is important to keep things short — only ten to fifteen minutes. Try to be regular and consistent.

How to Start

Read a story, ask the questions, and then have a short prayer. If you have very young children, the questions might be a bit difficult and you may want to rephrase them so that the answer is in the question.

When it comes to the prayer time, on some occasions my children liked to have short sentence prayers so that each in turn could pray out loud. Remember to teach your children to praise and thank the Lord first. Always be specific in your requests as well as in your praise. Perhaps you could begin a family prayer list that would include friends, teachers, schoolwork, pastor, missionaries, world leaders, and others. When you see answers, be sure to point them out to your children and talk about how good God is to answer our prayers.

Expect to have squirming and wiggling and a few interruptions when you are reading to little ones. It is surprising how much they absorb even while squirming. Don't make it a pressured competitive situation. If a child cannot answer a question or has not listened very well, don't worry. They will probably get it the next time around. With 100 stories, you have many opportunities!

Family Togetherness Time

I loved reading to my children! It was a wonderful way to spend concentrated time with them. Now I love reading to my grandchildren.

Reading can also spark marvelous conversations with your children. Learn to listen to them. Try to make your reading exciting by voice inflection. (My mother had such a wonderful Scottish burr and my grandmother could imitate all the Southern accents.) Use this book as a catalyst for family time; gather together for a time of fun and spiritual training. You will be surprised what this will lead to in family growth and closeness. God will bless your efforts.

A Book to Enjoy

Step into the Bible does not have to be limited to family devotions. Place it where it can be easily reached for little ones to look at on their own or for older children to read to themselves. Let them use it as a reference book for Sunday school lessons. It is a book to be used and enjoyed by the whole family. While your aim as parents may be to use it for spiritual development in your children, they don't have to be aware of that!

My mother used to say, "The best way to get a child to eat his vegetables is to have him see his parents enjoying their vegetables." So it is with spiritual training!

And so, you have in your hands a unique book that has stood the test of time and that has contributed to the nurturing of four generations of Grahams. It is a classic. The revisions have enhanced Charles Foster's original book by adding:

- Over 300 photographs and illustrations that will expand a child's world through color, art, geography, and culture. Contemporary photographs portraying biblical themes and situations will help bring the Bible closer to your children.
- Memory Verses to help establish a lifelong habit.
- Bible references so that you can refer back to Scripture.
- Questions at the end of each story to bring the Bible into your child's world.
- Parent Notes to offer you tips or advice.

It is my prayer that you enjoy this book together, and that God will use it to draw your family not only closer to each other but also closer to Him.

Ruth Graham

The LORD God Made It All

Genesis 1:3 – 5; 14 – 19; 26 – 27

A family once had a pet bird named Red. They chose this as a silly name, since the bird was yellow! It lived in a birdcage that hung in the kitchen. A man made the cage so he could sell it—it was very shiny and shaped like a Japanese house.

People can make cages, chairs, cars, houses, and many, many other things, but they can't make birds or things that are alive. Only God can create life.

Did you know God made you too? You are very special to Him; He loves you and cares about you.

God lives in heaven. We cannot see Him, but He sees us all the time, even at night.

You get ready for bed and get tucked in. Pretty soon the moon comes out and the stars twinkle up in the sky.

God made the sun and the moon and the stars up in the sky. He made everything. God is amazing.

Questions

- Who made the birdcage?
- Who makes things that are alive?
- Where does God live?
- When can God see us?
- Who else did God make?

Memory Verse

In the beginning God created the heavens and the earth.

Genesis 1:1

Seasons Change

Genesis 1:6 – 8

On a pretty spring day, it's fun to lie down in the grass and watch the clouds float along in the sky. Or in the summer, maybe you like to walk along the beach and imagine shapes in the clouds.

God made those clouds.

Sometimes they bring rain, and up north, the clouds sometimes bring snow in the wintertime. If you live up north, it can be fun to get up in the morning and discover that during the night, snow has covered the ground. You bundle up in coats, hats, boots, and gloves to go play outside.

God made the rain and snow that comes down from the clouds.

After the snow melts in springtime, maybe you like to go out to places where flowers grow. They start growing up from the dirt even under the snow. They know when to begin to grow. God tells them when.

The birds sing again; the grass grows green; the leaves come back on the trees; and the big, fat robins hop about on the lawn, snatching up worms.

Questions

- What do we see up in the sky besides the sun, moon, and stars?
- What do clouds bring?
- After the snow melts, what comes up from the ground?
- Who tells them to come up?

Memory Verse

In the beginning God created the heavens and the earth.

Genesis 1:1

God Made the Fish

Genesis 1:20 – 23

Have you ever gone fishing and caught a wriggling trout? Or maybe you have gone swimming in a lake or ocean and seen the shadow of a salmon flitting by your leg. Maybe you have an aquarium at home where you can watch many brightly colored fish swim all day long.

Fish can also live in streams and rivers, oceans and lakes, or maybe a pond in your backyard.

Fish don't have feet or legs like animals or wings like birds. Instead, God gave them fins so they could swim in the water. And they have a special way of breathing — through gills — so they can stay in the water all the time.

There are many kinds of fish. Some are very small and some have beautiful colors. Some are very big and some are funny looking. Some fish we never see because they live at the bottom of the ocean.

God made all the fish and everything in the oceans. He made the beautiful goldfish and also the huge whale. He takes care of them and has given them all they need to live in the water.

Questions

- Who made the fish?
- Do they have wings and feet?
- What do they have instead?
- Where did God make the fish live?
- Did God give them all they need to live in the water?

Memory Verse

In the beginning God created the heavens and the earth.

Genesis 1:1

God Made the Animals

Genesis 1:24 – 25

Have you ever visited a zoo? If so, you saw all kinds of animals! You may have seen baby white tigers, baboons, or flamingoes. Maybe you even got to ride an elephant!

Did you know that God made all the animals? Some animals are wild and live out in the woods or jungle, like wolves, bears, and tigers. Other animals are tame, like sheep and cows.

A zoo has lots of different kinds of animals. The wild animals are usually in a cage so they can't hurt us and people can't hurt them. The tame animals are put in a special area so we can pet them.

Tigers and lions are very fierce and could hurt us, but they live in countries very far from here. Or maybe you live where they do! We can go see them in the zoo where we are protected from them, or we can look at pictures of them. They are very beautiful and strong animals.

God also made zebras with their black and white stripes. He made giraffes with their long legs and blue tongues! He made cheetahs that run very fast and sloths that move very slowly. God made all kinds of different animals.

Questions

- Who made the animals?
- Have you been to a zoo?
- Which animal do you like best?

Memory Verse

In the beginning God created the heavens and the earth.

Genesis 1:1

God Made People

Genesis 1:26 – 28

You live on a big planet called earth. And lots and lots of other people live on it too.

Perhaps there are many families on your street. There are streets like yours all over the earth. There are so many people we can hardly count them.

God made this earth and gave us everything we need to live. He provided it for us to enjoy because He loves us. He does not want us to destroy or neglect it. We must take care of what He gave us and try to keep it beautiful.

Where you live, there are probably some people whose skin is a different color from yours or whose eyes are a different shape. And maybe you know someone who speaks in a language that sounds strange to you because you don't understand it. They might come from another country — maybe even the other side of the earth! The earth is a big place.

Even with all the people on the earth, no one is the same.

We are all special in the way we act, look, and think. Do you know what? God made us that way, and He loves us all.

Parent Note: This step establishes an awareness of the need to care for this beautiful world God gave us to enjoy. You may want to further the discussion on practical matters such as recycling and not littering.

Questions

- What is the name of the planet where we live?
- Are there lots of people on the earth who live here?
- Who made us?
- What are some things you can do to help keep God's earth beautiful?

Memory Verse

In the beginning God created the heavens and the earth.

Genesis 1:1

23

Paradise Lost

Genesis 2:15 – 3:19

Sometimes, especially in the spring or fall, when you are riding in the car with your mother or father, you may see a squirrel or bird that has been hit by a car and killed. It might make you sad to see it.

Maybe you think, *God will make it all better*, or you might ask, "Wasn't God taking care of that bird or squirrel?"

The answer to that started a long time ago. After God made the world and all the birds, animals, and fish, He looked around and said something was missing. He decided to make a man named Adam. And he did. Then God gave Adam a woman to be his wife—Eve—so he wouldn't be all by himself.

God was very happy that He made Adam and Eve. He put them in a beautiful garden, called Eden, that was full of wonderful plants and animals and birds. Everything was perfect. No one cried or was sick or died. All the animals loved each other. No one hurt anyone else. God gave Adam and Eve everything they needed and told them to enjoy everything. But He told them that there was fruit on one tree in the garden that they could not eat. He told them that they must obey Him.

Everything was great until one day when Eve began to talk with a slithery snake—who was Satan in disguise. He questioned what God had told her about the fruit tree. Soon Satan convinced Eve that it wasn't as important to obey God as she had thought, so she ate some of the fruit from that tree. She gave some to Adam too.

All of a sudden they realized that they had been very wrong to listen to Satan and disobey God. When they heard God coming in the garden, they hid.

God was very sad because He knew what Adam and Eve had done. They had disobeyed Him—that is called a sin. God knew He would have to punish them.

Part of the punishment was death. Sin always ends in death.

Ever since Adam and Eve sinned by disobeying God, people, animals, birds, and fish have died.

But God has provided a wonderful way for us to live forever. He sent His Son Jesus to take the punishment for us so that when we die, we can go to heaven and live forever. All we have to do is ask Jesus into our hearts to wash away our sins, and He does! Isn't that exciting?

Parent Note: This story contains many concepts—the reality of Satan and his efforts to get us to disobey God, penalty for disobeying God, the atoning death of Christ, eternal life, and original sin. Children accept these things so easily and do not question them. Do not make it complicated, just present it as it is.

Questions

- Was the garden of Eden perfect?
- What did God tell Adam and Eve not to do?
- Who did Satan make Eve doubt?
- What did Adam and Eve do?
- What was the punishment?
- When Adam and Eve sinned, why did that affect everyone else ever to be born?

Memory Verse

I will maintain my love to him forever,
and my covenant with him will never fail.

Psalm 89:28

New Hope

Genesis 6–9

Noah was a good man who loved and obeyed God. He obeyed God even though those around him did not. God decided to punish the people who did not obey Him.

God said He was going to send a flood to cover the whole world to destroy the ones who did not obey God. He told Noah to build an ark, how big He wanted it to be, and what animals to bring into the ark. God was going to send a lot of rain, but He promised Noah that He would take care of them.

Even though it might not have rained in a long time and his neighbors might have made fun of him, Noah did obey. He built the ark.

When the ark was finished, God told Noah to go inside with his family and all the animals. God shut the door and then it began to rain. It rained for forty days and forty nights. Water covered everything. Soon the ark began to float—higher than the houses, trees, and mountains! Everything that had been living on land died underwater.

But Noah, his family, and all the animals were safe inside the ark.

Then the rain stopped and the flood waters began to go down. The ark came to rest on a mountain. But they couldn't leave the ark because water was still everywhere.

After the flood waters dried up, God told Noah and his family they could leave the ark with the animals and stand once again on dry ground.

When they left the ark, the first thing Noah and his family did was worship God and thank Him for His protection. God promised that He would never again cause a flood to cover the whole earth. As a sign of this covenant, or promise, God made a rainbow.

This is Mt. Ararat in Armenia, near Turkey. Some people think the ark landed here.

Questions

- Who told Noah to build the ark? Why?
- How would you have felt to be on the ark with all those animals for so long?
- When Noah and his family left the ark, what was the first thing they did?
- What did God promise Noah?
- What was the sign that He would keep his promise?
- What was the difference between Noah and his neighbors?

Memory Verse

I will maintain my love to him forever, and my covenant with him will never fail.

Psalm 89:28

Abraham Follows God's Plan

Genesis 12:1–8

Have you ever moved from your home? If you did, you had to pack up your toys, clothes, books, and say good-bye to all your friends. That is hard to do. Can you imagine packing up your things and leaving but not knowing where you were going?

Ever since Adam and Eve had to leave the garden of Eden, God planned to redeem, or rescue, people from their sin. In this plan, He chose a group of people who would be the family of Jesus. This family began with a man named Abraham.

A long time ago, God told Abraham to leave his home, family, friends, and country to go to a place God would show him. Abraham had no idea where he would go. God promised to make Abraham's family into a great nation.

That must have been a hard decision for Abraham to leave everything, but he obeyed God. He did what God told him to do even though he didn't know where he was going. That is called faith—to obey even though you don't know what is going to happen to you.

Abraham took his wife, Sarah, his nephew, and everything he owned, including his cattle, sheep, camels, and servants. They traveled where God led. It wasn't an easy trip through the desert, but he was determined to follow God's plan.

When he got to the place called Canaan, God promised to give him the land—even though others were living there. Abraham believed God and took time to worship Him.

Abraham had no idea what his obedience would mean, but you and I have been affected by Abraham's decision to obey God. Through Abraham, all people would be saved. Jesus would be born into that family many years later. Because of Jesus, our sins can be forgiven and we can all know God. What a wonderful gift!

Questions

- What did God ask Abraham to do?
- What did God promise Abraham?
- What did Abraham do when he got to Canaan?
- Do you think it is important to obey God?
- What is faith?

Memory Verse

I will maintain my love to him forever,
and my covenant with him will never fail.

Psalm 89:28

A Mother and Son Are Saved

Genesis 21:14 – 21

God not only made all the different kinds of people and animals in the world to love, but He also made wonderful creatures called angels. Angels are God's messengers. Everyone who believes in Jesus has an angel all to himself or herself. Angels take care of us; they are often called guardian angels.

When angels are not busy taking care of people or doing other things for God, they live in heaven. Angels are very good and obey God when He tells them to do something.

A very long time ago, there was a little boy named Ishmael. His mother's name was Hagar. They lived in a tent (people lived in tents then), but they were told they had to leave their home. Ishmael and Hagar were sad, but they packed their things and left.

They had no place to go, so they wandered out into the desert. A desert is a place where practically no one lives. It isn't a very nice place. It can get very hot.

Pretty soon, little Ishmael got thirsty. His mother gave him some water, but then they ran out. They looked and looked for more water but couldn't find any. Ishmael got very weak. Hagar saw that he was going to die unless he got some water soon. She put him under a bush in the shade where it wasn't as hot.

She was very sad and walked away because she didn't want to see him die. She began to cry.

Then God sent an angel to her. The angel asked, "What is the matter, Hagar? Do not be afraid; God has heard the boy crying as he lies there. Lift the boy up and take him by the hand …" Then the angel brought her to a well.

So Hagar quickly got Ishmael some water to drink. He got strong and well. He grew up to be a man and lived in the desert.

Parent Note: If you would like to find out more about angels and what they do, a good resource is *Angels: God's Secret Agents* by Billy Graham.

Questions

- Where do angels live when they aren't taking care of people or doing other things for God?
- Why was Ishmael so sick?
- Why did Hagar walk away?
- What did the angel show Hagar?

Memory Verse

I will maintain my love to him forever, and my covenant with him will never fail.

Psalm 89:28

The Priestly Blessing

Numbers 6:22–27

The LORD said to Moses,
"Tell Aaron and his sons, 'This is how you are to bless the Israelites. Say to them:
"'"The LORD bless you and keep you;
the LORD make his face shine on you
and be gracious to you;
the LORD turn his face toward you
and give you peace."'
"So they will put my name on the Israelites,
and I will bless them."

Questions

- Who is speaking in this passage?
- Who is this a blessing for?
- What do you think this passage means?

Memory Verse

I will maintain my love to him forever,
and my covenant with him will never fail.

Psalm 89:28

Angels on a Stairway

Genesis 28:10 – 22

When you sleep, do you dream? In the morning when you wake up, do you remember what you dreamed? Sometimes dreams are very important.

Jacob was in trouble. He had lied to his father, Isaac, and stolen from his brother, Esau. His mother, Rebekah, told him to stay with his uncle until things cooled down. He was really afraid of what his brother might do to him. So he took off.

He traveled far away. The sun had gone down and he was tired. He found a good spot to lay down. He used a rock for his pillow! He was so tired, he went to sleep.

Jacob had a very important dream. He dreamed there was a stairway that went all the way up into heaven. He could see angels going up and down the stairway. At the very top was God. He told Jacob that he was the God of his grandfather Abraham and his father, Isaac. God promised to give Jacob and his children the land where he was sleeping. God said Jacob's family was going to be very big— hundreds and thousands of people. And because of his children, other people would be blessed. God promised to watch over him and bring him back home.

Jacob woke up early and remembered his dream. He realized that God had been there even though he had not been aware of it. He named the place where he had the dream "Bethel," which means "house of God."

This was going to be a very special place for Jacob and his family.

While he was there, he promised that he would serve God and give a tenth of all that God gave him.

Parent Note: The principle of tithing is very important. Introducing it early can make it a lifetime habit. God blesses us when we give him ten percent of what we have.

Questions

- Is there a time when you felt God very close to you?
- God made promises to Jacob. He has made promises in the Bible to you. Can you think of any?
- Do you give to God?

Memory Verse

Do not forget my teaching, but keep my commands in your heart.

Proverbs 3:1

Trouble with Brothers

Genesis 37

Sometimes brothers and sisters fight a lot. Have you ever wished one of your brothers or sisters would just go away?

Joseph's brothers hated him because he was their father's favorite son. One day their father gave Joseph a bright, colorful coat. None of the other boys had a coat like that. They hated Joseph even more.

Joseph liked to tell his brothers about his dreams. One of his dreams was about them bundling up grain in the field. His bundle stood up tall while his brothers' bundles bowed down to his.

They asked if he was going to rule over them. In another, he dreamed that the sun, moon, and stars all bowed down to him. This made his brothers angry.

One day his brothers were tending sheep in the fields. Jacob sent Joseph out to find out if they were okay. When his brothers saw him coming, they planned to kill him. But the oldest brother didn't think that was a good idea. Instead he told them to put Joseph in a deep pit, thinking he could rescue Joseph later. When Joseph arrived, his brothers took his clothes and the beautiful coat their father had given him. They threw him in the pit. Then they sat down to eat.

Soon they saw a group of travelers coming by on their way to Egypt. The brothers decided to sell Joseph to the travelers.

They took Joseph's beautiful coat and dipped it in blood so it would look like Joseph had been attacked by an animal. They took the bloody coat to their father and lied about what happened to Joseph.

Jacob was very sad because he thought Joseph was dead. No one could comfort him.

But that isn't the end of the story! You'll find out what happened to Joseph in the next story.

Parent Note: Favoritism in a family is harmful. It creates insecurities, bitterness, jealousy, and unhealthy attitudes. We see this in the stories of Jacob and Joseph. Ask God to help you love and treat your children equally. They are special in their own right and God has unique plans for them.

Questions

- Why were Joseph's brothers so mad at him?
- What did they do to Joseph?
- What did they say happened to Joseph?
- What was the truth?

Memory Verse

Do not forget my teaching, but keep my commands in your heart.

Proverbs 3:1

Dreams Come True

Genesis 39–50

In the last story, you learned that Joseph's brothers sold him to some travelers going far away to Egypt. His brothers thought they had seen the last of him.

Although Joseph may have been afraid and alone without his family, God was with him. His new boss saw that God was helping Joseph, so he put Joseph in charge of his house. It was a big job but Joseph did it well.

His new boss's wife was not a nice lady. She said terrible things about Joseph, so he was put in jail. But God was still with Joseph. He soon was put in charge of all the other prisoners.

One day, two new men were put into the jail. They had made the king very mad. He put them in the same jail as Joseph.

While in jail, the two men had two different dreams on the same night. They told Joseph about them. With God's help, Joseph told them what their dreams meant. Both of the men's dreams came true.

One night, the king had two dreams. He didn't know what they meant. He asked all the smartest men what his dreams meant. But they didn't know. Then one of the men who had known Joseph in jail remembered something. Joseph could tell them what the dreams meant.

With God's help, Joseph told the king what they meant. Joseph told the king that there were going to be seven good years and seven bad years. He advised the king to use the good years to prepare for the bad years. When the bad years came they were

ready. Soon people came from all over to ask the Egyptians for food. It all happened just as Joseph said it would.

The king was so happy with Joseph that he put him in charge of his whole kingdom. The only person more important than Joseph was the king himself.

Guess who came to Egypt looking for food. Joseph's brothers! They had no idea that their little brother was now so powerful in Egypt. They didn't recognize him. They came and bowed down to him. (Remember the dream he told his brothers about them bowing down to him? Joseph's dream had come true after all these years.)

He knew who they were but didn't let them know. He wanted to see if they had changed. They were afraid of this important man.

After many days, he showed them who he was. Joseph told them not to be afraid, because God had sent him to Egypt so that they would have food during the bad years. God was making sure their family would survive. He said, "You intended to harm me, but God intended it for good to accomplish what is now being done, the saving of many lives."

Joseph forgave his brothers and hugged them.

When the king heard that Joseph's brothers had come, he was very happy. The king told them to bring the rest of Jacob's family to Egypt so that he could take care of them.

And that is what they did!

Questions

- During everything that happened to Joseph, who was with him?
- Who gave Joseph wisdom?
- What would you have done to your brothers?
- What did Joseph do?
- Why is forgiveness so important?

Memory Verse

Do not forget my teaching, but keep my commands in your heart.

Proverbs 3:1

Baby for the Princess

Exodus 2:1-10

When a mother has a new baby, she prepares a cradle or crib for him. She lines it with clean padding and places a little mattress in the bottom so it will be soft. She wants it to be just right.

Remember the story about Joseph becoming an important man in Egypt? After Joseph died, a new king, called a pharaoh, came to power—a king that did not know Joseph. He saw that Joseph's family, the Hebrews, had become too large and powerful so he ordered all Hebrew baby boys to be killed. He was a very wicked pharaoh.

Moses was a baby boy. His mom loved him very much and didn't want him to be killed. She decided to hide him from the pharaoh. When she couldn't hide Moses at home any longer, she prepared a basket for him. She made sure it was tight and safe. She then put him in the basket and hid the basket in the reeds along the Nile River. It was like a little basket boat! Moses' mother asked his sister to stay nearby to watch over him in the basket.

One day, the pharaoh's daughter, a princess, went down to take a bath in the river. She discovered the basket in the reeds and peeked inside. There was baby Moses! He was crying and she felt sorry for him. "He must be a Hebrew baby," she said.

Moses' sister quickly asked the princess, "Shall I go and get one of the Hebrew women to take care of the baby for you?"

"Yes, go," the princess agreed. Moses' sister went home and got her mom. Moses' mom took him home with her and watched over him until he grew a little older. Then she took him back to the princess. He grew up in the palace as if he were a member of Pharaoh's own family.

Questions

- Why did the pharaoh want all the baby boys killed?
- What did Moses' mom do?
- Who watched over Moses?
- What happened to Moses?

Memory Verse

Do not forget my teaching, but keep my commands in your heart.

Proverbs 3:1

God's Message to Moses

Exodus 3–14

After Moses grew up, he had to move away from Egypt. While Moses was tending sheep, he noticed a nearby bush was on fire. This was a strange thing—a bush that was on fire but wasn't burning up! Moses went closer to see what was happening. As he got closer, God called his name from out of the bush, "Moses! Moses!"

Moses answered God and said, "Here I am."

God told Moses not to come any closer and to take off his shoes because this was a special place. Moses hid his face because he was afraid to look at God. He told Moses that He was the God of Abraham, Isaac, and Jacob. He had seen the troubles of His people, who were now slaves in Egypt. He had heard their prayers.

Then God said that He was going to send Moses back to Egypt to tell the new pharaoh to let the slaves go.

The slaves finally escaped, and Moses led them out of Egypt. But they weren't safe yet. Pharaoh and his army were close behind Moses and God's people.

Soon they came to the Red Sea. To get away from the army, God told Moses to raise his staff over the Red Sea to make a path. Moses and God's people followed the path through the sea safely to the other shore.

Moses raised his staff again, and the rushing water swept away Pharaoh and his army. Then Moses led the people to a place that God had promised Abraham so long ago. God keeps His promises.

Questions

- What did Moses see in the desert?
- Whose voice did he hear?
- How does God talk to you?
- Had God heard and seen the troubles of His people?
- What did God tell Moses he was going to do?

Memory Verse

Do not forget my teaching, but keep my commands in your heart.

Proverbs 3:1

We're So Hungry

Exodus 16 – 17:6

Moses and the Israelites walked through the desert for days and days. They were traveling to the Promised Land—and they couldn't wait to get there. The desert was dry and dusty. They were getting tired of this trip. They began to long for the things they had in Egypt.

Soon, people started grumbling that there wasn't anything to eat in the desert.

"If we were still in Egypt, we'd have plenty to eat!" someone complained.

"We are starving!" another whined.

God heard their cries. That night, He sent meat in the form of birds called quail for the Israelites to eat. The next morning, he sent bread called manna so that they wouldn't starve.

Moses and the Israelites kept walking and walking, farther and farther. They probably asked Moses, "Are we there yet?" They wondered if they'd ever get there. They started grumbling again. "Why did you bring us up out of Egypt to make us and our children and livestock die of thirst?" someone complained.

"We are thirsty!" another whined.

Moses didn't know what to do. The Israelites and their animals were so thirsty! There was no water in the desert. Moses prayed to God for help.

God led Moses to a rock. God said, "Strike the rock, and water will come out of it for the people to drink."

Whooooosh!

Water gushed from the dry rock like a fountain. The Israelites all had enough drink. Isn't God amazing?

Questions

- Where were the Israelites traveling to?
- What did the Israelites eat at night?
- What did the Israelites eat in the morning?
- Who sent them their food?
- What came out of a rock?
- Who made the water come out of the rock?

Memory Verse

Trust in the LORD with all your heart and lean not on your own understanding.

Proverbs 3:5

STEP 17

God's Rules

Exodus 20:3-17

God made all the people in the world, and He loves us. He also wants us to love and help each other. But because we want to do bad things more than we want to do good things, He gave us a set of rules. These rules are called the Ten Commandments.

1. You should only worship God.
2. Do not create anything that you think looks like God or that you treat like God.
3. Always behave in a way that shows love and respect for God.
4. One day of the week should be saved for rest and worshiping God.
5. Obey your parents and be respectful toward them.
6. You must respect life.
7. When people get married, husbands and wives must only share that special love with each other.
8. Do not take anything that does not belong to you.
9. Do not lie—always tell the truth.
10. You shouldn't want something so badly that you would do anything to get it.

These are very important rules. But no one can keep them perfectly. Only one person ever kept them and that was Jesus. When we read these rules, we see that we need forgiveness for what we have done wrong. To ask forgiveness means to say we are sorry and will try not to do them anymore. God is always ready to forgive us.

Parent Note: It may seem strange to include a story on the Ten Commandments in a book for children. But in this age in which relativity and situational ethics have been enthroned, it is vital for a child to know that there are some absolutes and a moral code given by God. Consider the illustration of a highway without painted traffic lanes. No one would know where they were supposed to drive—and all would be havoc. So it is morally and spiritually. God has given us a standard of excellence.

Questions

- Why did God give us these rules?
- Can you name a few?
- Who kept all these commandments perfectly?
- If we know we have done wrong,
  what can we ask God to do?
- Will He forgive us?
- How can you tell if you love something or
  someone more than God?

Memory Verse

Trust in the LORD with all your heart and lean not on
your own understanding.

Proverbs 3:5

Spying the New Land

Numbers 13:1–14:9

Do you like to explore? If there are woods near your house, maybe you like to find all kinds of treasures in the forest.

The children of Israel were at the border of the new land—the new home God had promised Abraham a long time ago. God told Moses to send explorers into the land to see what it was like. Moses sent out twelve explorers.

He told them to go throughout the country to see if the land was good and what the people were like there. Were they strong? Weak? Few? Many? He wanted to know what kind of towns the people

lived in and if they were well guarded. He wanted to know if the land was good for farming. Moses wanted to know all the details before sending the Israelites.

He asked the explorers to bring back samples of the fruits of the land. It was grape season so he asked them to bring back some grapes. Maybe grapes were Moses' favorite fruit!

The explorers traveled all over the country. When they reached the vineyard, they cut off a big bunch of grapes—it was so big that two men had to carry it on a pole between them! They went back to Moses to report what they had seen.

The explorers told him it was a good place where they could grow plenty of food. They also told Moses that the people who lived there were very strong and the towns were well protected. Ten of the twelve explorers were afraid and did not want the Israelites to go into the land. The Israelites listened to these explorers. The people became afraid and very upset. But two of the explorers, Joshua and Caleb, disagreed with the others and said that they should make the land their home.

Joshua stood up and told them that the land was very good. He reminded the Israelites that God had promised to give them the land. Joshua and Caleb urged the Israelites not to disobey God and not be afraid. God was on their side.

Parent Note: This story can give you the opportunity to talk about peer pressure. "My son, if sinful men entice you, do not give in to them." (Proverbs 1:10) Another way to say that verse is: "If bad boys tell you to do bad things, say no." It seems really hard at first, but God helps us just say no.

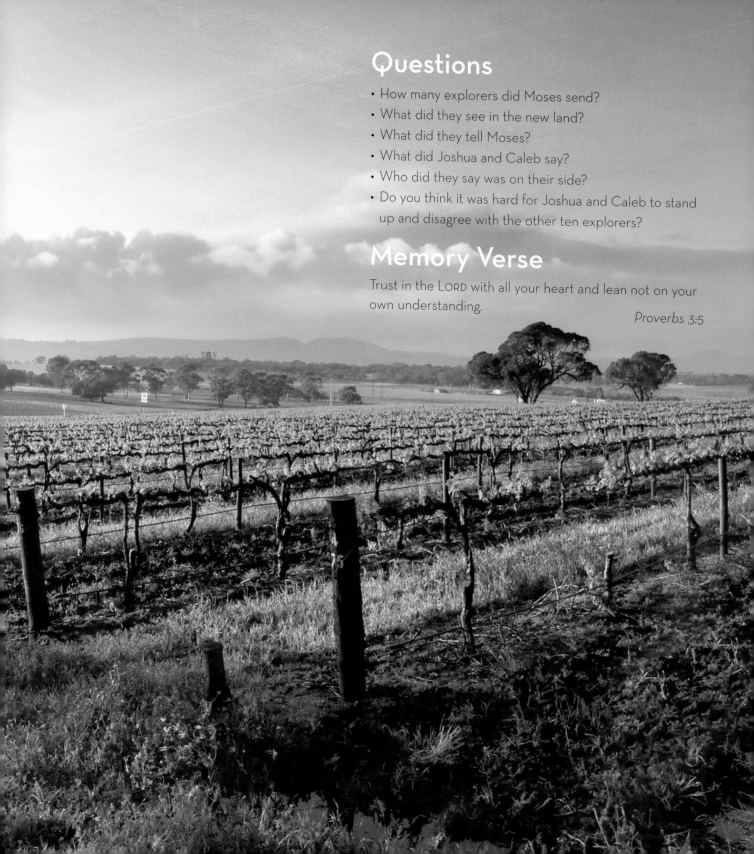

Questions

- How many explorers did Moses send?
- What did they see in the new land?
- What did they tell Moses?
- What did Joshua and Caleb say?
- Who did they say was on their side?
- Do you think it was hard for Joshua and Caleb to stand up and disagree with the other ten explorers?

Memory Verse

Trust in the LORD with all your heart and lean not on your own understanding.

Proverbs 3:5

A New Home

Joshua 1:1–9; 21:44–45; 23:1–11

In school when it's time for a test, sometimes you might get nervous. Maybe you worry that you haven't studied enough or that you won't know the right answers. But the Bible says we are not to worry about anything, because God is with us. If you keep thinking about it, God's promise will help.

The same was true for Joshua. God had a big job for him. Moses had died, so God told Joshua that he was now the new leader. There would be many hard days ahead, but God said to remember His Word.

God promised He would give them the land just as He had promised Moses. Although they must have been very sad that Moses was dead, it was also an exciting time. They had traveled in the desert a long time and now they were about to enter the land that God had promised Abraham years before. God keeps His promises!

God said He would be with Joshua just as He had been with Moses. He would never leave Joshua's side and Joshua would never have to be afraid. God told Joshua to be strong and brave and obey Him.

He told Joshua to think about God's Word all the time—to talk about God's Word. He was to make God's Word the most important thing in his life.

Just as God promised, He was with Joshua as they moved into their new home. It was not easy. Joshua, with God's help, was able to start a new home for the people of Israel.

Questions

- What did God promise Joshua?
- What did God tell Joshua to do?
- Did Joshua obey?
- What happened to Joshua and the people of Israel?

Memory Verse

Trust in the LORD with all your heart and lean not on your own understanding.

Proverbs 3:5

50

Fly Like an Eagle

Isaiah 40:28 – 31

The LORD is the everlasting God,
    the Creator of the ends of the earth.
He will not grow tired or weary,
    and his understanding no one can fathom.
He gives strength to the weary
    and increases the power of the weak.
Even youths grow tired and weary,
    and young men stumble and fall;
but those who hope in the LORD
    will renew their strength.
They will soar on wings like eagles;
    they will run and not grow weary,
    they will walk and not be faint.

Questions

- What did God create?
- Who does God help?
- What can we do if we need help?

Memory Verse

Trust in the LORD with all your heart
and lean not on your own understanding.

Proverbs 3:5

53

Called by Name

1 Samuel 1–3

Have you ever had to give back something you loved very much?

Maybe you were taking care of your friend's dog for the summer, but when your friend came home you didn't want to give him back. You had fallen in love with the dog and wanted to keep him. But you gave him back.

Hannah was very sad because she wanted a baby more than anything in the world. She went to God's house to pray very hard that God would give her a little boy. She promised God that if He gave her a little boy, she would give him back to God. That meant he would grow up in God's house learning to serve God and others.

God answered her prayer! She had a baby boy and named him Samuel. Hannah remembered the promise she made to God. Hannah and her husband took Samuel to God's house. He lived with Eli, the priest. His parents visited him each year. Hannah loved Samuel very much.

As Samuel grew up, he did what was right and pleased God. He was getting tall, and everyone liked him.

One night Samuel heard someone call his name, "Samuel." He thought it must be Eli so he went to Eli, but Eli told him he had not called. Eli sent him back to bed. Again, Samuel heard someone call his name, "Samuel." Samuel went to Eli and said, "Here I am. You called me." But again Eli told him that he had not called. Once more Samuel heard a voice call, "Samuel." Again he went to Eli. This time Eli knew God was calling Samuel. Eli said to go back to bed and when it happened again say, "Speak, LORD, for your servant is listening."

This time when Samuel heard, "Samuel! Samuel!" instead of going to Eli, Samuel said, "Speak, for your servant is listening."

Samuel learned to listen for God's voice and obey God. He grew up to become a great prophet. God was with him.

Questions

- What did Hannah ask God for?
- What did she promise God?
- Did she keep her promise?
- Even though Samuel was a little boy, who called him?

Memory Verse

" ... Set an example for the believers in speech, in conduct, in love, in faith and in purity."

1 Timothy 4:12

An Unlikely Cure

2 Kings 5:1–14

While Israel and another country were fighting, a little girl was taken from her home. She missed her home, but she loved and trusted God. Eventually she learned to love the new people and her new home, even though the people did not know or love God.

This little girl worked for a family. Naaman, the father, was very sick. The little girl was sad that he was so sick. She told Naaman's wife that back in her old home, a great prophet named Elisha could cure him.

Naaman went to the king to get permission to go to Israel. The king said yes.

So Naaman went to Elisha's house, but Elisha

didn't come out to meet him. He sent a messenger who told Naaman to go wash seven times in the Jordan River, and he would be healed. Naaman got angry. He thought that Elisha was rude not to come

out to talk to him. He didn't want to go wash in the Jordan River—it was very muddy. He decided to go back home instead of doing what Elisha said.

On their way back home, one of Naaman's servants asked, "If the prophet had told you to do some great thing, would you not have done it? How much more reason, then, to obey him when he tells you, 'wash and be cleansed.'" Naaman realized that this man was right. He turned around and went to the Jordan River.

When he got to the river, it was just as muddy as ever. But he went in and washed once. Nothing happened. Twice, three times—even after he had washed six times—nothing happened. But when he washed the seventh time, he was healed. He was completely well, just as Elisha had said. Naaman was very happy.

Naaman went back to Elisha's house to thank him. This time Elisha met him in person. Naaman told Elisha that now he believed in God.

Naaman began to love God because that one little girl, even though she was far from home, still loved and obeyed God.

It took a lot of courage for her to tell Naaman's wife about Elisha and her God, but she trusted God and was not afraid to talk about Him to others.

Parent Note: It is important for children to learn early to talk about Jesus with others. Witnessing is such an important key to our growth as Christians. Children who grow up seeing it done naturally will follow the example. It will lead to a more vital faith.

Questions

- Who told Naaman to go see Elisha?
- What did Elisha tell Naaman to do?
- Why did Naaman go back to Elisha's house?
- What did he tell Elisha?
- Why do you think Naaman was mad that Elisha didn't come to the door the first time?

Memory Verse

" … Set an example for the believers in speech, in conduct, in love, in faith and in purity."

1 Timothy 4:12

Young King

2 Kings 22–23

Josiah was only eight years old when he became king. Can you imagine? Though he was very young, he was a good king who wanted to obey God.

Many in his country had been worshiping idols, or fake gods, instead of the one true God. When Josiah was still young, he began to change the way things were done. Even as a child, Josiah lived differently than the kings who had come before him. He tried to live in a way that pleased God. Then as a young man, he began to rebuild the great temple in Jerusalem where the one true God was to be worshiped.

While workers were rebuilding the temple, they found God's Law, which had been written long before by God through Moses. As they read it, they realized that they had been sinning against God by putting other things in God's place.

King Josiah was upset and decided that he and the other people would make a covenant, or promise, to God to obey and do all that God had told them to do. Josiah was going to do this with all his heart and soul. He set an example even though he was young.

The king destroyed all the places where people worshiped other things instead of God. Then they celebrated with a feast in Jerusalem.

Even though King Josiah was young, the Bible tells us, "Neither before nor after Josiah was there a king like him who turned to the LORD as he did—with all his heart and with all his soul and with all his strength."

Questions

- How old was Josiah when he became king?
- Was he a good king?
- What did they find while rebuilding the temple?
- How had the people been sinning against God?
- What did they do about it?

Memory Verse

" ... Set an example for the believers in speech, in conduct, in love, in faith and in purity."

1 Timothy 4:12

Memory Verse

" … Set an example for the believers in speech, in conduct, in love, in faith and in purity."

Timothy 4:12

Daniel Faces the Lions

Daniel 6

There are many wonderful stories about angels and how God sends them to help us. Here is one story about angels.

Once there was a very important king. Everyone had to obey him. He had many lions that he kept in a den.

In the same country, there was a very good and wise man named Daniel. But some bad men didn't like Daniel.

These bad men talked the king into making a rule that no one could pray to anyone except him. Anyone who did would be thrown into the lions' den.

But Daniel loved and obeyed God. Daniel prayed to Him three times a day with his window open. Daniel would not pray to the king. The bad men peeked into Daniel's window, then ran to tell the king that Daniel was praying to God.

That made the king sad. The king liked Daniel, but he had made the rule. Now Daniel would have to be thrown into the lions' den. The bad men took Daniel and put him in with the lions. They placed a big stone at the entrance so no one could get out.

The king was worried about Daniel. When the king went to bed later, he couldn't sleep. Early the next morning, he ran to the lions' den to see what had happened to Daniel. He cried out, "Daniel, servant of the living God, has your God, whom you serve continually, been able to rescue you from the lions?" Daniel answered him, "O king, live forever! My God sent his angel, and he shut the mouths of the lions. They have not hurt me."

The king was so happy! He ordered his men to help Daniel out of the den. Everyone could see that Daniel was not hurt at all. God had taken care of him.

Questions

- What did the bad men want to do to Daniel?
- What rule did these men talk the king into making?
- Who did Daniel pray to?
- Who was sent to help Daniel?
- Is God able to rescue us from danger?
- Why didn't Daniel pray secretly to try to stay out of trouble?

Jonah Runs Away

Jonah 1–3

Have you ever run away from doing something you were told to do? Maybe your mom asked you to pick up your toys but you went outside to play with a friend. Did she punish you for disobeying her? Did you have to pick up your toys anyway?

There was a man named Jonah who lived a long time ago. God told him to go to a city named Nineveh to tell the people that they had been doing bad things. God was very unhappy with them. They needed to repent, or say they were sorry, and quit doing bad things.

Jonah didn't want to do that so he ran away from God. He got on a boat that was going far away.

The boat went way out into the ocean. God made a big storm with wind and waves to toss the boat around. The men on the boat were afraid that it would turn over and they would die.

Jonah was asleep. The captain woke him up and asked him to pray to God to stop the storm. The men on the boat found out the storm was Jonah's fault. They asked him who he was and what he was doing. They became afraid because they realized he was running away from God.

The storm was getting worse. They asked Jonah what they should do. Jonah told them to throw him overboard—then the sea would calm down. The men prayed that they would not die and tossed Jonah into the water. The storm quit and the ocean became calm.

Jonah was in the ocean when a big fish came and swallowed him up! It must have been very dark and smelly inside that fish.

He prayed inside the fish for three days. Jonah did not like it there, so he called out to God for help.

The big fish threw up and out came Jonah onto the beach!

Once again, God told Jonah to go tell the people of Nineveh that they had done bad things and needed to repent.

This time Jonah obeyed. He went to Nineveh and told the people to stop doing bad things. Many people believed in God because of what Jonah told them.

Questions

- Can we run away from God?
- Can you imagine what it must have been like inside the big fish?
- What did Jonah learn while he was in the fish?
- Did he obey God?
- What happened because he obeyed?
- What are some things you need to do to obey God?

Memory Verse

" … Set an example for the believers in speech, in conduct, in love, in faith and in purity."

1 Timothy 4:12

400 Years of Silence

Micah 6:8; Zechariah 1:3

Do you like to get mail? Isn't it fun to go to the mailbox and find something with your name on it? Sometimes you get a letter from someone very special, and that's even better! But maybe there are times when you don't hear from that special someone and it makes you sad. You wait and wait but nothing comes. You wonder, *Has something happened to her? Did he go away? Did she quit liking me? Did he forget me?*

God had given many messages to prophets to tell God's people what He wanted them to do. He wanted them to rebuild the temple and worship Him only. He wanted them to love mercy and walk humbly with Him. There were a few obedient people, but most of them continued to disobey God. They did bad things. He warned them that they would be punished.

And then God got quiet. He didn't say anything to His people for 400 years. That's a long time!

The ones who were obedient were hopeful that God would speak to them again. They hoped year after year. They had no idea how He would speak and He surprised them. He didn't shout from heaven. He didn't stand on a street corner preaching. He didn't send a letter.

He sent a baby—His very own Son!

Questions

- What were God's messengers called?
- Did the people obey?
- How long was God quiet?
- How did God surprise His people?
- Whom did He send?

Memory Verse

"For God so loved the world that he gave his one and only Son, that whoever believes in him shall not perish but have eternal life."

John 3:16

A Special Baby Is Born

Luke 2:1–12

Far away, there is a town called Bethlehem. It is a very old town. Many years ago a young woman named Mary came to Bethlehem with her husband, Joseph. They did not live in Bethlehem, so they had to find an inn where they could spend the night.

The inn was full of people, so Mary and Joseph would not have a room to spend the night. Joseph was sad and a little worried. He needed to find a place for Mary to be comfortable. She was going to have a baby. A special baby.

They went to the stable where the donkeys and cows stayed. In the stable, there was a place for the donkeys and cows to lie down and a place where they ate.

We eat meals from plates, but donkeys and cows eat out of mangers. There was a manger in the stable where Mary and Joseph stayed.

When the baby was born, Mary had no nice cradle to put Him in. She wrapped Him up in some cloths and put Him into the manger as His cradle.

Do you know what that baby's name was? His name was Jesus. And He was and is the Son of God.

Parent Note: This story begins to establish the deity of Christ, which is essential in understanding the Bible. Children readily accept this. They may ask why they cannot see Him or what He looks like or how He can live in heaven as well as in their hearts. Keep your answers simple and honest, such as: we can't see Him because He does not have a body like ours, but one time He did. He was seen by people; He ate with them and talked with them. We do not know what He looks like, and we don't know how He can be in two places at one time, but He can because the Bible *tells us so. The Bible is true.*

Questions

- Where did Mary and Joseph find a place to sleep?
- Why was Joseph worried?
- What usually sleeps in a stable?
- What was the baby's name?
- Who had sent Jesus to Mary?
- Since Jesus was God's own Son, why wasn't He born in a palace or at least a comfortable, clean place?

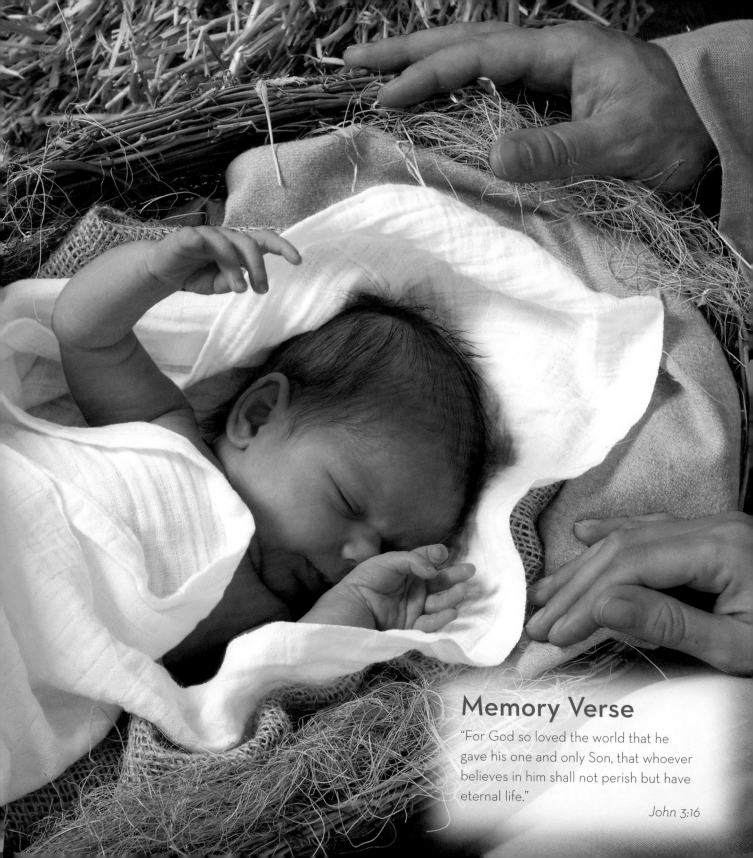

Memory Verse

"For God so loved the world that he gave his one and only Son, that whoever believes in him shall not perish but have eternal life."

John 3:16

The Shepherds Meet Jesus

Luke 2:8 – 16

Sheep are very useful creatures. If a farmer owns sheep, he can keep them fenced in the yard and let the sheep eat the grass. That is one way to keep the lawn mowed!

In the country where Jesus was born, people used to have many sheep. The sheep stayed out in the fields to eat the grass. But since there were wolves and bears in the country, men had to stay in the fields to protect them. These men were called shepherds.

On the night Jesus was born, an angel came from heaven and spoke to some shepherds. They were frightened, perhaps because they had never seen an angel before. The angel had a very important message from God.

The angel told them, "Do not be afraid. I bring you good news of great joy that will be for all people. Today in the town of David a Savior has been born to you." The angel called Jesus the Savior.

The shepherds found Jesus in the manger, as the angel said.

Questions

- Why did shepherds have to take care of their sheep?
- On the night Jesus was born, who came to talk to the shepherds?
- What did the angel tell them?
- What did the angel call Jesus?

Memory Verse

"For God so loved the world that he gave his one and only Son, that whoever believes in him shall not perish but have eternal life."

John 3:16

Wise Men Meet Jesus

Matthew 2:1–12

Some other men came to see Jesus in Bethlehem. They were called Wise Men. These Wise Men knew a lot about the stars. They used to stay up all night sometimes, looking at the stars, trying to learn all about them.

One night when they were looking up at the sky, they saw a new star that was different from all the

stars they had ever seen before. God had sent the Wise Men star for the Wise Men to see, so they would know His Son was born.

As soon as they knew that Jesus was born, the wise men wanted to see Him. But they lived far

away from Bethlehem. They did not know where Bethlehem was. How would they find the way? God made the star to lead them to Bethlehem.

The WIse Men followed the star to Bethlehem and found Jesus. They knelt down on the ground in front of Him and worshiped Him. Then they took out some presents and gave them to Him. Then they traveled home.

Questions

- Who else came to see Jesus in Bethlehem?
- Why did God send the star for the Wise Men to see?
- After they had seen the star, what did they want to do?
- What did they do when they saw Jesus?
- Where did they go afterward?
- How does God lead us to Jesus today?

Memory Verse

"For God so loved the world that he gave his one and only Son, that whoever believes in him shall not perish but have eternal life."

John 3:16

God Is Good

Psalm 100

Shout for joy to the LORD, all the earth.
    Worship the LORD with gladness;
    come before him with joyful songs.
Know that the LORD is God.
    It is he who made us, and we are his;
    we are his people, the sheep of his pasture.

Enter his gates with thanksgiving
    and his courts with praise;
    give thanks to him and praise his name.
For the LORD is good and his love endures forever;
    his faithfulness continues through all generations.

Questions

- Who made us?
- Who should we thank?
- Who will love us forever?

Memory Verse

"For God so loved the world that
he gave his one and only Son, that
whoever believes in him shall not
perish but have eternal life."

John 3:16

A Safe Place

Matthew 1:21

Suppose you were playing ball in your yard and the ball rolled out into the street. You ran into the street after the ball and did not look to see that a car was coming. The car was coming closer to you at a fast speed and would soon hit you. It would kill you.

But suppose that just then, a strong man picked you up and carried you to a safe place so that you would not be hit by the car. That man would have saved you from the car, right?

Jesus came from heaven to save us—not from a speeding car but to save us from our sins.

Our sins are the bad things we do, like disobeying Mom and Dad or lying or being selfish. Jesus came to save us from being punished for them after we die. That is the reason we call Him our Savior.

If you ask Jesus to forgive you and live in your heart, he promises to do it. Would you like to ask Jesus into your heart?

Parent Note: If your children indicate that they understand what sin is, take advantage of their open hearts and lead them to Jesus through a simple prayer. For a parent, there is no greater joy. Children are the only things we can take to heaven!

Questions

- What did the angel call Jesus?
- What are our sins? Can you think of some things you have done that have made God unhappy?
- Did Jesus come to help us stop doing these wrong things?
- If we love Jesus and ask for forgiveness, will we be punished for these things after we die?
- How can Jesus, who died so many years ago, take the punishment for our sins, when we're alive now?

Memory Verse

"Believe in the Lord Jesus, and you will be saved."

Acts 16:31

John the Baptist

Matthew 3:1–6; John 3:35–36

Jesus' cousin's name was John the Baptist. He was a very good man. He lived in the desert. Even though a desert is a place where practically no one lives, John the Baptist lived there.

John wore a scratchy coat and ate unusual food. He did not eat meat and bread as you probably do.

He ate locusts and wild honey. Locusts are insects that look like grasshoppers.

John loved and obeyed God. God told John to leave the desert to tell people about Jesus. He would teach them that Jesus was God's Son and to get ready for Him.

John went to a place near the Jordan River. Many people came to hear what he had to say. John told them that very soon they would see their Savior and that they must get ready.

How would they get ready? Should they put on their best clothes? No, that is not what John meant. The way to get ready for Jesus was to stop doing wrong things and to ask God to take away their sins.

Parent Note: This story introduces a bit of biblical cultural background. This helps establish the Bible as a historical book with real people and real events.

Questions

- Where did John the Baptist live?
- What did he eat?
- What did God tell him to do?
- What was the name of the big river John went to?
- What did John tell the people?

Memory Verse

"Believe in the Lord Jesus, and you will be saved."

Acts 16:31

Wash Away Our Sins

Matthew 3:13–17

Some of the people who listened to John obeyed him and stopped doing bad things. But others didn't pay any attention to him. John took everyone who listened and obeyed down to the Jordan River to be baptized.

Do you know what it means to be baptized? It means that we are sorry for all the wrong things we have done and have asked Jesus to forgive us. It is a way of saying to everyone that Jesus has washed away our sins and we have asked Him to live in our hearts.

While John was baptizing all the people at the river, Jesus came and asked John to baptize Him. And so John did.

Jesus' baptism was very special. Because Jesus never did anything wrong, there was no need to wash away any sins. He asked to be baptized to show that John's preaching was right and that soon Jesus was going to truly wash away our sins by His blood.

When Jesus came up out of the water after being baptized, a wonderful thing happened. He heard a voice from heaven speaking. It was God's voice, and He said, "This is my Son, whom I love; with him I am well pleased."

At the same time, a beautiful bird flew down and rested on Jesus. It looked like a dove, but it wasn't. It was the Holy Spirit.

Parent Note: The Holy Spirit is introduced here. If your child shows interest, a discussion as to who the Holy Spirit is and what He does (we cannot see Him but He lives in us and helps us obey God) will help your child in an important but often neglected subject. If you are unclear in your own mind about the Holy Spirit, *The Holy Spirit* by Billy Graham is an excellent resource.

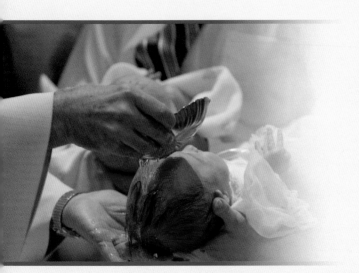

Questions

- What did John do at the river?
- What is baptism?
- Who came to John to be baptized?
- When Jesus came out of the water, whose voice did He hear?
- What did God say?

Memory Verse

"Believe in the Lord Jesus, and you will be saved."

Acts 16:31

Memory Verse

"Believe in the Lord Jesus, and you
will be saved."

Acts 16:31

Jesus Is Tempted

Matthew 4:1-11

The Holy Spirit led Jesus into the desert. Jesus stayed in the desert a long time—forty days and nights. He didn't have anything to eat the whole time.

While Jesus was out in the desert, Satan came to Him; he tried to get Jesus to do wrong. Remember, Satan was the one who convinced Adam and Eve to disobey God. He tries very hard to make us do wrong. He is very bad.

Satan knew that Jesus was hungry, so Satan told Jesus to change the stones lying on the ground into bread. Jesus could easily have changed the stones into bread, but He would not do it because that would be obeying Satan.

Then Satan took Jesus away from the desert to the top of a beautiful church called a temple. Satan told Jesus to throw Himself down from that high place. Satan said that God would send angels to catch Him. Then He would not be hurt when He fell.

But Jesus would not do this either, because to obey Satan would be wrong.

Then Satan took Jesus to a high mountain, and he showed Jesus many beautiful countries and cities. Jesus could see them all at the same time.

Satan told Jesus that if He would only bow down and worship Satan, jesus could have all those beautiful countries and cities for His own. But Jesus said only God should be worshiped, not Satan. In the Bible, God says that we should not worship any other gods.

When Satan realized that Jesus would not obey him, Satan went away and left Jesus. Then some good angels came and took care of Jesus.

Sometimes Satan comes to us and tries to make us do bad things. We cannot see him when he comes, but we can tell he is near, for he makes us feel like we want to do wrong.

When you feel like doing wrong things, ask Jesus to tell Satan to leave you alone. Then he will go away from us, as he went away from Jesus.

Questions

- How long did Jesus stay in the desert?
- Did He eat or drink anything?
- Who came to try to make Jesus do wrong?
- Whom did Jesus say he would obey?
- What should we say to Satan when he tries to make us do wrong?
- Is temptation a sin?

81

Jesus' First Miracle

John 2:1–10

Do you like to go to parties? Some of the nicest parties are weddings. The best part is when they cut the cake and you get a thick slice with lots of frosting!

Jesus liked parties too. He went to weddings. In Jesus' day, weddings lasted up to seven days. The Bible says that Jesus and some of His disciples went to a wedding in a town named Cana.

At the wedding, there was laughter and celebrating. Maybe some dancing. A party with all kinds of good food for the guests to eat.

So many people came to this wedding that soon there was nothing left to drink.

Jesus' mother, Mary, was also at the wedding. She noticed that the wine was all gone. She was concerned and took the groom's servants to Jesus and told them, "Do whatever He tells you."

Jesus told the servants to bring in some water and pour it into six tall jars in the dining room. The servants obeyed Jesus and filled the jars up to the very top.

Then Jesus told them, "Now draw some out and take it to the master of the banquet." The man tasted it and found out that it wasn't water now. It was very fine wine.

Jesus hadn't put anything into the water. He had changed the water into wine. This was His first miracle.

Jesus had authority over nature—including the jars of plain water. He could change it because He is the Son of God, and He can do the same things God can do.

Parent Note: This Scripture helps us see that Jesus wasn't always encumbered by His ministry. Here He was at a wedding party. He enjoyed people and laughed. He is not always stern, always demanding of us. He accepts us just the way we are. He does not hand us a list of don'ts—He gave us all things to enjoy, under authority of Scripture.

Questions

- What was the name of the town Jesus went to?
- What did Mary tell the servants to do?
- What did Jesus tell them to do?
- Who had changed the water into wine?
- Why could Jesus do miracles?
- Do you think that while Jesus was here on earth, He laughed and had fun?

Memory Verse

"Believe in the Lord Jesus, and you will be saved."

Acts 16:31

Jesus Says the Word

John 4:46-53

Jesus was in the town of Cana where He had attended the wedding party. There was a very important man who lived there. The man's son was very sick, and he was afraid his little boy would die.

When the man heard that Jesus was in Cana, he hurried to see Him. He begged Jesus to make his son well. He told Jesus, "Sir, come down before my child dies." The man thought that Jesus would have to see his son to make him well.

But Jesus told the man, "You may go. Your son will live." The man believed what Jesus said. On his way home, his servants came out to meet the man to tell him that his son was already well.

Jesus had made the boy well by just saying so. As soon as Jesus said that, the sickness left the boy and he was well. Jesus has authority over everything, including sickness. This was a miracle!

Questions

• What did the important man ask Jesus to do?
• Could the doctors make his son well?
• What did the important man think Jesus had to do to make his son well?
• What did Jesus say to him?
• What does Jesus have authority over?
• Why didn't Jesus go to the man's house and heal the young boy there?

Memory Verse

"Come, follow me," Jesus said, "and I will send you out to fish for people."

Matthew 4:19

Time Alone with God

Luke 4:42 – 44, 5:16

God had given Jesus a big, important job. His job was to show us what God was like and how much God loves us. Jesus knew that He had to spend time alone with God in order to do this job well.

The same is true for us. Our job is to show people what God is like. We need to spend time with God getting to know Him, reading about Him in the Bible, and talking to Him.

Early one morning before the sun was up, Jesus woke up and went out to the desert so that He could pray to God. Jesus knew how important it was to spend time alone with God.

Soon, the people from the town came to look for Jesus. When they heard that He had gone to the desert, they went out to find Him. When they found Him, they begged Him to stay in their town.

But Jesus said, "I must preach the good news of the kingdom of God to the other towns also, because that is why I was sent."

God had sent Jesus to our world to show us that God loves us. But because we have done many wrong things, called sins, we have to repent and ask God to forgive us. To repent is to be sorry for something you have done—so sorry that you never want to do it again. Jesus went to many towns telling the people these things.

Parent Note: This step introduces the doctrines of repentance and forgiveness. At this point you might want to ask your children if they want to ask Jesus to forgive them for their sins and come into their hearts.

Questions

- Where did Jesus go very early in the morning?
- Who went out to find Jesus?
- Why did God send Jesus?
- What are sins?
- What does *repent* mean?
- Do you follow Jesus 'example of being alone to pray?

Memory Verse

"Come, follow me," Jesus said, "and I will send you out to fish for people."

Matthew 4:19

Fishermen Follow Jesus

Luke 5:1–11

Jesus came to another town named Capernaum, near the Sea of Galilee. He liked to walk on the beach.

He saw two fishing boats on the beach; they belonged to fishermen who were washing the nets.

If you have ever gone fishing, you probably took a fishing pole and some worms. But you could only catch one fish at a time. If you used a net, you would catch a lot of fish at a time.

Jesus told one of the fishermen, "Put out into deep water, and let down the nets for a catch."

They let down their nets. But when the men tried to pull them back up into the boat, the nets were so full of fish the men couldn't lift them!

The men called out to their friends to come help. They came alongside and helped pull up the net. When they emptied the fish into the boats, they were so full of fish that they were close to sinking.

This was a miracle! No one else but God could do this.

Jesus told the men to follow Him. They immediately left their boats and nets to follow Jesus.

Questions

- What happened when Jesus walked on the beach?
- Who did Jesus see on the beach?
- Why weren't they fishing?
- When Jesus told them to fish, what happened?
- Did these men follow Jesus?
- Pretend you were in Peter's boat that day. How would you have felt and acted?

Memory Verse

"Come, follow me," Jesus said, "and I will send you out to fish for people."

Matthew 4:19

But the man, named Peter, said, "Master, we've worked hard all night and haven't caught anything. But because you say so, I will let down the nets."

Peter, James, and John rowed out on the sea.

Make Me Clean

Luke 5:12-15

Leprosy was a horrible disease that no one knew how to cure back in Bible times. This sickness made sores show up on a person's skin. Sometimes the sickness was so bad that people would lose feeling in their fingers or hands or feet or nose or ears.

Leprosy made people sick in the country where Jesus lived. When people got leprosy, they would have to leave their homes so that their families would not get sick. They didn't have nice hospitals to go to. They could not come back home until they were well. But no one could make them well except God.

When Jesus was walking in a town one day, a man who had leprosy came to Jesus. He knelt down on the ground in front of Jesus and said, "Lord, if you are willing, you can make me clean." Jesus didn't like to see such a sick man suffering. Jesus said, "I am willing." He then told the man, "Be clean." As soon as Jesus said that, the leprosy went away and the man was well. This was another miracle!

People couldn't help talking about how Jesus healed the man. After that, so many people came to Jesus and crowded around Him that He could not stay there anymore.

Parent Note: Leprosy shouldn't upset a child. Explain that it is not common today and medicines are available to help cure it. Emphasize the love and power of Jesus.

Questions

- What bad sickness did you learn about in this story?
- When a person got leprosy, could he stay with the family?
- Who can make a person well?
- What did the man with leprosy ask Jesus to do for him?
- Did Jesus do it?
- What did the people do?
- Why was Jesus willing to heal the man with leprosy?

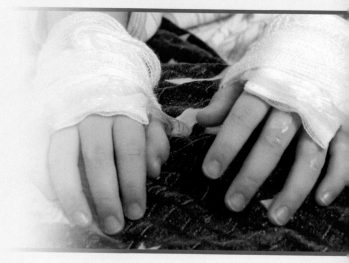

Memory Verse

"Come, follow me," Jesus said, "and I will send you out to fish for people."

Matthew 4:19

Rejoice and Be Glad

Matthew 5:1–12 (NIrV)

Jesus saw the crowds. So he went up on a mountainside and sat down. His disciples came to him Then he began to teach them.

He said,

"Blessed are those who are spiritually needy. The kingdom of heaven belongs to them.

Blessed are those who are sad. They will be comforted.

Blessed are those who are free of pride. They will be given the earth.

Blessed are those who are hungry and thirsty for what is right. They will be filled.

Blessed are those who show mercy. They will be shown mercy.

Blessed are those whose hearts are pure. They will see God.

Blessed are those who make peace. They will be called sons of God.

Blessed are those who suffer for doing what is right. The kingdom of heaven belongs to them.

"Blessed are you when people make fun of you and hurt you because of me. You are also blessed when they tell all kinds of evil lies about you because of me. Be joyful and glad. Your reward in heaven is great. In the same way, people hurt the prophets who lived long ago."

Questions

- Who taught this passage to a crowd?
- Do you know someone that is like the person Jesus is describing?
- How could you become like the person Jesus is describing?
- What is our reward for believing in and following God?

Memory Verse

"Come, follow me," Jesus said, "and I will send you out to fish for people."

Matthew 4:19

A Hole in the Roof

Luke 5:18–26; Mark 2:1–12

Jesus went back to Capernaum, where he had helped Peter and his brother catch so many fish. There were many houses there, but they did not look like the houses people live in today. They were small, square houses that were only one story high. The roofs on these houses were flat—a person could go up and walk around on them.

Jesus was teaching in one of these houses. In the same town, some friends of a sick man wanted to take him to Jesus, but the man was so sick and weak he could not walk.

So they decided to pick up the bed he was lying on and carry him to Jesus. When they got to the house where Jesus was, there was such a big crowd that they could not get through the door.

The man's friends carried him up onto the flat roof of the house and opened up a hole in the roof. Then they carefully lowered the man on his bed into the room where Jesus was.

Jesus saw how much faith the men had and the effort they put in bringing their sick friend to Him. He said to the sick man, "I tell you, get up, take your bed and go home." Just by saying that, Jesus made the man well. This was another miracle!

The man had been so sick that he couldn't even stand up. But now Jesus had made him well, and he could walk, run, and carry his bed all the way home!

When the people who had crowded the house and doorway saw this, they were surprised and said to each other, "We have never seen anything like this!"

Questions

- When Jesus was teaching in one of these houses, who was brought to Him?
- Why couldn't the man's friends get him through the door?
- How did they get the sick man into the house?
- What did Jesus do?
- Would you go to a lot of trouble to bring a friend to Jesus?

Memory Verse

"Do to others as you would have them do to you."

Luke 6:31

Read the Bible

2 Timothy 3:14–17

The Bible is a wonderful book, full of exciting stories. It is important to read the Bible every day. It tells us about Jesus and how to live like Him.

Suppose you had done something wrong, and one of your parents was angry and about to punish you. But your brother felt sorry for you and asked to be punished in your place. Wouldn't that show how much your brother loved you?

This is what Jesus did for us. We have done wrong, and God was going to punish us; but Jesus loved us so much that He came down from heaven to be punished in our place. The Bible is the book that tells us about this.

Parent Note: This step underlines the substitutionary atonement of Jesus (Jesus took over our punishment for sin so that we could have a relationship with God). This is a key ingredient to the Christian faith.

Questions

- Who does the Bible tell us about?
- If you were going to be punished for doing something wrong and your brother asked to be punished in your place, what would that show?
- When God was going to punish us, what did Jesus do?
- Why did He do this?
- Do you have a time each day when you read the Bible?

Memory Verse

"Do to others as you would have them do to you."

Luke 6:31

Jesus Says to Be Kind

Matthew 6:26

Jesus walked up the side of a mountain to pray to God, His Father. Jesus prayed all night long. In the morning, twelve men had come to listen to Him. He chose these men to stay with Him all the time. They were to listen to Him and learn from Him. They were to go wherever He sent them and do whatever He said. Jesus made these twelve men *disciples*.

Jesus came down the mountain to be with a crowd of people. He was soon surrounded by people wanting to listen to Him and be healed of their diseases. Jesus taught them how to live a holy

He told the people that they must be kind to each other. Just as Jesus taught us, it is important to be kind to people and also to all animals. It pleases God when we take care of His creatures.

Jesus said that the people should not fight with each other or be angry at each other. And when someone is unkind to them, they must not be unkind back. They should be kind to the people who are unkind and pray for them.

Parent Note: Children have very tender hearts but can be very unkind to each other. You might point out a practical application of this story by saying, "Remember when Allison took your ball, and you hit her? What do you think Jesus would have told you to do?"

Questions

- How many men did Jesus ask to stay with Him all the time and learn from Him?
- What did Jesus tell the people on the mountain?
- How should people treat each other?
- If someone is unkind to us, what should we do?
- What are some things you can do this week to help your family or friends?

life. He told them that they should not be proud and think that they are better than other people. Instead, they must think of ways to help other people. When they realize they have done something wrong, they must be sorry for doing it.

Memory Verse

"Do to others as you would have them do to you."

Luke 6:31

The Lord's Prayer

Matthew 6:9–13

Prayer is like talking to your best friend. One day, there was a boy who couldn't find his glasses. He was very upset and began to cry. In the midst of his tears he remembered that even if the glasses were lost, God still knew where they were. The boy asked Jesus to help him find the missing glasses. A few minutes later, the boy remembered that he had left his glasses on the kitchen table beside his book. Then he thanked God for helping him.

Jesus taught the people what they should say when they were praying to God. He said, "This, then, is how you should pray:

'Our Father in heaven,
hallowed be your name,
your kingdom come,
your will be done
on earth as it is in heaven.
Give us today our daily bread.
And forgive us our debts,
as we also have forgiven our debtors.
And lead us not into temptation,
but deliver us from the evil one.'"

This is called the Lord's Prayer, because the Lord Jesus teaches us to say it. But whenever we say this prayer, we must remember that we are speaking to God—we must think about what we are saying.

Jesus not only teaches us to say the Lord's Prayer, but He tells us to pray to God for everything we need. For God is our Father who lives in heaven, and He loves to give His children the things they pray for, and the things that are best for them.

Parent Note: Prayer cannot be overemphasized. But we can't expect our children to learn to pray if we don't pray ourselves. Children's prayers are so precious as they learn to talk to Jesus. What a privilege to teach them how! Remind them that they can pray about anything: needs, wisdom, guidance, protection, friends, sick pets, sore fingers, help in finding a lost toy—everything! As you pray with them, keep it simple and short. And don't forget to thank God for His answers.

Questions

- What did Jesus teach the people?
- What is the name of the prayer that Jesus taught them?
- When we say this prayer, to whom are we speaking?
- What should we pray to God for?
- Does God love to give His children the things that are best for them?
- Can you think of something you prayed about and how God answered?

Memory Verse

"Do to others as you would have them do to you."

Luke 6:31

Pray to Please God

Matthew 6:5-8

There were some men in Jesus' country called Pharisees.

They were the leaders and teachers. But they thought they were better than other people. They didn't like Jesus or what He taught.

The Pharisees used to say their prayers out in the open where people could hear them. They wanted people to think they were good — and they wanted praise from other people. Whenever anyone looked at them, they were very careful to do good things; but when no one looked at them, they weren't as concerned about what God thought of their prayers.

The Pharisees didn't want people to listen to Jesus, but to them instead. They were always trying to find something wrong with Jesus and His disciples. They were not nice.

Jesus said we shouldn't be like the Pharisees. We should not say our prayers for other people to hear, but for God to hear.

We shouldn't do good things because we want other people to think we are good. We should do good things because we want to please God.

Questions

- Why did the Pharisees want people to hear them?
- When anyone looked at them, what were the Pharisees very careful to do?
- When no one saw them, did it matter as much?
- Who do we want to hear our prayers?
- Is it easier to be good when someone is watching you?

Memory Verse

"Do to others as you would have them do to you."

Luke 6:31

Have Faith

Matthew 8:5–13

Jesus traveled back to the town of Capernaum. A man who was a Roman commander lived there. This soldier had a servant who was very sick. The soldier was afraid that the man would die.

So the soldier said to Jesus, "Lord, my servant lies at home paralyzed, suffering terribly." Jesus told the soldier, "I will go and heal him." But the soldier replied, "Lord, I do not deserve to have you come under my roof. But just say the word, and my servant will be healed. For I myself am a man under authority, with soldiers under me. I tell this one, 'Go,' and he goes; and that one, 'Come,' and he comes. I say to my servant, 'Do this,' and he does it."

Jesus was happy that the soldier believed that just by saying so, Jesus could make the man well. Jesus told the soldier, "Go! It will be done just as you believed it would."

The soldier went home and found that the sickness had left his servant. He was already well, just as Jesus had said. This was a miracle.

Questions

- Who came to Jesus when He was in Capernaum?
- What was wrong with the man who worked for the soldier?
- What did the soldier want Jesus to do?
- Why was Jesus happy?
- When the soldier got home, what did he find?

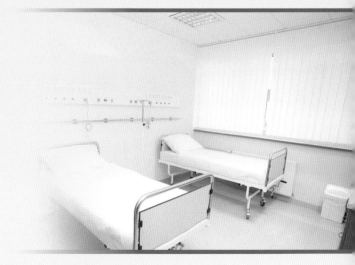

Memory Verse

"Be strong and courageous.
For the LORD your God goes with you;
he will never leave you nor forsake you."

Deuteronomy 31:6

Back to Life

Luke 7:11–16

Jesus traveled to another town called Nain. Nain had a wall with gates so the town officials knew who was going in and out.

As Jesus came closer to the town, He met some men coming out of the gate. They were carrying a dead man in a coffin. They were going to bury him in a grave.

The dead man was his mother's only son, and her husband had already died, so now she lived alone. She was walking in front of the coffin. She was crying. She did not think she would ever see her son again.

When Jesus saw this woman crying, His heart went out to her. He said, "Don't cry." Then He went over to the coffin where her son lay and touched it. The men carrying the coffin stopped.

Jesus spoke to the dead man. He said, "Young man, I say to you, get up!" As soon as Jesus said these words, the dead man came to life again. He got out of the coffin and began to talk. The man and his mother were together again!

This was a miracle.

When the people saw the dead man come to life, they were filled with wonder and praised God. They said that surely Jesus had been sent from God because no one else could bring a man back to life.

Questions

- When Jesus got close to the town of Nain, who was being carried out of the gates?
- What did Jesus say to the man's mother?
- What did He say to the dead man? What happened then?
- How did the people feel when they saw the dead man come to life?
- Who did they say had sent Jesus to them?

Memory Verse

"Be strong and courageous.
For the LORD your God goes with you;
he will never leave you nor forsake you."

Deuteronomy 31:6

Lotion in a Jar

Luke 7:36-50

In the country where Jesus lived, people used to buy a very nice lotion called ointment. They would use the ointment to rub on their hair and skin to make them soft and smooth. The ointment smelled like flowers.

One day a man invited Jesus to his house for dinner. In that country, instead of having tables and chairs in the dining room, they had a very low table and sat on big pillows. Also, instead of shoes, the people usually wore sandals because it was very hot.

While Jesus was eating dinner at the man's house, a woman came inside where they were eating. She held a little jar of ointment. It had cost her a lot of money. She went over to Jesus, knelt at his feet, and began crying. Then she dried Jesus' feet with her hair, kissed them, and poured the ointment on them. The woman did this to show how much she loved Jesus. She loved Him because He forgave her.

Do you remember what a Savior is? We learned a story to help us understand it. In the story, you imagined that you were running after a ball and ran into the street. You didn't see a car coming—but a man nearby did. He ran out and pulled you out of the street just in time. That man saved you from being killed by the car.

In the same way, Jesus came from heaven to forgive this woman and save her from her sins. She had disobeyed God by doing many bad things, but now she was truly sorry for her sins and didn't want to do bad things anymore. Jesus told her that because she was sorry for her sins, God would forgive her and would not punish her. When Jesus did this, He was her Savior. She loved Him for coming from heaven to save her from her sins.

Questions

- What did the woman pour the ointment out of?
- What did she do to Jesus' feet?
- Was she sorry for the bad things she had done?
- When Jesus forgave her, what was He to her?
- Did she love Him because He forgave her?
- Would you give something to Jesus that cost you a lot of money?

Memory Verse

"Be strong and courageous.
For the LORD your God goes with you;
he will never leave you nor forsake you."

Deuteronomy 31:6

Good Soil

Matthew 13:3–8; Mark 4:3–8; Luke 8:5–8

One day, Jesus told a parable to some people. A parable is a story that teaches something very important.

This parable was about a man who planted wheat in his field. Wheat grows up like tall grass and is harvested, and the little kernels of wheat are collected to grind into flour. Bread is made out of wheat flour.

The man in this parable scattered some seeds on the ground as he walked along. But some wheat fell outside the field. Birds flew down and ate the seeds.

Some of the seeds fell in the wrong place — where there were lots of rocks and no soil. The wheat could not form roots and grow up. Some seeds fell where there were lots of thorny bushes. The bushes would not give the wheat any room to grow. The wheat was crowded out.

But the rest of the seeds fell where the ground was soft and fertile. The wheat got plenty of rain and sun, so they formed roots that went deep into the ground. The plants grew tall. Soon it was harvested and the man had much more than he had at the beginning.

This same thing happens when pastors and teachers tell people the things that Jesus wants them to know. Some people do not listen. It is as if birds come and take away every word they speak, because the people do not remember.

But some people remember, and they let words and teachings about Jesus speak to their hearts.

Questions

- What was this man planting in his field?
- Could the seeds that fell in the wrong place grow up?
- What flew down and ate the seeds?
- Did the seeds that fell in the right place grow up and produce more plants?
- What kind of soil are you?

Memory Verse

"Be strong and courageous.
For the LORD your God goes with you;
he will never leave you nor forsake you."

Deuteronomy 31:6

The Fruit of the Spirit

Galatians 5:22–26

The fruit of the Spirit is love, joy, peace, forbearance, kindness, goodness, faithfulness, gentleness and self-control. Against such things there is no law. Those who belong to Christ Jesus have crucified the flesh with its passions and desires. Since we live by the Spirit, let us keep in step with the Spirit. Let us not become conceited, provoking and envying each other.

Questions

- What are the fruits of the Spirit?
- What is your favorite one?
- Are any of them hard to follow or appreciate?

Memory Verse

"Be strong and courageous.
For the LORD your God goes with you;
he will never leave you nor forsake you."

Deuteronomy 31:6

Seeds of Faith

Matthew 13:31–32

If you plant seeds in a garden and make sure that they have enough water and sunlight, eventually they will sprout up into beautiful plants. From the tiniest seed comes the biggest oak tree!

Jesus told people about the mustard seed. It is a very small seed—about the size of a pinhead. If you held one in your hand, you would hardly be able to see it. But if you planted it in the ground, it would sprout and grow into a plant big enough for birds to sit on its branches.

The words Jesus speaks to your heart are like a tiny mustard seed planted in the ground. As you water it by obeying Jesus, learning all you can about Him, and talking to Him, the seed grows bigger and bigger—like the mustard seed when it grows to be a beautiful plant.

Questions

- What kind of seed did Jesus tell the people about?
- How big is a mustard seed?
- When it is planted, what does it grow into?

Memory Verse

For it is by grace you have been saved, through faith—and this is not from yourselves, it is the gift of God...

Ephesians 2:8

Bread with Yeast

Matthew 13:33

Do you like to make bread? The whole house smells good when it bakes in the oven. Then it tastes so delicious when you slice it and spread butter on a warm piece.

Jesus told a story about a woman making bread. First she takes some wheat flour and adds some salt. Then, before adding water, she adds something called yeast. She begins to mix it all together to make dough.

She works it some more with her hands—this is called kneading the dough. She kneads the dough so that the yeast will be mixed in. Then she sets the dough aside, covers it up, and leaves it for several hours. When the dough is baked later in the oven, the yeast makes nicer, softer bread.

When we keep our hearts soft to Jesus and allow His Word to be "kneaded" in our lives, we begin to love Him more and more. And as we love Him more, we begin to live like Jesus. We actually want to obey Him and be just like Him!

Questions

- What was the woman in the story making?
- After she had mixed the flour, salt, and water, what else did she add?
- What does the yeast do?
- How is Jesus' word like the yeast?

Memory Verse

For it is by grace you have been saved, through faith—and this is not from yourselves, it is the gift of God. . .

Ephesians 2:8

Beautiful Pearls

Matthew 13:45–46

Jesus told the people a story about a man who wanted to buy some pearls. Pearls are beautiful, little, round, white stones created inside an oyster, which lives deep in the ocean. Sometimes we see pearls in rings or bracelets or necklaces.

The man in Jesus' story did not want to dive down under the water to find them. He wanted to buy them from the divers who had found them already. So he went to the divers and asked to see all the pearls they had to sell.

At last one of the men showed him a very beautiful pearl. It was larger and prettier than any pearl he had ever seen before, but it would cost more money than he had. So he told the diver to keep that pearl until he came back again. Then he sold everything he had so he could get enough money to buy that pearl.

Perhaps he had horses, cows, sheep, and land. He sold everything to get money to buy the pearl. When he finally had the pearl, he was very pleased because he had the thing that he wanted more than anything else in the world.

What should we want more than anything else in this world? It isn't a pearl, for that will not make us happy.

We should want to please God and ask Him to come into our heart to live. We should obey Him, learn about Him, and talk to Him. And when we do wrong, we should ask God to forgive us.

We should want this as much as the man wanted the beautiful pearl.

Questions

- What did the man in the story want?
- What did he do to get enough money to buy it?
- How did he feel when he got the pearl?
- Did he want it more than anything else?
- Should we want to please God that much?

Memory Verse

For it is by grace you have been saved, through faith— and this is not from yourselves, it is the gift of God. . .

Ephesians 2:8

Jesus Calms the Storm

Matthew 8:23 – 27; Mark 4:35 – 41; Luke 8:22 – 25

Jesus got into a boat with His disciples to go across a lake. While they were sailing, He lay down and went to sleep. He was tired.

Pretty soon a storm came up that made the waves very high and rough. The little boat was being tossed around on the lake, and water was spilling into the boat. The disciples were scared. They were afraid the boat would sink.

They woke up Jesus and said, "Lord, save us, we're going to drown!"

Jesus asked the disciples why they were afraid of the wind and the waves while He was there to keep them safe. Then Jesus stood up and told the wind not to blow and the waves to be calm.

Do you know what? The wind and waves obeyed Him. The wind quit blowing and the waves became calm.

The disciples were amazed when they saw this happen. They said to each other, "Even the winds and the waves obey Him!"

Questions

- What was Jesus doing in the boat during the storm?
- How did the disciples feel? What did they do?
- What did they say to Jesus?
- Should they have been afraid while He was with them?
- If you had been in the boat that day, what would you have done?

Memory Verse

For it is by grace you have been saved, through faith — and this is not from yourselves, it is the gift of God. . .

Ephesians 2:8

Jesus Has Power Over Demons

Mark 5:1–20; Luke 8:26–38

Remember the good angels that live in heaven? Well, there are some very bad angels that do not live in heaven. No one bad or wicked lives in heaven. We call these bad angels "demons."

Demons do not have bodies like ours with hands and feet. We cannot see demons and they can go into places where we cannot go. Sometimes demons torment people and make them very sad. Sometimes they even make people do bad things. Fortunately, Jesus is stronger than the most powerful demon.

When Jesus got out of the boat and walked on the shore, a man came to Him who was tormented by demons. The demons made this man very angry and mean so that he acted like a wild animal.

Not even chains around his hands and feet could hold him down. He would just break the

chains and roam around in the mountains. All the people were afraid of him and stayed very far away.

There were caves in these mountains. Sometimes wild animals lived in the caves. This poor man with the demons lived in the caves.

Day and night, the man would cry out and cut himself with sharp rocks.

The poor man could do nothing to make the demons go away. But Jesus could make them leave. Jesus spoke to the demons and told them to come out of the man.

Nearby was a big herd of pigs that were eating. The demons asked if they could live in the pigs instead, and Jesus said, "Yes." The demons went into the pigs—then ran them into the lake where the pigs all drowned.

The man was changed. He was not mean and angry; instead he was quiet and calm like other people. Jesus had made the demons leave the man alone. He thanked Jesus and wanted to follow Him. But Jesus told him to go home and tell his friends how he had been made well.

Parent Note: The existence of demons should not scare a child. The emphasis should be put on Jesus' power and His love and care for us. We must not confuse mental illness with demon possession or vice versa. They are very different. We have authority over demons by the power of Jesus' blood and name.

Questions

- Can we see demons?
- Who is more powerful—the demons or Jesus?
- When Jesus was walking on the shore, who came to Him?
- What had the demons made this man act like?
- What did Jesus make the demons do?
- Was the man well after that?

Memory Verse

For it is by grace you have been saved, through faith—and this is not from your-selves, it is the gift of God. . .

Ephesians 2:8

One Touch: Healed

Luke 8:43 – 48; Matthew 9:18 – 26; Mark 5:24 – 33

A big crowd was following Jesus. Some people were very close to Him and bumping against Him.

In the crowd there was a woman who had been sick for a long time. She had gone to see many doctors, hoping they could make her well. She had spent all her money but they could not cure her.

As soon as she saw Him, she said to herself, "If I could just touch His clothes, I will be healed." So she crept up quietly behind Jesus, reached out, and touched His clothes. Immediately, she knew that she was well.

Jesus stopped and looked around. He asked who touched His clothes. The disciples had not seen the woman do it, and with so many people crowded around, they wondered why Jesus asked who had touched Him. But Jesus said, "Someone touched me; I know that power has gone out from me."

When the woman saw that Jesus knew she had touched Him and that she could not hide, she was afraid. She came to Jesus and knelt in front of Him. She told Him that she had touched His clothes and had been made well.

Jesus spoke to her very kindly. He said, "Daughter, your faith has healed you. Go in peace."

Questions

- What had the woman spent all her money doing?
- What happened when she touched Jesus' clothes?
- Was Jesus kind to her? What did He call her?
- What did He tell her?
- If you are being pushed and shoved in a crowd, do you get angry and scared? How did Jesus act?

Memory Verse

"Ask and it will be given to you; seek and you will find; knock and the door will be opened to you."

Matthew 7:7

Just Believe

Mark 5:22-24,35-42

A man who was very upset because his little girl was sick and he was afraid she was going to die came to Jesus. The man said, "Please come and put your hands on her so that she will be healed and live."

So Jesus went with the man, but as they were approaching the house, someone came to them and told the man, "Your daughter is dead; why bother the teacher anymore?" But Jesus told the man, "Don't be afraid; just believe."

When they went inside the house, people were crying; they were very sad that the little girl was dead. Jesus told them to leave the house. He took three of His disciples and the little girl's parents into the room where she lay.

Jesus went to the side of her bed, held her hand, and said, "Little girl, I say to you, get up!" As soon as He said that, the little girl opened her eyes, sat up, and began to walk. She was alive again!

Jesus told her parents to get her something to eat. They were so happy because their little girl was alive. She would live with them and love them just as before.

Parent Note: When we experience difficulties or are faced with change or the unknown, we often feel afraid. Jesus does not want us to feel that way. He wants us to trust Him. The opposite of fear is faith and trust.

Questions

- What did the man ask Jesus to do?
- When they were approaching the man's house, what did someone tell them?
- What did Jesus tell the man?
- When Jesus went beside the little girl's bed, what did He say to her?
- What happened to the little girl?
- How did her mom and dad feel when they saw she was alive?
- When you are sad, is it okay to cry?

Memory Verse

"Ask and it will be given to you; seek and you will find; knock and the door will be opened to you."

Matthew 7:7

God Watches Over Us

Matthew 10:29 – 32

We learn all about Jesus and His Father, God, in the Bible. The Bible is what God wants us to know and learn about Him. One thing God wants us to know is that He loves us very much.

He doesn't say life will be easy or that He will give us everything we want. But He promised He will always watch over us.

God gives the birds food to eat. He cares about the birds. Did you know that if one tiny, little bird falls out of its nest, God knows about it? But God cares a lot more about you and those who love Him than He does about the birds. He cares so much about you that He even knows how many hairs are on your head!

If we love and obey God, we don't ever need to be afraid. God promises He will always watch over us. He keeps His promises.

Questions

- Who does the Bible tell us about?
- Does God care about the birds?
- But who does God care about more?
- If we love and obey Jesus, what will God always do?
- Does God give us everything we want?

Memory Verse

"Ask and it will be given to you; seek and you will find; knock and the door will be opened to you."

Matthew 7:7

Riches in Heaven

Matthew 8:19 – 20

One day a man came to Jesus and told Him that he wanted to live with Him all the time. But Jesus told the man that He had no home to live in.

Jesus said that the little birds had homes—they had their nests up in the trees. And the wild animals had homes—they had caves and holes in the ground where they could go. But when Jesus was tired, He had no place to sleep. He had no home.

Jesus was even poorer than the birds and animals. But Jesus was not always poor. He used to live in heaven—He was not poor there. He had everything to make Him happy.

Why, then, did He come to this world where He would be poor and have problems? Because He loved us, and He wanted to make us God's children—then we can go to heaven after we die. If Jesus loves us so much He came from heaven for us, we ought to love Him and ask Him to live in our hearts.

Questions

- What did Jesus say about the birds and the animals?
- What was the reason Jesus had no home?
- Was Jesus always poor?
- Why did He come to this world to be poor and have problems?
- What should we do?

Memory Verse

"Ask and it will be given to you; seek and you will find; knock and the door will be opened to you."

Matthew 7:7

The Good Shepherd

Psalm 23

The LORD is my shepherd, I lack nothing.
    He makes me lie down in green pastures,
he leads me beside quiet waters,
    he refreshes my soul.
He guides me along the right paths
    for his name's sake.
Even though I walk
    through the darkest valley,
I will fear no evil,
    for you are with me;
your rod and your staff,
    they comfort me.

You prepare a table before me
    in the presence of my enemies.
You anoint my head with oil;
    my cup overflows.
Surely your goodness and love will follow me
    all the days of my life,
    and I will dwell in the house of the LORD
    forever.

Questions

• Who is the Good Shepherd?
• Are we ever really alone? Why or why not?
• Do you pray when you get scared?
• Do you feel better after you pray?

Memory Verse

"Ask and it will be given to you;
seek and you will find; knock and
the door will be opened to you."

Matthew 7:7

Barns of Food

Luke 12:16 – 21

It is important to save your money, even when you don't have very much of it. Maybe you have a special box where you keep your money for spending. Maybe you have another box where you keep the money you will give to God.

It is wise to save money for the future. But sometimes people try to save money to feel safe rather than trust God for the future. Some do it so they can buy whatever they want. Some do it to feel very important.

Jesus loved to tell the people stories to teach them things. One day, He told them a story about a man who had many good things to eat and drink. The man built great big barns to store all the food for himself.

When he had put all of it away, he said to himself, "You have plenty of good things laid up for many years. Take life easy; eat, drink and be merry."

But as soon as the man said that, God spoke to him and told him, "You fool! This very night your life will be demanded from you." Then all the things that he had saved for himself would not do him any good. Instead, someone else would get it all.

Jesus told that story to teach us to not be like that man, because all the man cared about was getting rich and doing what would please himself. Instead, we should always care about others and do what will make God happy.

God gives us all things to enjoy, but the important thing is not the things we have, but our relationship with God.

Parent Note: Selfishness is a very real problem in our society. Emphasize this step by pointing out a recent family incident. Perhaps one of your children has a difficult time sharing toys.

Questions

- Why did Jesus like to tell stories?
- What did the man in this story have?
- When he had put everything away in the barns, what did he say to himself?
- But what did God say to him?
- Who should we try to make happy?

Memory Verse

"Therefore go and make disciples of all nations, baptizing them in the name of the Father and of the Son and of the Holy Spirit."

Matthew 28:19

If You Have Faith

Matthew 9:27 – 31

Two blind men were following Jesus one day. They could not see Him, but someone had told them that Jesus was there. Perhaps they had heard how He brought the little girl to life again and thought He could make their eyes well.

So the blind men followed Jesus and called after Him, saying, "Have mercy on us, Son of David!" Jesus stopped and asked them, "Do you believe that I am able to do this?" They said, "Yes, Lord." "According to your faith will it be done to you," He said. Then He put out His hand and touched their eyes. By only touching them, He made their eyes well. They were so happy!

They went away and told all the people who lived in that country how Jesus had cured them in one moment.

Parent Note: A blind person cannot see all the beautiful things God has made, like the sun and sky. Have your children close their eyes and imagine what it would be like to be blind. What senses do they need to be more aware of when their eyes are closed?

Questions

- As Jesus was leaving, who followed Him?
- What did the blind men want Him to do?
- What did Jesus ask the blind men?
- How did He make their eyes well?
- What did the blind men do after they were made well?

Memory Verse

"Therefore go and make disciples of all nations, baptizing them in the name of the Father and of the Son and of the Holy Spirit."

Matthew 28:19

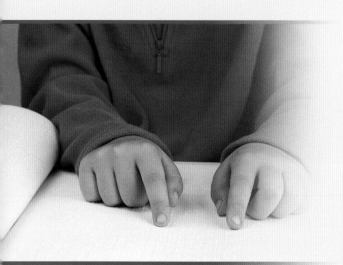

Disciples Teach Too

Mark 6:7-15

Jesus went to many other towns to teach the people. But He could not teach all the people by Himself: there were too many of them. So He sent out His twelve disciples to teach the people who lived in the cities where He could not go.

The disciples went to the towns and taught the people about Jesus. They told the people how He had come down from heaven to take away their sins and make them God's children. Jesus enabled the disciples to make sick people well and dead people alive, just as Jesus did Himself. He let the disciples do these wonderful miracles so that the people would listen to what the disciples said and believe that God had sent them.

After the disciples had taught the people, they came back to Jesus. They told Him where they had been and what they had done.

Questions

- Did Jesus go to other towns to teach people?
- Why could He not teach all the people in other cities by Himself?
- Who did He send to teach the people in the towns where He could not go?
- What did the disciples do while they were teaching the people?
- Could they have done this without Jesus' help?

Memory Verse

"Therefore go and make disciples of all nations, baptizing them in the name of the Father and of the Son and of the Holy Spirit."

Matthew 28:19

139

Lots of Leftovers

John 6:1-15; Mark 6:32-44

Jesus and His disciples wanted to be alone, so they got into a boat and went to the other side of the lake. But when the crowd saw where Jesus was going, they followed Him.

Jesus was kind to them; He taught them about God and heaven. He healed the sick people.

As it got dark, the disciples went to Jesus and said, "This is a remote place … and it's already very late. Send the people away so they can go to the surrounding countryside and villages and buy themselves something to eat." Jesus told the disciples, "You give them something to eat."

The disciples told Jesus that would cost a lot of money! They wanted to know if they should spend that much on bread and give it to the crowd to eat. Jesus asked them, "How many loaves do you have?" They told Him they had only five loaves of bread and two small fish.

Jesus took the loaves of bread and fish in His hands and thanked God for the food. He divided the bread and fish into pieces and gave them to the disciples to serve the people.

They didn't run out of food because as they gave out a piece of fish or bread, another piece would come. They had plenty. The crowd had all they wanted to eat.

After everyone was finished, Jesus told the disciples to pick up the leftovers. They had twelve baskets of food left over! This was a lot more food than when they began feeding the people.

Jesus made the bread and fish keep coming until all the people had enough to eat. This was a miracle. We cannot do miracles, but Jesus can, because He is the Son of God. He can do the same things that God can do.

Questions

- Was Jesus kind to the crowd that followed Him?
- What did Jesus tell the disciples to do?
- How many loaves of bread and fish did they have?
- Whom did Jesus thank for the food?
- How many baskets full of food were left over?
- Why could Jesus do miracles?

Memory Verse

"Therefore go and make disciples of all nations, baptizing them in the name of the Father and of the Son and of the Holy Spirit."

Matthew 28:19

Walk on Water

Matthew 14:22 – 33

One day, Jesus told the disciples to get into a boat and sail to the other side of the lake. He wanted to be alone, so He stayed by Himself on the beach. When the disciples left, He went up on a mountainside to pray. He knelt down on the ground and prayed to God. Remember how important Jesus thought it was to spend time with God?

During the night Jesus came down from the mountainside to the beach. He saw the disciples out in the middle of the lake struggling with their boat because the wind was blowing hard against them. The waves were high and rough.

So Jesus went out to them, walking on the water. He walked on the water as if it were dry land! When the disciples saw Him coming toward them on the water, they were afraid. They did not know who or what it was.

But Jesus called out to them, "Take courage! It is I. Don't be afraid." One of the disciples named Peter asked Jesus if he could come out on the water. Jesus answered, "Come."

So Peter got out of the boat and began to walk on the water toward Jesus.

But when he heard the loud wind and saw the rough waves all around him, he got scared and began to sink. He called out to Jesus, "Lord, save me!" Jesus reached out His hand and caught Peter so that he would not sink into the water. Jesus asked Peter why he was afraid. Jesus would take care of him and keep him safe.

As soon as Jesus and Peter got into the boat with the other disciples, the wind and waves calmed down. They were able to cross the lake.

Questions

- Where did Jesus see the disciples?
- How did He get to them?
- What happened to Peter as he was walking on the water?
- What did Jesus ask Peter?
- When they got into the boat, what happened?
- If you had been in the boat that night, would you have gotten out like Peter did?

Questions

- What did Jesus say people must do if they loved Him?
- What did He call this?
- If you do something right even if it may hurt you or cause you to lose friends, is this taking up the cross?
- Is thinking of other people, rather than yourself and what you want, taking up the cross?
- Does it make Jesus happy?
- Can you think of a time when you helped someone even though it caused you to lose friends?

Memory Verse

"The time has come … The kingdom of God has come near. Repent and believe the good news!"

Mark 1:15

Helping Others

Luke 9:23 – 26

Many people came to listen to Jesus and learn what He would teach them. Jesus told the people that if they loved Him, they would always obey Him, even if they had to do something they didn't want to do. He said, "If anyone would come after me, he must deny himself and take up his cross daily and follow me."

Suppose one day you were playing outside with some other boys and girls, but one of the boys got too rough and started to hurt one of your friends. Your friend was smaller than the others and could not play as well as they could, so the children made fun of him.

You didn't like them making fun of him and hurting his feelings, so you protected him. You made a special effort to play with him, even though the other children now wouldn't play with you.

That is taking up your cross. It is doing something that is right even though it might cause you to lose some friends or to be hurt. You do it because Jesus would have done the same thing.

Or suppose someone gave you a ten-dollar bill and told you to go buy something you wanted. You thought of a toy you wanted and your mother took you to the store to buy it.

But as you were going into the store, you saw a friend of yours who didn't have very many toys. You stopped and thought about the nice toys you had at home that this friend had admired. So you went over to him, gave him the money, and told him to go buy something he wanted. He smiled, thanked you, and hurried off to the toy section.

This is taking up the cross: putting someone else first instead of only thinking about yourself and what you want.

This makes you happy because in your heart, you know you have done the right thing. But the best part is you have made Jesus happy.

Two Visitors

Matthew 17:1–8

Jesus and three of His disciples, John, Peter, and James, went up a mountain to pray. While there, Jesus' face was changed so that it looked bright and shining, like the sun. And His clothes looked as white as light.

All of a sudden, two men joined them. They were Moses and Elijah, two great prophets who had lived a long time ago and now lived in heaven. They did not look like other men: they looked more beautiful.

Moses and Elijah had come back to the world where we live to talk to Jesus a little while.

Soon a bright cloud came onto the mountain; it covered the three disciples. They heard a voice speaking out of the cloud. It was God's voice. He said that Jesus was His Son, whom He loved very much. God told the disciples to listen to Jesus.

When the disciples heard God's voice, they were afraid. They fell face down on the ground. But Jesus told them to stand up and not be afraid.

They stood up and looked around, but Moses and Elijah were not there now: they had gone back to heaven.

Questions

- What happened while Jesus was praying?
- Where do Moses and Elijah live?
- Whose voice spoke out of the cloud?
- What did God say?
- When the disciples got up from the ground, where had Moses and Elijah gone?
- How does God speak to us today?

Memory Verse

"The time has come. … The kingdom of God has come near. Repent and believe the good news!"

Mark 1:15

Who's the Best?

Luke 9:46 - 48

One day when the disciples were walking along together, they began to argue. Each of them wanted to be the greatest.

When they reached Jesus, He asked what they had been arguing about. They were ashamed and did not want to tell Him; they did not think that He had heard them. But He knows everything we say, and He knew what they had said when they were arguing.

Jesus told the disciples that they should not want to be the greatest, but they must be willing to serve other people. Jesus is not happy with us when we are proud and think we are better than other people. He is happy with us when we are humble and think of other people first instead of ourselves.

Questions

- What did the disciples argue about?
- Does Jesus know everything we say?
- When He asked the disciples why they had been arguing, how did they feel?
- What did He say?
- Does he tell us to think of other people first?
- Do you like to be first?

Memory Verse

"The time has come. ... The kingdom of God has come near. Repent and believe the good news!"

Mark 1:15

Show Mercy

Matthew 18:23 - 34

One day Jesus told a parable about a king. A man who worked for the king had borrowed some money from him. Now the king wanted the money back, but the man did not have any money.

In that country, if you needed money and you had no other way of getting some, you could sell yourself, your wife, and your children to be slaves. This is what the king told the man he must do. This made the man very sad.

He begged the king to wait just a little longer so that he could find some other way to get the money. He didn't want to sell his family. He promised the king that as soon as he got the money, he would repay him.

The king felt sorry for him. The king said that the man should not sell his family to get the money. The king would forget all about the money the man owed him. The man would never have to pay back what he owed. This made the man so happy — he thanked the king for being so kind.

As the man left the king's palace, he met a man who owed him some money. He told this man to pay up. The poor man said he couldn't but promised to as soon as he could.

The man who worked for the king got very mad and demanded to be paid. He couldn't wait. Because the poor man couldn't pay him immediately, the man had him thrown into prison.

When the king heard about this, he got very angry. He sent for the man who worked for him and said, "You wicked servant … I canceled all that debt of yours because you begged me to. Shouldn't you have had mercy on your fellow servant just as I had on you?" The king had the man punished.

Jesus told this story to teach us that we must forgive others when they have said or done something unkind to us. Jesus said that we must forgive from our hearts.

Questions

- When the man could not pay the money, what did the king say must be done to him and his wife and children?
- When the man heard this, what did he ask the king to do?
- What did the king do?
- When the poor man asked the king's servant to wait until he could get some money to pay him, was he willing to wait?
- What did the king say when he heard how cruel the man who worked for him had been?

Memory Verse

"The time has come. … The kingdom of God has come near. Repent and believe the good news!"

Mark 1:15

Wonderful Counselor

Isaiah 9:6–7

For to us a child is born,
    to us a son is given,
    and the government will be on his shoulders.
And he will be called
    Wonderful Counselor, Mighty God,
    Everlasting Father, Prince of Peace.
Of the greatness of his government and peace
    there will be no end.
He will reign on David's throne
    and over his kingdom,
establishing and upholding it
    with justice and righteousness
    from that time on and forever.
The zeal of the LORD Almighty
    will accomplish this.

Questions

- What child is going to be born?
- Can you remember what Jesus is also called?
- Why was Jesus born?

Memory Verse

"The time has come. … The kingdom of God has come near. Repent and believe the good news!"

Mark 1:15

The Cost of Following Jesus

Luke 9:51–56

One day while Jesus and His disciples were walking together, they came close to a small town. Many people lived in this town. Jesus sent some of His disciples into town to ask if they would let Him stop there to rest and eat.

But the people in the town were very unfriendly and unkind to Jesus. They told Him that He could not stay. Two of the disciples, James and John, got very mad at the people and wanted to punish them. They asked Jesus if they could bring down fire from heaven to burn up the houses and all the people.

Jesus was sad that the people in the town did not want Him to stop, but He was unhappy with James and John for wanting to do such a mean thing. Jesus told James and John, "The Son of Man did not come to destroy men's lives, but to save them."

He did not punish the people who were so unfriendly and unkind but went on to another town to rest and eat.

Questions

- What did the people say to Jesus' disciples?
- How did James and John feel?
- Was Jesus displeased that James and John wanted to do this?
- What did Jesus say that he had come to this world for?
- Instead of punishing the people, what did He do?
- Do you sometimes like to get back at someone who's been mean to you, as James and John wanted to do?

Memory Verse

Forgive as the Lord forgave you.

Colossians 3:13

The Good Samaritan

Luke 10:30 - 37

Jesus told the people another story. A man was walking alone on a dangerous road. All of a sudden, robbers jumped out of their hiding place and beat up the poor man. They took everything he had—food, clothes, and money. Then they left him beside the road to die.

After a while another man came down the road. He was a priest in the temple who would tell people to be kind to each other. But he was not kind himself, so he passed the beaten man. He crossed over to the other side of the road, pretending not to see the man, and went away.

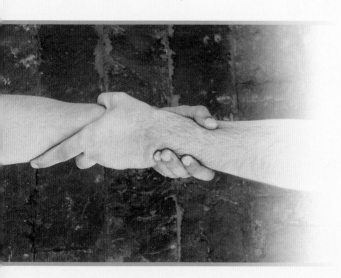

Soon another man came down the road, but he didn't help the poor man either. He kept on walking just as the priest had done and left the man on the ground.

But after these men had passed by without helping the poor man, someone else came by. He was called a Samaritan, and he was riding on a donkey. As soon as he saw the man lying beside the road, he stopped and got down to help him.

The Samaritan was very kind and gentle. He lifted the poor man and put him on the donkey; then he walked very carefully and slowly beside the donkey to be sure the man would not fall off. The Samaritan took the man to a nearby inn and stayed up all night to take care of him.

The next day, when the Samaritan had to go away, he gave some money to the man who owned the inn and asked him to take care of the hurt man until he got well.

Jesus told this story to teach us to be like the good Samaritan. We should be kind to everyone we meet but especially to those who need our help. We should always be helpful whenever we can be, even when it is a lot of work.

Parent Note: Our society is very selfish; we "don't want to get involved." But this is contrary to the teaching of Jesus. Begin to instill in your child thoughtfulness and unselfishness by example. Perhaps one Thanksgiving, your family could serve a local food kitchen. Get involved in local nonprofits that provide food or services to senior citizens, children, or the homeless.

Questions

- Who beat up the man in the story and took everything that he had?
- After this happened, who first came down the road?
- Instead of helping the wounded man, what did he do?
- What did Jesus want to teach us by telling this story?
- Which man in this story are you like?
- Can you think of a time you helped someone who needed you?

Memory Verse

Forgive as the Lord forgave you.

Colossians 3:13

Get to Work ... or Listen?

Luke 10:38 – 42

Jesus visited some friends who lived in a town called Bethany. His friends were sisters named Mary and Martha. They also had a brother named Lazarus.

When Jesus came to their house, Mary stopped what she was doing to sit near Jesus to listen to what He was saying. Perhaps he was telling her about how their sins could be forgiven and how she would go to heaven when she died.

But Martha was busy in the kitchen preparing the meal. It might have been hot. She was probably tired. How could she get it all done? She had cleaned the house and was busy trying to make her guests feel at home.

Martha asked Jesus to tell Mary to get busy and help with the housework. Jesus said, "Martha, Martha, you are worried and upset about many things, but only one thing is needed. Mary has chosen what is better." It was more important for Mary to learn what He had to say and the things He taught than to be busy doing other things.

Parent Note: Priorities are an important lesson, but perhaps more for us as parents than for children. They should be one of our priorities. Raising children is very hard work in our society. It is a huge responsibility, not to be taken lightly. In Mark 9:37, Jesus said, "Whoever welcomes one of these little children in my name welcomes me."

Questions

- When Jesus came to the house, what did Mary do?
- Why did she want to listen to Jesus?
- How did this make Martha feel?
- What did Martha ask Jesus to tell Mary?
- What did Jesus tell Martha?
- Is it easier for you to play with your toys or listen to what your parents tell you to do?

Memory Verse

Forgive as the Lord forgave you.

Colossians 3:13

STEP 74

Is It Really Him?

John 9:1–38

Jesus was walking along the street when He saw a man who had always been blind, even when he was a little boy. Now the man was grown up, but he couldn't work because he couldn't see.

So he would sit down in the street and beg the people who passed by to give him some money to buy food and clothes.

Jesus felt sorry for the man. He spit on the ground, stooped down, and made it into mud. He put it on the blind man's eyes. Then He told him, "Go wash in the Pool of Siloam."

The man went to the pool and washed his eyes. He could see! But it was not the mud or water in the pool that made his eyes well. Jesus made his eyes well.

When the people who knew the blind man saw him walking along like any other person who could see, they asked, "Isn't this the same man who used to sit and beg?" Some claimed that it was the man.

Others said, "No, he only looks like him." But the man himself said, "I am the man." Then they asked him, "How then were your eyes opened?" He told them, "The man they call Jesus made some mud and put it on my eyes. He told me to go to Siloam and wash. So I went and washed, and then I could see."

But the men who asked him were not happy with what he told them. They did not love Jesus and would not believe that He could make blind people well. So when the man said that it was Jesus who made him well, they were angry and would not talk to him.

Jesus heard how unkind they had been to the man. He found him and asked him if he believed in the Son of God.

"Who is he, sir?" the man asked. "Tell me so that I may believe in him."

Jesus said, "You have now seen him; in fact, he is the one speaking with you."

Then the man said, "Lord, I believe," and he worshiped Him.

Questions

- When the man washed his eyes in the pool, what happened?
- Who made his eyes well?
- How did the people treat the man when he told them Jesus cured him?
- When Jesus told the man that God's Son had made him well, what did the man do?
- Do you have friends who don't think Jesus can help them?

Memory Verse

Forgive as the Lord forgave you.

Colossians 3:13

The Lost Sheep

Luke 15:3–7

Have you ever lost something important to you? Maybe it was your favorite stuffed animal. You looked everywhere for it. In the closet. Under the bed. In the car. Behind the sofa. Outside. You asked your friends and family to help you look. You looked and looked until you found it.

Jesus told a story about a shepherd who had one hundred sheep. One day when he counted his sheep, he found that there was one missing. There were only ninety-nine. The missing sheep was important to him, so he wanted to find the sheep and bring it back. He went to go look for the lost sheep.

Maybe, like you, he looked everywhere he could think of where a sheep might get lost: behind bushes and rocks, up on a mountain, down in a valley. Maybe he even looked in a dark, scary cave! The sheep was nowhere to be found! But the shepherd didn't give up.

He kept looking until—finally—he found the one sheep. He was so happy to finally find it! He tenderly picked up his woolly friend, put it on his shoulders, and took it back home.

The shepherd called his friends and neighbors and told them he had found his sheep. "Rejoice with me; I have found my lost sheep."

When Jesus told this story, he said that when one person repents, there is a party in heaven! Each person who repents is very important to Him. He said, "There will be more rejoicing in heaven over one sinner who repents than over ninety-nine righteous persons who do not need to repent."

Questions

- How many sheep did the shepherd have?
- Why did he look for the one sheep?
- What did he do when he found it?
- When someone repents, what happens in heaven?

Memory Verse

Forgive as the Lord forgave you.

Colossians 3:13

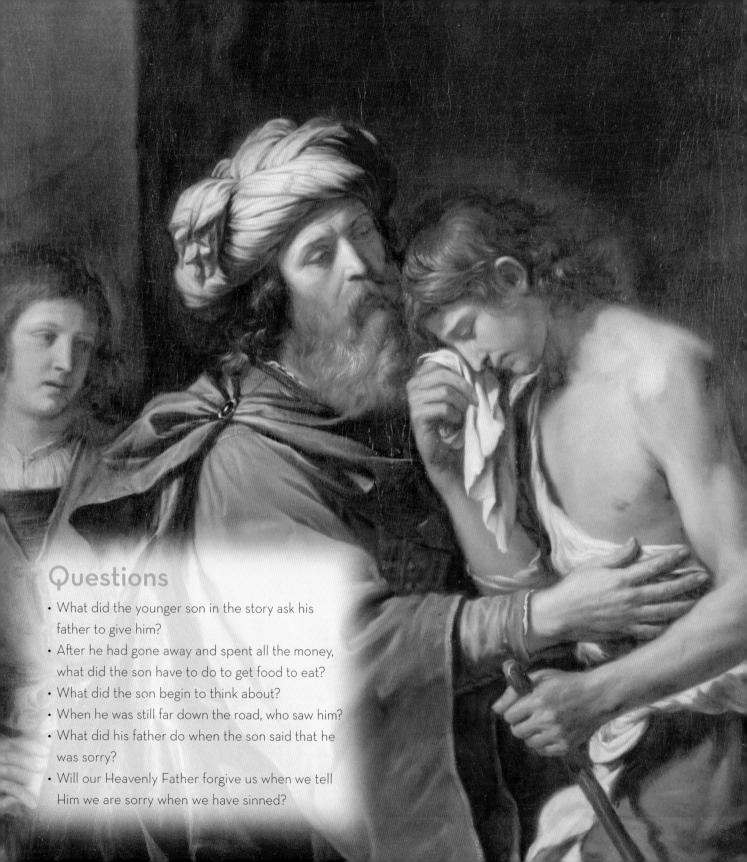

Questions

- What did the younger son in the story ask his father to give him?
- After he had gone away and spent all the money, what did the son have to do to get food to eat?
- What did the son begin to think about?
- When he was still far down the road, who saw him?
- What did his father do when the son said that he was sorry?
- Will our Heavenly Father forgive us when we tell Him we are sorry when we have sinned?

The Lost Son

Luke 15:11 – 32

Jesus told the people a story about a man who had a son. One day the son said to his father, "Give me my share of the estate." He said that he wanted it right away. So his father gave him all the money set aside for him.

The son took the money and left his father's house. He went far away to another country. He wasted all the money his father had given him until it was all gone.

So he had to go to work just to get something to eat. The man he worked for sent him out to feed the pigs but never gave this young man enough to eat. He got very hungry.

He began to think about his father's house. No one was hungry, not even the people who worked for his father. He always had plenty to eat, a bed to sleep in, clean clothes, and a father who loved and cared for him. So he said to himself, "I will set out and go back to my father and say to him, 'Father, I have sinned against heaven and against you. I am no longer worthy to be called your son; make me like one of your hired men.' Perhaps he will forgive me and let me work for him."

So he headed home. When he was still far away, his father saw him. His father did not wait for him to come any closer but ran down the road to meet him. His father was so happy! He gave his son a big hug and kiss.

The son began to tell his father how sorry he was to have been so selfish and greedy to take the money and leave home. But his father told the people who worked for him to go get new clothes for his son to wear and put a ring on his finger and shoes on his feet.

His father told them to prepare a big meal to celebrate. "I thought he was lost but now he is found," said the father.

Jesus told this story to teach us something. If we have sinned but are sorry for it, we can go to our Heavenly Father to tell Him we are sorry. He will always forgive us, just as the father in this story forgave his son.

Memory Verse

"For the Son of Man came to seek and to save the lost."

Luke 19:10

Calling All Angels

Luke 16:19–28

Jesus told the people about a rich man who had a lot of money and could buy everything he wanted. He wore beautiful clothes and had good things to eat every day. In the same city where the rich man lived, there was a poor man. His name was Lazarus.

Lazarus was not only poor, he was also sick and weak—his body was covered with sores. Because he was so poor and sick, his friends used to carry him to the rich man's house and lay him down just outside the gate, so that he might be able to get the pieces of leftover bread from the rich man's dinner.

The dogs that roamed the street seemed to pity Lazarus because they came to lick his sores.

Finally, Lazarus died. God sent some of His angels to get him.

The angels came and carried Lazarus up to heaven. He was not sick or poor anymore when he got to heaven, for God loved him and made him well. He had a new body. God gave him everything to make him happy.

After a while, the rich man died too. But the angels did not come for him. He went to the place where wicked people go, called hell. Hell is a place where God is not. And while the rich man was being punished for his sins, he could see Lazarus far away in heaven.

The rich man wanted to go to heaven where Lazarus was, but he could not. He had done many bad things and had not asked God to forgive him. We can ask God to forgive our sins because Jesus paid for them. We can ask Him to teach us how to love and obey Jesus. Then when we die, God will send His angels to take us up to heaven too.

Parent Note: This chapter mentions hell. Children can understand this concept because they have to deal almost daily with being punished for doing something wrong! But again, you need not dwell on the idea of punishment and hell but instead emphasize heaven and the anticipation of it if we have Jesus in our hearts.

Questions

- When Lazarus died, who carried him to heaven?
- Was he sick and poor in heaven?
- Where was the rich man sent?
- Why couldn't the rich man go to be with Lazarus in heaven?
- If we ask God to forgive us and ask Jesus to live in our hearts, where will we go when we die?
- Have you ever not helped someone because you thought you were too good?

Memory Verse

"For the Son of Man came to seek and to save the lost."

Luke 19:10

Being Sorry

Luke 17:3–5

Sometimes people would rather do bad things; sometimes it seems easier and more fun than doing what is right. But Jesus said we should not do what is wrong — like telling lies, taking things that don't belong to us, fighting with brothers or sisters, or talking back to Mom or Dad. If we are doing these things, we must stop doing them. If we ask Jesus to help us stop doing wrong, He promises to help us.

We should obey God and do what is right — that makes God happy. He won't stop loving us for doing bad things, but there will be consequences.

Perhaps someone hurts you and you get very angry. Jesus said, "If he sins against you seven times in a day, and seven times comes back to you and says 'I repent,' forgive him." If he is unkind many times, but later says he is sorry, we must forgive him every time — no matter how often it may be.

Questions

- Do people sometimes like to do wrong more than they like to do right?
- If we go on doing wrong and will not stop, what will happen to us?
- What should we do?
- If any person is unkind to us and later says he is sorry, what must we do?
- Is it easy to forgive someone who has hurt you?

Memory Verse

"For the Son of Man came to seek and to save the lost."

Luke 19:10

He's Alive!

John 11:1-6, 17-44

One day, Lazarus, a good friend of Jesus, got very sick. This was a different Lazarus than the one Jesus described who died and went to heaven. Mary and Martha sent someone to tell Jesus that their brother Lazarus was sick.

When Jesus heard, He went to their house because He loved them very much. By the time Jesus got there, Lazarus was already dead and had been buried. The people who were with Mary and Martha were so sad that Lazarus was dead. When Martha heard that Jesus was coming, she went to meet him. Mary stayed home.

Martha told Jesus that if He had been here, her brother would not have died. But she knew God would give Jesus whatever He asked.

Jesus said that her brother would rise again. "I am the resurrection and the life," He said. People who believe in Jesus would live even though their bodies die.

Jesus asked them where they had buried Lazarus. It was a cave with a big stone rolled over the front of the opening. Jesus told the people to roll the stone out of the way.

In a loud voice, Jesus said, "Lazarus, come out!" As soon as He said that, Lazarus came out! He was alive! Jesus had made Lazarus alive again by just saying those words. Jesus did this miracle so that people would honor God.

Jesus told the people nearby to undo the clothes that had been tightly wrapped around Lazarus for burial so that he could walk. Then Lazarus went home with his sisters, Mary and Martha. They were very happy.

Parent Note: Death is a matter to be discussed with children. It is more frightening to children to be uncertain about such a subject than to learn about it. But here the emphasis should always be not on death but on the promise of a wonderful heavenly home with Jesus if we have asked Him into our hearts. No doubt a member of the family—perhaps a great-grandparent or grandparent—has died during your child's lifetime. While there were sorrow and tears because of the loss, there was also great joy knowing the loved one was in heaven with Jesus.

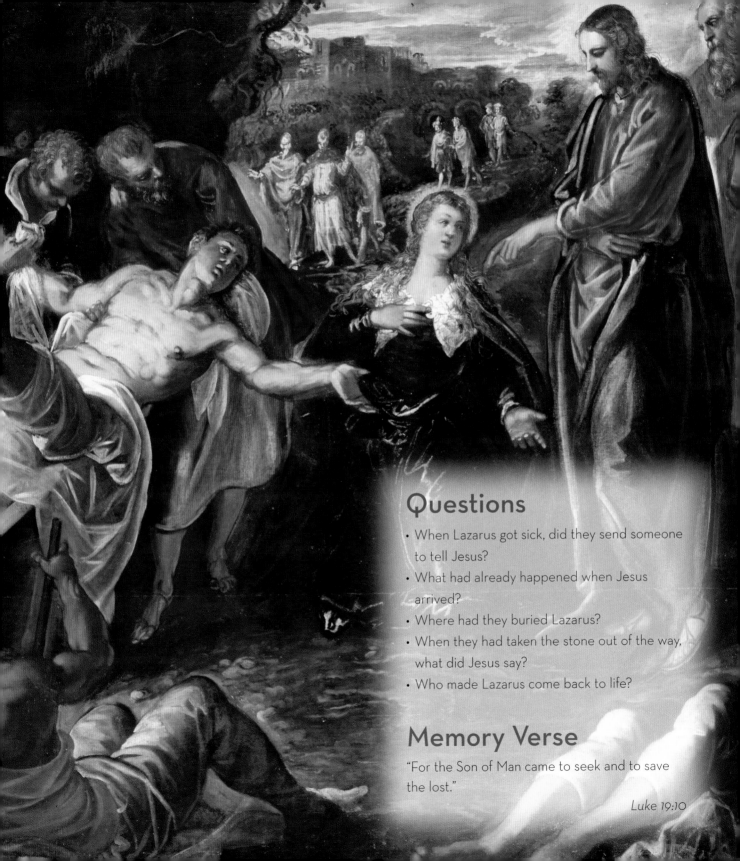

Questions

- When Lazarus got sick, did they send someone to tell Jesus?
- What had already happened when Jesus arrived?
- Where had they buried Lazarus?
- When they had taken the stone out of the way, what did Jesus say?
- Who made Lazarus come back to life?

Memory Verse

"For the Son of Man came to seek and to save the lost."

Luke 19:10

Secure in Jesus Christ

Romans 8:35 – 39

Who shall separate us from the love of Christ? Shall trouble or hardship or persecution or famine or nakedness or danger or sword?…

No, in all these things we are more than conquerors through him who loved us. For I am convinced that neither death nor life, neither angels nor demons, neither the present nor the future, nor any powers, neither height nor depth, nor anything else in all creation, will be able to separate us from the love of God that is in Christ Jesus our Lord.

Questions

- Can you think of anything that can separate you from God's love?
- What is the one thing that separates us from God?

Memory Verse

"For the Son of Man came to seek and to save the lost."

Luke 19:10

Jesus Loves Children

Matthew 19:13–15; Mark 10:13–16; Luke 18:15–17

Jesus was very busy. Many people were crowding around Him. They wanted Him to help them with their problems.

Some people brought children to Jesus. They wanted Him to hug and pray for them. But Jesus' disciples thought this would bother Him. They told the people to take the children away.

This made Jesus unhappy because He loved children (and still does!). Perhaps He liked to play games with them, tickle them, and laugh with them.

He told the disciples, "Let the children come to me, and do not hinder them, for the kingdom of God belongs to such as these." He gathered them to Him and blessed them.

Jesus loves children. If they love Him and ask Him to live in their hearts, He will always be with them.

Questions

- Who were brought to Jesus?
- Why did the disciples try to send the children away?
- What did Jesus tell the disciples?
- If children love Jesus and ask Him to live in their hearts, what will He do?

Memory Verse

"What is impossible with man is possible with God."

Luke 18:27

Big Wooden Cross

Matthew 20:17 – 19; Mark 10:32 – 34; Luke 18:31 – 33

Jesus told His disciples that people were going to betray and kill Him when they went to Jerusalem. (This was predicted years before by prophets.) The soldiers were going to be very cruel to Jesus. They would beat Him, call Him names, spit on Him — and then kill Him. But He told the disciples that He would be brought back to life again!

The way that Jesus would be killed would be very painful and cruel. He was going to be nailed to a cross. A cross was made of two large pieces of wood fastened together. The Roman soldiers were going to put Jesus on the cross by pounding big nails through His hands and feet into the wood of the cross. They would leave Him there until He died.

Many people in Jerusalem wanted this to happen. They were angry that Jesus told them about their sins and for saying that He was God's Son. They did not want to be told about their sins, and they did not believe that Jesus was God's only Son.

Questions

- How did Jesus say the people would treat Him when He came to Jerusalem?
- What would they nail Him to?
- What was a cross made of?
- Why were the people so angry at Jesus?
- Would they believe that Jesus was God's Son?

Memory Verse

"What is impossible with man is possible with God."

Luke 18:27

Short Man

Luke 19:1–10

Governments are very large and do many things for people. But to do those things, they need money. Adults pay the government some money out of what they earn at work. This is called a tax.

Long ago, a king would send out men called tax collectors to get the taxes from the people. One of these tax collectors was named Zacchaeus.

When Jesus came to his town, Zacchaeus tried very hard to see Jesus. Zacchaeus was a short man and couldn't see over the heads of the other people. So he ran ahead of the crowd and climbed a tree.

When Jesus reached the tree, He looked up and said, "Zacchaeus, come down immediately. I must stay at your house today." Zacchaeus hurried down from the tree. He was very glad that Jesus was coming to his house.

Zacchaeus listened to all the things Jesus told him. He promised to do all the things Jesus told him to do. He said, "Look, Lord! Here and now I give half of my possessions to the poor, and if I have cheated anybody out of anything, I will pay back four times that amount."

This made Jesus happy. He told Zacchaeus, "Today salvation has come to this house … For the Son of Man came to seek and to save what was lost."

Jesus wants us to be kind to other people and help them when we can. He wants us to be very careful never to take anything that belongs to another person. If we have, we must give it back. This makes Jesus happy.

Questions

- Why couldn't Zacchaeus see Jesus?
- Where did he go so he could see Jesus?
- When Jesus came to the tree, what did He say to Zacchaeus?
- What did Zacchaeus say he would do?
- What can we do to make Jesus happy?

Memory Verse

"What is impossible with man is possible with God."

Luke 18:27

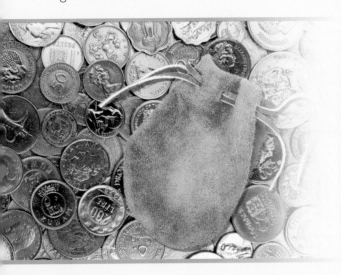

Hosanna in the Highest

Matthew 21:1-17; Mark 11:1-11; Luke 19:28-40

Before Jesus reached Jerusalem, He told two disciples to go to a nearby town. They would find a donkey and her young colt tied there. Jesus told the disciples to untie the animals and bring them to Him.

If anyone asked what they were doing, they would say that Jesus needed them. Then, Jesus said, the people would send the donkey and colt to Jesus.

So the disciples did what Jesus told them to do. They went to the town and found the donkey and colt. As they untied the animals, some men asked what they were doing. The disciples said that Jesus needed the animals. Then the men let the disciples take the animals to Jesus.

Jesus rode into Jerusalem on the colt's back. A big crowd of people followed Him. They all cried out, "Hosanna to the Son of David!"

"Blessed is he who comes in the name of the Lord!"

"Hosanna in the highest!"

Some people took off their coats and laid them on the ground. Other people cut off branches from the trees and laid them on the ground for Jesus to ride over them. That was what they used to do when a king rode through the streets.

They did this to show how glad they were to have Jesus come into their city.

Then Jesus went up to the Temple. Some blind people came to Him. People who couldn't walk were brought to Him. Jesus healed them.

Questions

- What did Jesus send two of His disciples into town to find?
- What did some of the people put down on the ground for Him to ride over?
- Why did they do this?
- Who did Jesus heal?

Memory Verse

"What is impossible with man is possible with God."

Luke 18:27

Time-out

Luke 19:45–47

When Mom or Dad says you have done a bad thing, how do you feel? Do you get mad at them? Do you want them to go away? It is hard to have someone tell us we are wrong. It hurts us. And you might get a time-out or be grounded for doing wrong. But learning when we're wrong is for our own good, to make us better people.

Jesus tried to tell some people they were doing wrong too.

After Jesus was praised in the streets of Jerusalem, He found some people selling cattle and doves in the temple. He got very angry because this was against God's rules. He told the people to stop.

He told them about their sins. Remember what sins are? They are all the bad things we do, like disobeying our parents or telling lies. In this story, people were making lots of money in the temple — where they were only supposed to pray and worship God.

The people who didn't love Jesus were getting more and more unhappy with Him. They did not want to be told about their sins. In fact, they wanted to get rid of Jesus. They wanted to kill Him.

But they were foolish to think they could try to stop God's plan.

Questions

- Do you ever get mad at someone who tells you that you're doing wrong things?
- What were the people selling in the temple?
- Why was it wrong to sell things in the temple?
- Did Jesus know that people wanted to kill Him?

Memory Verse

"What is impossible with man is possible with God."

Luke 18:27

A Wedding Feast

Matthew 22:1–13

Jesus told some people a story. It was about a king who decided to have a big party because his son was getting married. When the party was ready, the king sent his servants out to say that it was time to come. But no one would come.

Then the king sent his servants out again to tell the people that it was time to come to the party. Good food was waiting on the table. But some people would not listen to the servants. And others listened but did not obey.

When the king heard what they had done, he was very angry and sent his soldiers out to punish them.

The king then called more servants and told them that although the party was ready, no one was there to eat the food. The people who had been asked to come would not be allowed to come now, since they had refused. It was too late for them.

So the king sent his servants out into the streets and found people who would come to the feast. The servants brought many people. These people got to enjoy the party.

God has given us many good things. He has prepared heaven for us and has given us everything we need to enjoy it. We just have to accept his invitation. Asking Jesus into our hearts is the only way to go to heaven.

Questions

- When the king sent his servants to tell the people to come to his feast, what did the people do?
- Did the people who came in from the streets get to enjoy the party at the king's house?
- What do we have to do to go to heaven?

Memory Verse

Come near to God and he will come near to you.

James 4:8

Love Each Other

Matthew 22:35 – 40

One day, a man came to Jesus and asked Him what God wanted us to do most. Jesus said, "Love the Lord your God with all your heart and with all your soul and with all your mind."

We cannot see God, but He is so good to us that we can love Him without seeing Him. Sometimes people we love go away where we cannot see them. But we keep on loving them. We want to see them.

We will see God when we die, but He wants us to love Him now before we die. We should love God more than we love anyone else.

And Jesus said there was another thing God wants us to do besides love Him: He wants us to love each other. If we love each other, we will be kind to each other. When we are kind to each other, we show God that we love Him.

Questions

- What does God want us to do more than anything else?
- Who else does God want us to love besides Him?
- If we love each other, how will we treat each other?
- Why does God want us to love each other?

Memory Verse

Come near to God and he will come near to you.

James 4:8

Give God the Best

Mark 12:41-44; Luke 21:1-4

During church service, an offering plate is passed. We give money to God through the offering.

In Jesus' time, the church, or temple, had a box. People dropped in just as much as they wanted to. After a lot had been dropped in, the priests opened the box and took the money out. They bought things with it for the temple. This money was the same as if it were given to God; it was God's money.

One day Jesus watched people drop their money into the box. Rich people dropped in a lot. Soon, a poor woman dropped in only a little bit of money.

Jesus said to the disciples, "I tell you the truth, this poor [woman] has put in more than all the others. All these people gave their gifts out of their wealth; but she out of her poverty put in all she had to live on."

The rich people had plenty of money left over for themselves. But the poor woman had nothing left for herself, because she gave all that she had. This showed how much she loved God.

God isn't impressed by how much we give. What matters most is that we love Him by giving Him the best that we have.

Parent Note: Giving is an important concept in the Christian faith. It should be taught early and made a lifelong habit—to give back a portion of what God has so bountifully given to us. Help your children see that it is a joy to give to God.

Questions

- Who gave the large amount of money to the temple?
- Who came and dropped in a little bit of money?
- Did the poor woman have any left for herself?
- How can we show God that we love Him?

Memory Verse

Come near to God and he will come near to you.

James 4:8

The Second Coming

Matthew 24:36–46; 1 Corinthians 15:52; 1 Thessalonians 4:16–18

Jesus told the disciples about when He was going to come back to earth again. This is called the second coming. The second coming will happen in the last days in history. Jesus will come from heaven on that day, and all the angels will come with Him. Everyone will be able to see Him.

No one knows when Jesus will come back. It could happen while you're at school or while you're sleeping. Jesus will come back when many people don't care about Him. And we know it will happen when we least expect it. That's why it's important to always be ready.

Don't wait to give your heart to Jesus or to share Him with your friends. He could come back next week or tomorrow — or even today. That's why it's best to be ready for Jesus to return.

Questions

- Who is coming to this world at the second coming?
- Who will come with Jesus?
- When will Jesus come back?
- Who can you tell about Jesus?

Memory Verse

Come near to God and he will come near to you.

James 4:8

The First Communion

Luke 22:19 – 20

[Jesus] took bread, gave thanks and broke it, and gave it to [the disciples], saying, "This is my body given for you; do this in remembrance of me."

In the same way, after the supper he took the cup, saying, "This cup is the new covenant in my blood, which is poured out for you …"

Questions

- Who gave thanks and broke bread?
- Who did Jesus give the bread to?
- What is another word for *covenant*?
  (Hint: think Noah's rainbow)

Memory Verse

Come near to God and he will come near to you.

James 4:8

Be Prepared

Matthew 25:1–13

Jesus told a story about some young women who went out in the night, carrying lamps with them. They went out to meet a man who had just been married. This man was called a bridegroom.

When the women reached the bridegroom's house, they sat down to wait until he came home. Soon they all fell asleep.

In the middle of the night, somebody called out, "The bridegroom is coming; go out to meet him." They all got up quickly and began to get ready. But they found that while they were asleep their lamps had run low on oil.

Some of the young women were wise and had brought extra oil with them; they poured this oil into their lamps so the flame wouldn't go out. These women were ready when the bridegroom came home. He invited them into his house and gave them a feast.

But the other women were foolish. They did not bring any extra oil with them, so they had to go buy some. By the time they came back, it was too late. The bridegroom had gone inside and shut the door. They could not get in.

This is the way it will be when Jesus comes again. Some people will be ready like the wise young women who had their lamps burning. And Jesus will take those who are ready up to heaven. He said, "Therefore keep watch, because you do not know the day or the hour."

But some people, like the foolish young women whose lamps were low on oil, will not be ready. If we want to be ready for Jesus when He comes back, we must love Him and ask Him to be our Savior.

Questions

- Were the wise young women ready when the bridegroom came?
- What did the foolish young women have to do?
- Where will Jesus take the people who are ready to meet Him when He comes back?
- If we want to be ready when Jesus comes back, what must we do?

Memory Verse

Now faith is confidence in what we hope for and assurance about what we do not see.

Hebrews 11:1

Money Hungry

Matthew 26:14–16; Mark 14:10–11; Luke 22:3–6

The disciples followed Jesus wherever He went and listened to what He taught. All the disciples loved Jesus—except one. Judas loved money more than anything else. He did not love Jesus.

Remember the people who wanted to kill Jesus? Judas went to these men and asked how much money they would give him if he showed them where Jesus was. The men told Judas they would pay him thirty pieces of silver.

Judas decided that as soon as he could find Jesus in a place by Himself, he would show these wicked men where Jesus was. Then they could take Jesus away to kill Him.

Questions

- Did the disciples all love Jesus?
- Which one didn't love Jesus?
- What did Judas love more than anything else?
- How much money did the men promise to give Judas?
- What did Judas decide to do?

Memory Verse

Now faith is confidence in what we hope for and assurance about what we do not see.

Hebrews 11:1

The Last Supper

Matthew 26:17 – 46; Mark 14:12 – 42; Luke 22:7 – 46; John 14:2 – 3

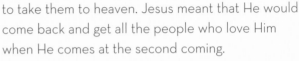

The people who lived in Jesus' country used to have a feast for a special reason. They had this feast once a year; in fact, Jewish people still have this feast. It is called the feast of the Passover.

Jesus and His disciples wanted to eat a feast of the Passover together. But they did not have a place to have the feast. Jesus told the disciples to go into town and ask where they could hold their feast.

The disciples did exactly what Jesus told them to do. A man showed them a room to use. So the disciples prepared the feast.

That evening around suppertime, Jesus told the disciples this would be the last time He would eat the Passover feast with them. He knew that He would die soon.

Jesus told the disciples that He was going to get heaven ready for them. Later He would come back to take them to heaven. Jesus meant that He would come back and get all the people who love Him when He comes at the second coming.

After Jesus and the disciples had eaten, they sang a song together. Then they went to a garden. Jesus went off by Himself to pray.

Jesus knew that some men were going to take Him that night and kill Him. He told God, "My Father, if it is possible, may this cup be taken from me. Yet not as I will, but as you will." He didn't want to die, but He knew that was the only way we could have our sins forgiven and go to heaven.

Questions

- What was the name of the feast that was celebrated by Jesus and His disciples?
- What did Jesus tell the disciples as they were eating?
- Is Jesus going to come back and take all those who love Him to heaven?
- What did Jesus do in the garden?
- Why was Jesus willing to die?
- What kind of place do you think heaven will be?

Memory Verse

Now faith is confidence in what we hope for and assurance about what we do not see.

Hebrews 11:1

Betrayed

Matthew 26:47 – 50

After the Passover feast, Jesus went to the garden to pray. Most of the disciples went with Him.

Judas, the disciple who didn't love Jesus, did not go to the garden with Jesus. Instead, he found the men who wanted Jesus killed and told them where Jesus was.

The men paid Judas the thirty pieces of silver they had promised him.

Some men went with Judas to get Jesus. The men carried sticks and swords to fight with. They carried lanterns too, so they could see in the dark. Judas showed the way to the garden.

On the way, Judas told them how they would know which one was Jesus. Judas would go up to Jesus and kiss Him.

Soon they reached the garden. Judas went up to Jesus and said, "Greetings, Rabbi," and kissed Him. (Rabbi means teacher and was something the diciples called Jesus.) The men grabbed Jesus and took Him away.

Questions

- Did Judas go with Jesus to the garden?
- Who did Judas tell where Jesus had gone?
- What did those men give Judas?
- How did Judas say they would know which one was Jesus?
- After Judas had kissed Jesus, what did the men do to Jesus?

Memory Verse

Now faith is confidence in what we hope for and assurance about what we do not see.

Hebrews 11:1

Taken Away

Matthew 26:47–56; Mark 14:43–52; Luke 22:47–53; John 18:1–12

The disciples didn't want the soldiers to take Jesus away from the garden. A disciple named Peter took a sword and cut off one soldier's ear. But Jesus told Peter to put his sword away. Then Jesus touched the man's ear and made it well again.

Jesus did not want the disciples to fight. He said that God would send down many angels from heaven to fight for Him if He asked. But Jesus would not ask them to come. He was willing to let the men take Him. And He was willing to let them kill Him.

Why was Jesus willing to let the men do these things? Because that was the way He was going to be punished for all our sins.

The men took Jesus out of the garden. The disciples were afraid that they would be taken too, so they ran away.

Questions

- How do you think the disciples felt when they saw the men taking Jesus away?
- What did the disciple named Peter do?
- What did Jesus do to the man's ear?
- Why was Jesus willing to let the men kill Him?
- When the disciples saw the men taking Jesus away, what did they do?
- What would you have done?

Memory Verse

Now faith is confidence in what we hope for and assurance about what we do not see.

Hebrews 11:1

Death of a King

Matthew 26:57 – 68; 27:1 – 2, 11 – 50

In our country, when people do wrong things, only judges can tell them what their punishment will be. But in Jesus' day, the governor could also punish people for doing wrong things. The men who wanted to kill Jesus took Him to the governor.

These men were very bad because they lied when they said Jesus had done wrong. Jesus had not done wrong. He claimed to be God's Son, which was true, but it made the men mad. They said that Jesus was bad and should be killed.

The governor did not care about Jesus. He told his soldiers they could beat Jesus with a whip. It hurt Jesus and cut His skin. They got some thorny branches, wound them together into a crown, and pushed them down on Jesus' head. They spit on Him and called Him names.

Then the soldiers made Jesus carry the wooden cross through the crowds that had gathered to watch. They hammered great big nails through His hands and feet to hold Him to the cross.

It hurt Jesus very much, but He did all of that for you and me. He loves us that much.

The soldiers then raised the cross with Jesus nailed to it, and they stayed there to watch Him die.

Questions

- What did the men who wanted to kill Jesus tell the governor about Him?
- Were they lying?
- What did the governor do?
- What did the soldiers do?
- What did the soldiers nail Him to?
- Why did Jesus go through all of that?

Memory Verse

. . . We wait for the blessed hope – the appearing of the glory of our great God and Savior, Jesus Christ . . .

Titus 2:13

Jesus Is Alive!

Matthew 28:1 – 8; Mark 15:42 – 47; 16:1 – 7; Luke 23:53

There was a rich man named Joseph who loved Jesus. When Joseph saw that Jesus was dead, he went to the governor and asked if he could take Jesus down from the cross and bury Him. The governor said okay.

Joseph had Jesus taken down from the cross. Joseph wrapped up Jesus' dead body in some new, clean cloth. A good man named Nicodemus helped Joseph. They buried Jesus in a tomb that was hollowed out of a rock. Some women who loved Jesus saw where they buried Him.

The governor sent some soldiers to roll a big stone in front of the tomb and then stand guard to keep Jesus' disciples away.

Early in the morning when it was still dark, God sent an angel from heaven. The angel's face was bright like lightning, and his clothes were as white as snow. When the soldiers saw the angel, they were so afraid that they fell to the ground. They could not move.

The same women had come back to look at Jesus' tomb. They wondered who would roll away the big stone for them. But when they got there, the stone was already rolled away, and an angel was there. The women were afraid. But the angel said, "Do not be afraid, for I know that you are looking for Jesus, who was crucified. He is not here; He has risen."

Jesus had told His disciples that He would be killed. He had also said that He would come back to life and leave the grave after He had been buried. Jesus did what He had said He would do!

Questions

- Where did Joseph and Nicodemus bury Jesus?
- Whom did the governor send to watch over the tomb?
- What did they roll over the door of the tomb?
- What did the women see when they arrived?
- What did the angel tell them?
- Had Jesus told the disciples that He would come to life again?

Memory Verse

. . . We wait for the blessed hope — the appearing of the glory of our great God and Savior, Jesus Christ . . .

Titus 2:13

Go and Tell Others

John 20:19 – 21; Matthew 28:18 – 19

After He came out of the grave, Jesus went to the disciples. They were in a room together. Even though the door of the room was closed, Jesus came into the room. The disciples were afraid when they saw Him. They didn't think it was Jesus; they thought He was dead.

Jesus told them not to be afraid. He showed them His scarred hands and side, so they would know it was Him. He had really come to life again.

Later, Jesus told the disciples, "Therefore go and make disciples of all nations, baptizing them in the name of the Father and of the Son and of the Holy Spirit, and teaching them to obey everything I have commanded you." He wanted everyone to know how much He loved them and that He had been punished in their place by dying on the cross for them. Jesus wanted the disciples to go all over the world to tell every person about Him.

Questions

- How did the disciples feel when they saw Jesus?
- Why were they afraid?
- What did Jesus show them so they would know He had really come to life again?
- What did He tell the disciples to do?
- What does He want you and me to do?

Memory Verse

. . . We wait for the blessed hope — the appearing of the glory of our great God and Savior, Jesus Christ . . .

Titus 2:13

In Heaven

Luke 24:50; Acts 1:9-10

After Jesus talked to the disciples, He took them out of the city to a place by themselves. While He was speaking to them, all of a sudden He began to go up into the sky. He went up higher and higher, until they saw Him go into a cloud. Soon they could not see Him anymore.

Two angels appeared and told the disciples, "Men of Galilee, why do you stand here looking into the sky? This same Jesus, who has been taken from you into heaven, will come back the same way you have seen him go to heaven."

Jesus is in heaven now. But He sees all of us as if He were here with us—and He is!

We don't ever need be afraid, for Jesus does not forget us. He hears us when we pray to Him, and He helps us to be good. When He comes to this world again, at the second coming, He will call us all to Him. He will take us to heaven, and we will live there with Him always.

Questions

- While Jesus was speaking to the disciples, where did He begin to go?
- Who came and spoke to the disciples?
- Where did the angels say that Jesus had gone?
- Where is He now?
- Does He see everyone?
- How long will His followers stay with Jesus in heaven?

Memory Verse

. . . We wait for the blessed hope—the appearing of the glory of our great God and Savior, Jesus Christ . . .

Titus 2:13

Jesus Promises the Holy Spirit

John 14:12 – 18, 26 – 27

"Very truly I tell you whoever, believes in me will do the works I have been doing, and they will do even greater things than these, because I am going to the Father. And I will do whatever you ask in my name, so that the Father may be glorified in the Son. You may ask me for anything in my name, and I will do it.

"If you love me, keep my commands. And I will ask the Father, and he will give you another advocate to help you and be with you forever – the Spirit of truth. The world cannot accept him, because it neither sees him nor knows him. But you know him, for he lives with you and will be in you. I will not leave you as orphans; I will come to you.

"But the Advocate, the Holy Spirit, whom the Father will send in my name, will teach you all things and will remind you of everything I have said to you. Peace I leave with you; my peace I give you. I do not give to you as the world gives. Do not let your hearts be troubled and do not be afraid."

Questions

- When Jesus went back to heaven, who did He send?
- Where does the Holy Spirit live?
- How do we show that we love God?
- What does the Holy Spirit do?

Memory Verse

. . . We wait for the blessed hope – the appearing of the glory of our great God and Savior, Jesus Christ . . .

Titus 2:13

Life in New Testament Times

Many years ago there were trails that came into the land of Israel. They came from the east and from the north. Other trails came from Egypt in the south.

Traders used these trails to travel from one place to another. They traveled mostly by camel. They bought and sold goods along the way. The trails went *through* Israel. But they also *met* in Israel. It was almost as if Israel were the center of the world.

In a way, Israel *was* the center of the world. Jesus was born there. All of the things that happened in Bible times seemed to say, "Israel is a special land."

Places of Worship

The beautiful temple stood in the city of Jerusalem. It was the center of worship for the Jews. Herod built the temple again not long before Jesus was born. The temple was on a hill. Its shining, white marble walls could be seen all over the city. Large stone gateways opened on all four sides. Jesus called this temple his Father's house (John 2:16).

Each Jewish town also had a smaller meeting place. These were called "synagogues." The leader of the synagogue studied the Old Testament and the Jewish laws. He then could teach the people.

On the inside, synagogues looked much like some of our churches. The people sat on benches. The leader stood on a stage. A special box held the scrolls of the books of the Bible.

On the Sabbath day, the people came to the synagogue to worship. The leader read a verse to call the people to worship. Then there were readings of thanksgiving and praise. Someone would lead in prayer. After that, the leader might ask someone to read from the Bible. Any member who was able to teach could give the sermon. The service was closed with a blessing.

The Laws of God

God gave the Jews the Ten Commandments and many other laws at Mount Sinai. Many Jews thought that trying to keep the law was the only way to please God. They began to add many of their own laws to God's laws. They began to see themselves as very good people.

Jesus told the Jews that they were going in the wrong direction. They were so busy doing many little things that they were forgetting the more important big things. They were forgetting to love others. They were forgetting to take care of the poor. They were forgetting to love God.

The Sabbath Day

God gave the people of Israel the Sabbath as a day of rest. On the seventh day of every week, they rested from their work. They offered special sacrifices.

The scribes and Pharisees later added hundreds of laws about how people should keep the Sabbath holy. Then the people forgot that God gave the Sabbath to be a blessing. Instead they just worried about obeying all the rules.

On the Sabbath day, people could not travel very far. They could not carry anything from one place to another. They were not supposed to spit on the ground. If they did, they might be plowing a little row in the dirt. And that would be work! If a hen laid an egg on the Sabbath, they were not supposed to eat that egg. The hen had worked on the Sabbath to lay it.

When Jesus and his disciples picked some grain and ate it on the Sabbath, the Pharisees said they were working. When Jesus healed sick people on the Sabbath, the Pharisees got angry and wanted to kill him.

Religious Groups

The most important religious groups in New Testament times were the Pharisees and the Sadducees. The Sadducees were rich and powerful men. The high priest, the chief priests, and rich businessmen all belonged to the Sadducees. The Sadducees were against any new group that tried to change Jewish life. That's why they were against Jesus and his disciples. The Sadducees also rejected many of the teachings of the Pharisees. They did not believe that people would live again. They did not believe in angels or demons. They did not keep all the laws of the Pharisees. They only kept the law of Moses.

The Pharisees added hundreds of laws to the law God gave Moses. They were mostly interested in keeping all these laws. But many of the Pharisees forgot some of God's other laws. They were proud of how good they were, and they did not love other people. However, there were also Pharisees who truly loved God and tried to do what was right.

Seventy of the most important Pharisees and Sadducees made up the Jewish high court. This court was called the Sanhedrin. The high priest led the court. The Romans let this court decide what to do when someone had broken a Jewish law. But this court did not have the power to put anyone to death. If the Sanhedrin thought someone deserved to die, they had to bring the person to the Roman courts.

The Roman Empire

Rome had begun to grow larger and stronger before Christ was born. Wars were fought and many new lands were added to the Roman Empire. This empire was very large. It included Spain and Germany, North Africa, Asia Minor, Syria, and Israel.

Many good things happened because of Roman rule. There was peace between all of the different countries in the empire. The Romans also set up good government everywhere. They built roads for safe and easy travel. Many of the people were able to speak and understand the same language — Greek.

The Romans did not know that all these things would make it easier for the gospel to spread to many lands. They did not know that God had prepared the way for Jesus and the spread of the good news. Later Jesus' disciples traveled more easily to faraway lands because there was peace and because there were good roads. They could bring the gospel in the Greek language to many people in many areas.

The Jews hated the Romans. They believed the Romans had no right to rule over them. They believed the Romans had no right to take their money for taxes. They didn't like the soldiers who lived in their country. The Jews also hated the Romans because they tried to change the Jewish way of life. The Romans wanted everyone to act like Romans. The Jews were looking for the Messiah. They thought he would become their king and would free them from the Romans.

The Jews hated tax collectors even more than they hated the Romans. Many tax collectors were Jews who were working for Rome. Many tax collectors were dishonest. They took more money than they were supposed to take. They were cheating their own people to help the enemy.

Jesus often talked and ate with tax collectors. Matthew was a tax collector. So was Zacchaeus. Both became followers of Jesus.

Everyday Life

Life in New Testament times was much different from life today. It was a simple life. Most people did not have any extras. In fact, they often had just enough to live. The people worked hard, and children had to share the work.

Houses

The people built their houses of mud bricks that were hardened by laying them out in the sun. Sometimes the front part of the house had no roof over it. This part was like a small yard. Behind it was a living room with small bedrooms at the back. The floor of the house was of hard and smooth clay. Builders made the roof of heavy wooden beams with boards laid across them. They covered the boards with a mixture of mud and straw. This flat roof was a good place to work or sit. Sometimes people slept on the roof on hot nights. Usually a ladder or sometimes steps led up to the roof.

Most people had very little furniture—just some wooden stools, a low wooden table, and some sleeping mats. There was a place for fire and sometimes a small clay oven for baking bread. There was no chimney, so the smoke had to find its way out of the small, high window openings. Some houses had wooden doors. Others had doorways covered with grass mats or cloth.

Food

The people ate foods like milk and cheese, grapes, figs, olives, honey and barley cakes, eggs, chickens, fish and goat meat, beans, cucumbers, and onions.

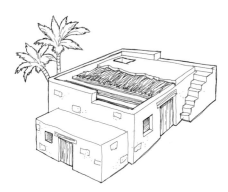

The first meal of the day was usually bread and cheese. Sometimes a family would eat a light meal at noon. Again, bread was the main part. The people had their large meal of the day in the evening. They usually ate bread and fish, fruit, and vegetables. The common people often ate meat only on very special days.

Clothing

The clothing of New Testament times was simple. Besides underclothing, the people wore robes with a belt tied around the waist. Over the robe, they often wore a cape. Children usually had shorter, knee-length clothing. They sometimes wore a kind of pullover shirt. Women decorated their clothing with brightly colored weaving and sewing.

The people wore sandals without socks. Their feet were often dusty from walking on their dirt streets and roads. They washed their feet often.

Work

The people did many different kinds of work. Some were farmers and builders and makers of pottery. Others were bakers and doctors and teachers. There were watchmen who guarded the cities. There were workers in leather and workers in metal. Jesus' father was a carpenter. He also knew about herding sheep. Peter, James, and John were fishermen. Matthew was a tax collector. There were scribes who wrote letters and copied the laws and the books of the Bible.

Women had to work hard in their homes. The first thing they would do in the morning was make the bread for the day. They would grind the grain into flour, then make dough into loaves of bread and bake them. The women also had to carry water from the well and get wood for the fire. They made all the clothes for the family, spinning and weaving their own cloth out of flax and wool.

Parents expected their children to help with the work. Girls helped their mothers with all the household work. Boys helped their fathers in their work and were expected to follow the same trade as their fathers.

Schools

Parents taught their children Bible verses when they were still very young. They learned verses from the law and stories from the Old Testament.

When boys were five or six years old, they went to school. The leader of the synagogue taught them. For the first four years, they studied mostly the first five books of the Bible. By then they knew the laws of God very well. They also learned how to read and write Hebrew. For the next several years, they studied other books of the Bible and other Jewish writings.

When a Jewish boy reached the age of twelve or thirteen, he was considered to be a man. The boy and his family and friends celebrated with a special ceremony and often a party. Most boys left school at this age.

Conclusion

The time of the New Testament was the best possible time for Jesus to come. The people were looking for the Messiah. The safe roads made it much easier for early Christians to travel to spread the good news of the Savior. The common language made it much easier for them to tell others about Jesus. People were eager to hear about him. God had everything planned and ready.

Acknowledgments

Someone once said, "If you see a turtle on a fence post, you know he didn't get there by himself." No book is the effort of one person. It takes a team! I am fortunate to have had an amazing team working together to develop the book you now hold.

I thank each one:

Catherine DeVries, Product Development Director, was the team leader and lead she did!

Sarah Drenth, National Accounts Manager for Sales, created a strategy for getting the word out about this book. You heard about it because of her work.

Kris Nelson, Creative Director, made this book beautiful and exciting with the creative design and myriad photographs and illustrations she doggedly hunted down.

Helen Schmitt, Associate Marketing Director. You hold this book in your hands as a direct response to her wonderful work getting the message out about it.

Kristen Tuinstra, Associate Editor, had to ride hard on me to get the book done. Her patience and her marvelous sense of humor were always an encouragement. She added her own creative writing talents to help me over the humps during my too busy schedule. All editors should be such fun to work with!

Wes Yoder, my literary agent and friend, who early on saw the vision for this timeless book.

And finally my children, their spouses, and my adorable grandchildren, who bring me such joy and pride as I watch them grow and mature in their understanding and application of God's Word to their daily lives.

Photography and Artwork Credits

Step 54 stormy sea © Bradley Mason; sailing with sunset © Eric Gevaert; detail of waves © Bradley Mason

Step 55 Caves of Matala (Crete), Greece, the Roman era cave cemetery © Marcin Szmyd; manacles © Shutterstock; cougar © John Pitcher

Step 56 crowd © Shutterstock; woman wearing scarf © Ragne Hanni; sleeves of garment © Carlos Santa Maria

Step 57 girl in bed © Ensa; father and daughter © Rosemarie Gearhart; open hand © cozyta

Step 58 girl holding pigeon in Venice © Eril Nisbet; Starling eating berries © Janet Forjan-Freedman; blue eggs in nest © Yitzchok Moully

Step 59 man pointing the way in the Sandia Mountains in Central New Mexico © Kevin Lange Photography; woman rests by campfire © Danny Warren; hungry baby birds © niknikon

Step 60 lambs © Photodisc; sheep grazing along the English countryside © Shawn Mulligan; Jesus statue with lamb © Bernd Klumpp

Step 61 Mediterranean bounty © Donald Gruener; boy with messy face © Alison Hausmann Conklin

Step 62 blind boy with canes © istockphoto; child reading Braille, detail © Carmen Martínez Banús

Step 63 Jesus Teaching Teachers. Dore, Paul Gustave (1832-1883) © Planet Art; girl and teacher © Shutterstock; girl looking up © Lisa F. Young

Step 64 bread products © Tomo Jesenicknik; two fish © Olena Kucherenko; offering bread © Gino Santa Maria

Step 65 dark stormy sea © Felix Möckel; sailboat at sunset © Fielding Piepereit; stormy sea detail © Bradley Mason

Step 66 happy sister with brothers © Marzanna Syncerz; boy helping younger boy © Sonyae

Step 67 Transfiguration. Sanzio, Raphael (1483-1520) © Planet Art; sun shining on mountain © Photodisc; Baltic Sea © Tomasz Resiak; landscape © Fotostock

Step 68 sibling rivalry, smiling sisters © Lisa Eastman; children shaking hands © H. Tuller

Step 69 coins in treasure chest © Achim Prill; handshake against sky background © Peter Elvidge

Step 70 baby in a manger © Paige Roberts; stained glass © Jim DeLillo; gold crown © istockphoto

Step 71 city wall of Jerusalem © Claudia Dewald; dirt road © Michel de Jijs

Step 72 man on donkey © istockphoto; helping hand © Alvaro Heinzen; profile of donkey © Jason Lugo

Step 73 Jesus in the House of Martha and Mary. Le Sueur, Eustache (1617-1655) © Erich Lessing / Art Resource, NY; pots of food cooking © Joseph Luoman; broom © Octavian Florentin Babusi

Step 74 Pool of Siloam © Todd Bolen; muddy hands © Sandramo; man's eyes © Amanda Rohde

Step 75 sheep on hillside © Willi Schmitz; lonely lamb © Tissa; statue of the Good Shepherd © Scott Anderson

Step 76 Return of the Prodigal Son. Guercino, Giovanni Francesco (1591-1666) © Alinari / Art Resource, NY; pigs © Ovidiu Iordachi

Step 77 old iron gate © Laila Røberg; stairs © Shutterstock; angel statue © Jacob Hellbach

Step 78 a boy in trouble © Rob Friedman; girl looking sad © shutterstock:girl with friend looking sorry © Galina Barskaya

Step 79 Resurrection of Lazarus. Tintoretto, Jacopo Robusti (1518-1594); Cameraphoto / Art Resource, NY (including both details); tomb © Zondervan

Step 80 stained glass of Jesus healing sick boy © Laura Clay-Ballard; father and child holding hands © Peter Galbraith

Step 81 children running across lawn © shutterstock; boy lying in grass © Cindy Minear; smiling girl © Andrea Gingerich

Step 82 wooden cross on the Rigi Mountain in Switzerland © Andreas Kaspar; railroad spike © Niki Crucillo; crown of thorns © istockphoto

Step 83 sycamore trees © Cecelia Henderson; pouch © Dan Fletcher; coins © Flat Earth

Step 84 Christ's Entry into Jerusalem. Flandrin, Hippolyte (1809-1864) © Erich Lessing / Art Resource, NY; donkey in field © Mike Morley; palm branch © Photodisc

Step 85 child in time-out © Lewis Jackson; child and mother © Nicholas Monu; time-out chair © Mary Gascho

Step 86 wedding bouquet © shutterstock; wedding buffet (large image) © istockphoto; wedding buffet © DIGIcal; rose © Kuzma

Step 87 walking through woods © Vladimir Ivanov; © mother and daughter holding hands © Michel de Nijs; happy girls © BananaStock

Step 88 collection plate © Sean Locke; shaking bank for money © Steve Snyder

Step 89 Last Judgment, ceiling fresco in the sacristy of the Abbey Church of the Vorau, 1716 © Erich Lessing / Art Resource, NY (includes details)

Step 90 Communion still life © Magdalena Kucova; receiving communion © Michael Blackburn; French bread varieties © Photodisc

Step 91 oil lamp and scarves from the Middle East © Lori Martin; flame © Odelia Cohen; burning Roman oil lamp © Christian Bernfeld

Step 92 leather pack and coins, detail © PMSI Design; stacks of coins © José Carlos Pires Pereira

Step 93 The Last Supper. Titian, Tiziano Vecelli (1485-1576) © Planet Art; Garden of Gethsemane © Alex Slobodkin; chalice and bread © M. Kucova

Step 94 Garden of Gethsemane © Alex Slobodkin; Jesus praying, stained glass © Michael Westhoff; Roman gladiator © Winter

Step 95 Roman soldier with shield, hand and sword © Sue Colvil; sword and shield © Henning Martens

Step 96 three crosses at sunset © PhotographerOlympus; spikes, crown of thorns © Jill Fromer

Step 97 inside empty tomb © Glenda Powers; angel statue © Charles Taylor; Jesus statue © Tiburon Studios

Step 98 antique map with compass © Valerie Loiselux; hands showing scars © Gino Santa Maria; Earth in starfield © Kativ

Step 99 sun breaking the clouds © Aleksej Kostin; boy and clouds © Judy McPhail; angel statue © Manuela Krause

Step 100 father and child walking on beach © Natalia Sinjushina; bird flying at sunset © Eric Gevaert; dove in flight on black background © Christine Balderas

page 2 lambs © Photodisc

page 5 vineyard © Benjamin Goode

page 9 fisherman © Timur Kulgrain

page 10 Ruth Graham and grandchildren © Todd Bauder/Contrast Photography

page 214-219 *Life in New Testament Times* illustrations © Zondervan

page 220 children running across lawn © Tom Horyn

Notes